Occult Botany

"Reading *Occult Botany* was a deeply enjoyable experience that satisfied many diverse urges for botanical knowledge, from within one compendium. This strange and intriguing miscellany leads one through portals and fields, gardens and celestial realms. The plant entries are helpful for quickly finding zodiac and planetary correspondences; they also include a peppering of rare pieces of occult lore alongside many medicinal applications of old. The footnotes are very helpful and detailed, as are the charts, tables, and appendices within the book. It is a must-have for students and seekers who wish to blend the magical with the medicinal, the earth with the stars."

Corinne Boyer, folk herbalist, teacher, and author of *Under the Witching Tree*

"A refreshing look at the treatises of occult herbalism. A fascinating, in-depth deep dive and understandable approach to the esoteric arts as they pertain to plant, animal, and mineral medicines with an emphasis on occult botany and the Hermetic arts, including lab alchemy. A one-of-a-kind, expansive dictionary of magical plants with special care given to the translation and annotations in the footnotes to further illustrate the understanding of these plants then and now. A must-have for any seeker of esoteric herbalism."

Catamara Rosarium, master herbalist, owner of Rosarium Blends LLC, and cofounder and convener of the Viridis Genii Symposium

"*Occult Botany* gives us a needed look at esoteric herbalism from 1902, when philosophical arts had not yet veered dangerously into the pseudoscience of the new age that we have today. We can see whole philosophies at work within these pages that should help modern readers navigate their way out of the philosophical cul-de-sacs that modern herbalism has been circling for too long."

MARCUS MCCOY, BLACKSMITH, HERBALIST, EDITOR OF *VERDANT GNOSIS,* AND COFOUNDER OF THE VIRIDIS GENII SYMPOSIUM

Occult Botany

Sédir's Concise Guide to Magical Plants

Paul Sédir

TRANSLATED, EDITED, AND ANNOTATED
BY R. BAILEY

Inner Traditions
Rochester, Vermont

Inner Traditions
One Park Street
Rochester, Vermont 05767
www.InnerTraditions.com

Originally published in French under the title *Les plantes magiques: Botanique occulte, constitution secrète des végétaux vertus des simples, médecine hermétique, philtres, onguents, breuvages magiques, teintures, arcanes, élixirs spagyriques* by Bibliothèque Chacornac, Paris, 1902

Note to the reader: *This book is intended as an informational guide. The remedies, approaches, and techniques described herein are meant to supplement, and not to be a substitute for, professional medical care or treatment. They should not be used to treat a serious ailment without prior consultation with a qualified health care professional.*

Cataloging-in-Publication Data for this title is available from the Library of Congress

ISBN 978-1-64411-260-1 (print)
ISBN 978-1-64411-261-8 (ebook)

Printed and bound in the United States by Versa Press, Inc.

10 9 8 7 6 5 4 3

Text design by Virginia Scott Bowman and layout by Debbie Glogover
This book was typeset in Garamond Premier Pro with Argent CF, Futura Std, Gill Sans MT Pro, and Gotham, used as display fonts

To send correspondence to the translator of this book, mail a first-class letter to the translator c/o Inner Traditions • Bear & Company, One Park Street, Rochester, VT 05767, and we will forward the communication.

To Papus,

I dedicate this humble work to you,

who first awakened my mind to the Occult Mysteries.

Over the past twelve years you have introduced me to the many branches of science and revealed to me both their perfect blossoms and their blights. I have long tasted the fruits of your arduous labor, and now that I finally understand the path you follow, it brings me joy to acknowledge publicly the great debt I have incurred to you. May the heaven following your example train up many new laborers in the employ of the Great Husbandman, to cultivate the earth from whence they came, until the Master of the Vineyard appears in all his glory.

Paul Sédir
Epiphany, 1901

Contents

PART THREE

A Concise Dictionary of Magical Plants

Symbols and Tables

SYMBOLS

Elements

🜁	AIR
🜂	FIRE
🜃	EARTH
🜄	WATER

Planets and Luminaries

♄	Saturn
♃	Jupiter
☉	Sun
♂	Mars
♀	Venus
☿	Mercury
☾	Moon

Signs of the Zodiac

♈	Aries
♉	Taurus
♊	Gemini
♋	Cancer
♌	Leo
♍	Virgo
♎	Libra
♏	Scorpio
♐	Sagittarius
♑	Capricorn
♒	Aquarius
♓	Pisces

TABLES

Translator's Foreword

Paul Sédir, pseudonym of Yvon Le Loup (1871–1926), was a prominent figure in occult societies both in France and abroad in the late nineteenth and early twentieth centuries. His entry onto the French occult scene has become the stuff of legend. He first encountered Papus* in 1889 at Lucien Chamuel's Librairie du Merveilleux, a famous occult bookshop on rue de Trévise in Paris (now the site of a charming café). The esoteric poet Victor-Émile Michelet, who happened to be there that day, saw a gaunt and ungainly young man walk into the shop and without any pretense of formality announce, "Voilà! I've come to take up occultism!" Michelet broke out in uproarious laughter. But Papus didn't blink an eye. He saw something in the young man that Michelet, in that moment, could not see. "Very well, my boy," Papus replied. "Come to my house Sunday morning." That Sunday the "genius of physicians" tasked the young neophyte with curating his personal library. There he gave him the name Sédir, an anagram of the French word *désire,* because he so embodied Louis Claude de Saint-Martin's concept of *l'homme de désir,* the "man of desire" or "man of aspiration."[1] This esoteric nom de plume, in truth, could not have been more fitting.

Sédir's occult curriculum vitae is impressive by any standards. There were

*Gérard Encausse (1865–1916), although still a medical student at the time, had already adopted his famous pseudonym, Papus. He took the name from Éliphas Lévi's commentary on an incomplete edition of an excerpt from the *Nuctemeron of Apollonius of Tyana.* Lévi lists Papus as the *génie de la première heure* in medicine, the first daemon of the first hour of the night (see Lévi, *Dogme et rituel de la haute magie,* 2:391). The name Papus, however, does not derive from the *Nuctemeron* itself, as so many authors have claimed.

few initiatic orders in France of which Sédir was not a high-ranking member. He sat on the Supreme Council of Papus's Martinist Order and served as the director of the Hermanubis Lodge. He became a doctor in kabbalah in Stanislas de Guaïta's Ordre kabbalistique de la Rose-Croix, for which he wrote theses on the divinatory Urim and Thummim (*thèse de licence*) and kabbalistic conceptions of the universe (*thèse de doctorat*). Under the name Tau Paul, he was consecrated bishop of Concorezzo (located twelve miles northeast of Milan) in Jules Doinel's Église gnostique de France. Other notable groups with which Sédir was affiliated include François Jollivet-Castelot's Société alchimique de France and the French chapter of the Hermetic Brotherhood of Luxor headed by François-Charles Barlet. As a full professor at the École hermétique and Faculté des sciences hermétiques, he taught courses and gave seminars on a variety of subjects from Hebrew language to homeopathy and Hermetic medicine.

As an author, Sédir was nothing short of prolific. He wrote widely on traditions of magic, alchemy, and mysticism, publishing more than twenty books, including *Les miroirs magiques* (1894), *La médecine occulte* (1900), *Le fakirisme hindou et les yogas* (1906), and *Histoire et doctrines des Rose-Croix* (1910). He also produced important French translations of Latin, German, and English esoterica, including Jacob Boehme's *De signatura rerum* (1622), Johann Georg Gichtel's *Theosophia practica* (1721), and Peter Davidson's *The Mistletoe and Its Philosophy* (1898).

Les plantes magiques is by far Sédir's most influential and enduring work. The sheer number of reprints, editions, and translations is testament to its far-reaching influence. To cite just two examples, *Магические растения,* the augmented Russian translation produced by the Martinist Alexander Valerianovich Troyanovsky in 1909, has remained in print to this day. The book *Botánica oculta: Las plantas mágicas según Paracelso,* translated by Rudolfo Putz (pseudonym of Rossendo Pons) and first published in Barcelona in 1932 by Librería Sintes, is actually little more than a redacted Spanish translation of Sédir's book, only its authorship is attributed not to Sédir but to Paracelsus! It was this Spanish translation published under the name of Paracelsus that came to form the basis of Samael Aun Weor's *Tratado de medicina oculta y magia práctica* (1977). To Aun Weor's credit, he expresses serious doubts over whether Paracelsus is the book's true author.

Sédir spent six years researching and writing *Les plantes magiques*. Papus had announced its plan and scope to readers of the journal *L'Initiation* in 1896, but the book was not to appear until 1902.[2] One year prior to its publication, when the project was nearing completion, Papus listed the book as required reading in an advertisement for a program of courses at the École hermétique.[3] This would suggest that Sédir had written the book principally for use in a classroom setting, as he had his previous book, an introductory grammar of the Hebrew language titled *Élemens d'hébreu* (1901). In fact, the Université libre des hautes études, one of several outgrowths of the École hermétique, had been offering a course titled "Occult Botany" as early as 1891, and the Faculté des sciences hermétiques had been offering a course with the same title at least as early as 1898.[4] In all likelihood, it was Sédir's participation in a course such as this—either as student or teacher, or both—that served as the catalyst for the book.

Sédir's frequent recommendations to students throughout the book confirm this *Sitz im Leben* (setting in life). However, because Sédir's book was designed for a classroom setting and intended to be read alongside other books and course materials, in several places Sédir glosses over important materials covered in other publications, most notably in Papus's heavy tomes, to which his students would most certainly have had access. Since English readers will have neither the luxury of such a setting nor ready access to Sédir's sources, the vast majority of which remain untranslated, the production of an augmented English edition was needed. In chapter 2, for example, Sédir presents Papus's esoteric investigations into plant anatomy, plant embryology, and plant morphology in the starkest of outlines, and throughout the book he makes reference to specific planetary days and hours but without ever explaining or tabulating them for his readers. Of course, such cursory treatments or sins of omission would not be problematic in a classroom setting. Nonetheless, to make Sédir's text more accessible to English readers, where necessary I have augmented derivative sections with more thorough descriptions based on his sources. All revisions and additions are clearly identified as such and fully cited in the endnotes.

Other editorial improvements include the use of the symbol ⋆ before plant names for which there are entries in Sédir's dictionary in part 3; the addition of bibliographical citations for the many works Sédir neglects to cite; full and corrected citations for the works he does cite; and a number of translator's notes to clarify obscure words and phrases or to inform readers of any significant changes or corrections made to Sédir's text. This augmented English edition also includes translations of three articles by Sédir on related topics as appendices.

USING SÉDIR'S DICTIONARY

Each entry in Sédir's dictionary serves as a plant meditation. Readers are invited to contemplate the astral characteristics of plants and connect or reconcile them with their terrestrial properties, herbal actions, botanical names, medico-magical applications, or mythological traditions. Even with the augmented materials, the entries are still far from comprehensive. They leave the door open for interested readers to draw their own parallels and make their own additions. The following user's guide offers a window into the poetics and mechanics of Sédir's pioneering system of plant correspondences.

Latin Binomials

All plants now have up-to-date scientific names in standard binomial nomenclature. When Sédir uses outdated binomials, outdated genus or species names, or pseudo-binomials in use among old herbalists or apothecaries, these have been relegated to the footnotes with references to the original text. Latin binomials consist of two parts: the genus name (that is, the Latin or latinized generic name—always capitalized) and the species name (that is, the Latin or latinized specific name—always lowercased). Each binomial is followed by an author citation, indicating the name(s) of the botanist(s)—usually in abbreviated form—who first published the now accepted botanical name according to the formal requirements set in the *International Code of Nomenclature for Algae, Fungi, and Plants* (ICN).[5] When the species name no longer belongs in its original generic placement, the author of the original generic placement always precedes the author of its current generic placement in parentheses.

Take *Amanita muscaria* (L.) Lam., the Latin binomial for fly agaric, for example. Carl Linnaeus (= L.), the father of modern taxonomy, first identified this muscimol mushroom as *Agaricus muscarius.*[6] The generic epithet

Agaricus is a latinization of the ancient Greek name for a variety of tree fungi (ἀγαρικόν), which Dioscorides claims to derive from *Agaria,* the name of a town in Sarmatia where these fungi abounded.[7]* The specific epithet *muscarius,* an adjective meaning "of or pertaining to flies," derives from the Latin noun *musca* (fly) and was so applied because this species was traditionally used for catching flies. The fly agaric gained its current scientific name in 1783 when Jean-Baptiste Lamarck (= Lam.) placed it in the genus *Amanita,* a latinization of the ancient Greek name for various species of mushroom (ἀμανῖται).[8] It is not essential that readers know each and every author abbreviation, and indeed, most publications on magical herbs omit them, but they are included here for the sake of completeness.[9]

In places where Sédir provides only the genus name or, a much more common occurrence, the French common name corresponding to the genus name, I supply the Latin binomial for the type species (abbreviated "type sp."). This means that the information provided applies to multiple species within the genus but especially to the type species. In botany, *type species* is an unofficial term for the most representative species in a genus, and it is most often the case that the type species is the species the author had chiefly in mind when the genus was established. I employ the botanical system of type species here to avoid the ambiguous usage of spp. (the Latin abbreviation for the plural form *species*). The advantage of using this system is that it allows for greater precision in identifying the particular species Sédir had in mind when he compiled his entries.

Take ⋆agaric again, for example, which appears in *Les plantes magiques* under the French common name *agaric.* First, it is more often than not the case that generic French common names like *acanthe* (⋆bear's breeches), *aigremoine* (⋆agrimony), and *aulne* (⋆alder) designate the type species of their respective genus. Although the French common name *agaric* has been used as an epithet for mushrooms of both *Agaricus* and *Amanita* genera, it is clear from Sédir's description of the agaric's occult properties that his entry concerns mushrooms of the latter genus. The typical way of relaying this information would be *Amanita* spp., which would indicate two or more species in the genus *Amanita* but without specifying any one species in particular.

*Among the ancient Greeks, fly agaric was known as black agaric (ἀγαρικὸν μέλαν).

But whereas *Amanita* spp. might give readers the false impression that all or most species in the *Amanita* genus possess the same oneirogenic properties Sédir describes, classifying Sédir's entry under *Amanita muscaria* alone would exclude other species in the genus.

Here, the formula "Latin binomial (type sp.): *Amanita muscaria* (L.) Lam." identifies *Amanita muscaria* and one or more species in the genus *Amanita* (for example, the species *A. pantherina, A. gemmata, A. regalis,* and *A. strobiliformis* also possess psychoactive properties). This allows for greater precision not only in identifying the most probable species Sédir had in mind when he compiled his lexical entries but also in identifying one definitive and representative specimen in the genus, which readers may use as a starting point for further research.

French Common Names

Each entry also includes a list of French common names. Often these are more comprehensive than Sédir's original lists of variant names. This has been done not only for practical purposes—namely, to avoid confusion—but also to add an additional layer to Sédir's system of herbal mysticism. For readers who may wish to consult the original French publication, the first name in each series, unless otherwise noted, is the name under which the entry appears in the French original (the rest proceed in alphabetical order).

In several places Sédir provides separate entries for different French common names for the same species. For example, the original French publication includes one entry for *capillaire,* which identifies its planetary signature as Saturn, and another entry for *cheveux de Vénus,* which identifies its zodiacal signature as Taurus. Both *capillaire* and *cheveux de Vénus* are, however, common names for ⋆maidenhair fern (*Adiantum capillus-veneris* L.). In such cases, entries have been merged to form a single entry. In other places, Sédir provides French common names for two or more different plant species in a single entry. Such conflations are usually attributable to one or another of Sédir's sources or to the application of a single French common name to multiple plant species. For example, Sédir's original French publication includes both *armoise* (⋆mugwort) and *herbe de Saint Jean* (⋆Saint John's wort)—two unrelated plant species—in a single entry because the common name *herbe de Saint Jean* was, and still is, frequently used as an epithet for mugwort. In such cases, entries have been separated to form two distinct entries.

French common names, to a greater extent than English common names, can also be revealing indicators of the astral or occult properties of plants. More so than their English counterparts, French plant names are replete with god names, saint names, and animal names. Just as Sédir recommends combing through Greek mythologies to discover the secret properties of plants, readers may find it equally rewarding to peruse the hagiographies of saints whose names are immortalized as plant names.* A knowledge of animal signatures, too, can be revealing with respect to French plant names and herbal correspondences. For example, ★greater burdock, which Sédir places under the rulership of Saturn, is known in French by the common name *chou d'âne,* literally "donkey's cabbage," the ass having long been regarded as a Saturnian animal.[10]

Moreover, and more importantly, due to the nature of French pronunciation, French common names lend themselves more easily to the system of occult linguistics known as Green Language. A great many French common names also have patois origins—which is to say, they are examples of argot, cant, or slang. According to Fulcanelli, who gave the now classic interpretation of *art gothique* (International Phonetic Alphabet [IPA]: aʁ.gɔ.tik)—that is, the "Gothic art" of the cathedrals—as an expression of esoteric slang or *argotique* (IPA: aʁ.gɔ.tik), the explanation of a word is to be sought not in its literal root but in its "cabalistic origin."[11]

Take the herb ★wormwood, *Artemisia absinthium* L., for example. The French common name *absinthe* derives from the Latin *absinthium,* itself a derivative of the Greek plant name ἀψίνθιον. Rabelais's *Gargantua and Pantagruel,* which Fulcanelli describes as an esoteric "novel in cant," includes a fascinating catalog of plant names according to their linguistic origins. Among plants whose names have their origins in contrariety and antiphrasis—linguistic jargon for the (usually ironic) use of a word in the opposite sense of the generally accepted meaning—Rabelais includes "*absynthe,* the contrary of *pynthe,* because it is bitter to drink" (*Abſynthe, au contraire de pynthe, car il eſt faſſheux a boyre*).[12] Although the literal root

*See, for example, Sédir's entries ★bitter cress, ★mignonette, ★mullein, ★speedwell, and ★valerian.

of ἀψίνθιον is presently unknown,* Rabelais adopts what was a popular etymology in the sixteenth century, deriving it from ἀπίνθιον, a nonlexical compound form comprised of the Greek α privative, expressing negation, and the ghost word πίνθιον, which is apparently supposed to be a derivative of the verb πίνειν, "to drink," if not a diminutive of the noun πίθος, "wine jar."† Rabelais's Old French *pynthe* is modern French *pinte,* meaning "pint." Taken as a whole, this pseudo-etymology means "undrinkable," "impotable," or "unpleasant to drink." Whereas this fanciful etymology is rightly rejected by modern linguists, it remains valid as a form of occult linguistics.

From the standpoint of biblical literature, wormwood could justly be characterized as an "unholy herb." The Hebrew name for wormwood, *la'anah* (לענה), derives from the unused root לען, which is cognate with Arabic *la'n* or *la'ana* (لعن), meaning "to curse" or "damnation." In the Hebrew Bible, wormwood is often used figuratively to describe anything bitter or poisonous, and the same is true of the New Testament usage of the *koinē* variant ἄψινθος: "The name of the star is Wormwood. And a third of the waters became wormwood, and many died from the water, because it was made bitter."[13]

In French argot, however, wormwood acquires a very different and even contrary meaning. According to modern linguists, French *absinthe* (IPA: ap.sɛ̃t) morphed into the common plant name *herbe-sainte* (IPA: ɛʁb.sɛ̃t)—that is, "holy-herb" or "sacred-herb"—through homophony and antiphrasis.[14] Whether the linguistic equation *absinthe* equals *herbe-sainte* is truly a case of antiphrasis; *herbe-sainte* exemplifies what Fulcanelli would describe as the "cabalistic origin" of *absinthe.* Moreover, the holy nature of wormwood is equally well established in ancient tradition. Pliny describes the ancient Roman custom of awarding a draught (not a pint) of wormwood to

*The cluster -ινθ-, which suggests a pre-Greek origin, can be found in several other ancient Greek plant names, such as τέρμινθος (terebinth), ἐρέβινθος (chickpea), μίνθη (mint), and ὑάκινθος (hyacinth).

†This pseudo-etymology was popularized by early botanists like Charles Estienne and Leonhart Fuchs; cf. Lăzar Şăineanu's characteristically excellent study, *L'histoire naturelle et les branches connexes dans l'œuvre de Rabelais,* 116. Although the origin of this pseudo-etymology remains something of a mystery, the same etymology also appears in Friedrich Sylburg's edition of the *Etymologicum magnum* (see Sylburg, *Etymologicon magnum seu Magnum grammaticae penu,* 83 s.v. ἀψίνθιον).

the winners of four-horsed chariot races, "doubtless," Pliny says, "because our forefathers were of the opinion that good health was the most valuable reward they could bestow on their skill."[15] Even so, for occult linguists like Fulcanelli, both *apinthion* and *herbe-sainte* would be equally valid and informative argotologies.

Numerous examples of comparable linguistic morphologies could be cited, such as *mandragore* (★mandrake) morphing into *main-de-gloire* (hand-of-glory—the dried hand of a hanged man) or *aigremoine* (★agrimony) into *grimoino* (grimoire—a handbook of magic). The possibilities for study are truly limitless. Although I supply modern Latin binomials for the purposes of identification, Green Language "rules" may be applied to these or to any outdated scientific names, never mind to any common name in any language, dead, artificial, or modern.*

Magical Plant Names

Sédir's dictionary is replete with magical, secret, Chaldean, alchemical, Paracelsian, and Hermetic plant names. Here, all of Sédir's references to obscure plant names appear under the lexical category "occult properties" to avoid confusion.† In the original French publication, such plant names appear alongside French, Latin, and Greek plant names without explanation. For example, Sédir's original entry for ★cleavers (*Galium aparine* L.) appears under the following sequence of names: "*Glouteron, Philadelphus,* or *Apparine* (*sic*)."[16] Historically, the French common name *glouteron* has been used to describe both ★greater burdock and ★cleavers, but Sédir's inclusion of the species name *aparine,* a derivative of the ancient Greek plant name ἀπαρίνη, makes the identification of the plant as cleavers a certainty. Sandwiched between these two plant names, without any explanation, is the Latin name *philadelphus,* which refers not to *Philadelphus,* the genus of ornamental

*In several places Sédir himself makes use of Green Language. See, for example, his comments on the French words *pur* (pure) and *sain* (healthy) in chapter 2 and the notes to Sédir's entries ★cyclamen and ★chinaberry.

†These names are also written without capitalization so that readers do not mistake them for genus names, and the same is true of all traditional, rather than scientific, binomials in use among old herbalists and apothecaries, such as *remora aratri,* the old pharmaceutical name for ★restharrow.

shrubs whose type species is the sweet mock orange (*P. coronarius* L.), but rather to a common name for cleavers that was popular among the old herbalists and spagyrists. This Latin variant may be traced back as far as the time of Pliny, who says that the Greeks gave cleavers the synonymous common name φιλάνθρωπον, meaning "lover of humanity," because it cleaves so eagerly to the clothes of passersby.[17]

Sédir's source for the vast majority of the alchemical or Hermetic plant names was Antoine-Joseph Pernety's *Dictionnaire mytho-hermétique,* first published in 1758. Pernety's dictionary remains a valuable resource even today, but it contains numerous errors, especially with respect to plant names. In fact, the vast majority of errors, typographical and otherwise, in Sédir's original publication are attributable to Pernety's voluminous reference work.

Pernety's Hermetic plant names should be taken with a grain of salt. A number of these names are merely latinized derivations of Greek, Hebrew, or Arabic plant names. Pernety never supplies their source languages, but Sédir, who was not only a gifted linguist, philologist, and translator but also one who had unfettered access to Stanislas de Guaïta's impressive occult library, would most certainly have recognized the vast majority of them. Whenever the source language of such a name is clearly identifiable, I have supplemented the entry with this information. Many of these names were in use among contemporary eighteenth-century herbalists, apothecaries, physicians, and iatrochemists (those who sought to provide chemical solutions for medical ailments), and so readers should not automatically assume that these names have mystic or cryptographic origins in the Hermetic sciences. For example, the alchemical or Hermetic plant name *philadelphus,* a latinized form of the Greek compound φιλάδελφος, meaning "filial love," appears in a number of eighteenth-century medical lexica, including those compiled by Dutch iatrochemist Steven Blankaart, Swiss naturalist Albrecht von Haller, and English physician Robert Hooper.[18] This, of course, is not to say that such names are argotologically insignificant, for occult linguistics is a holistic discipline and one that does not exclude words or names, barbarous or otherwise, on academic grounds. Recourse to such encyclopedic reference works, in fact, unearths a panoply of fascinating, bizarre, and very uncommon plant names.

TABLE 0.1. THE CHALDEAN HERBS

	Chaldean Herb	*Liber aggregationis* (1493)	*Albert le Grand* (1703)	*Les plantes magiques* (1902)
1.	⋆Heliotrope	*Ireos*	*Ireos*	**Ileos*
2.	⋆Nettle	*Royb*	*Royb*	**Roybra*
3.	⋆Shepherd's purse	*Loru(m)borot*	*Loromberot*	om.
4.	⋆Greater celandine	*Aq(ui)laris*	*Aquilaire*	*Aquilaris*
5.	⋆Periwinkle	*Iterisi*	*Vetisi* or *iterisi*	**Herisi*
6.	⋆Catnip	*Bieith*	*Bieith*	*Bieith*
7.	⋆Hound's tongue	*Algeil*	*Ageil*	*Algeil*
8.	⋆Henbane	*Ma(n)sela*	*Mansesa*	**Mansera*
9.	⋆Lily	*Augo*	*Ango*	**Augoeides*
10.	⋆Mistletoe	*Luperax*	*Luperax*	*Luperax*
11.	⋆Centaury	*Isiphilon*	*Isiphilon*	**Siphilon*
12.	⋆Sage	*Colorio* or *coloricon*	*Colorio* or *coloricon*	*Coloricon*
13.	⋆Vervain	*Olphauas*	*Olphanas*	om.
14.	⋆Lemon balm	*Celayos*	*Celeyos*	**Celeivos*
15.	⋆Rose	*Eglerisa*	*Eglerisa*	*Eglerisa*
16.	⋆Bistort	*Cartulin*	*Cartulin*	om.

In addition to plant names from Pernety's dictionary, Sédir also supplies many of the so-called Chaldean* plant names from the *Grand Albert,*

*Throughout this book the appellation Chaldean refers not to the ancient Chaldeans, Assyrians, or Babylonians but more generally to astrologers, magicians, or persons versed in the occult arts. The appellation Chaldean lost its meaning as a reference to a specific ethnicity or land after the fall of the Babylonian Empire but survived as a technical term to describe a societal class of astrologers and astronomers in southern Mesopotamia. Among ancient Greek and Roman authors, any foreign person or group with a flair for the astral could be labeled Chaldean, and this is especially true among medieval writers, who often use the appellations Chaldeans and magi interchangeably. Although Pseudo-Albertus provides Chaldean names for sixteen herbs (see Table 0.1), there is no evidence that any of these names have Semitic origins.

a compendium of the magical properties of herbs, stones, and animals based in part on the authentic writings of Albertus Magnus. The work appears to have begun circulating in manuscript form sometime shortly before or after Albertus's death in 1280. Most manuscript copies of the Pseudo-Albertan grimoire bear the title *Experimenta Alberti* (Experiments of Albertus) or *Secreti Alberti* (Secrets of Albertus), but the earliest printed editions, upon which all other editions and translations are based, bear the title *Liber aggregationis* (Book of Collections). In several places, however, Sédir gives variants that are otherwise unattested in Latin or French editions of the *Grand Albert.* These *hapax legomena*—variants that occur only in this plant guide—appear to be Sédir's own word formations; which is to say, he appears to have deliberately altered the spellings of some of these names (those marked with asterisks in Table 0.1, p. xxiii) to convey their underlying, hidden meaning.

For example, as if to more clearly elucidate the *lily as an "herb of manifestation," Sédir gives its so-called Chaldean name in the form *augoeides.* In Latin editions of the *Grand Albert,* the name of Chaldean herb no. 9 is *augo,* but French editions read *ango.* No Latin or French edition of the *Grand Albert,* nor any herbal tradition derived therefrom, so far as I am aware, gives *augoeides* as a plant name, either for the lily or for any other plant. The ancient Greek adjective αὐγοειδής literally means "light-formed"—a compound of αὐγή, "light (of the Sun)," and εἶδος, "form." The term was employed chiefly by Neoplatonists like Porphyry, Iamblichus, and Proclus to describe the luminous body (σῶμα), soul (ψυχή), spirit (πνεῦμα), or vehicle (ὄχημα), and it is now used by modern occultists in the form of the neuter substantive αὐγοειδές, meaning "luminous body" or "luminous vehicle," to refer to the Higher Genius or Holy Guardian Angel. Papus similarly equated the luminous body with the Egyptian *khu,* the part of the spirit or soul that survives the body after death.

> The luminous body, which presides over the vital functions and is at the same time the immortal spirit's *means of manifestation* within the order of universal life, is no different from the light of life that circulates in the interzodiacal spaces. It is composed of the same ethereal matter that constitutes the inbreathing and outbreathing of all that exists.[19]

At any rate, Sédir's inclusion of *augoeides* as a plant name suggests that his choice of variant for each of the Chaldean herbs, and especially when the formation given is otherwise unattested in Latin or French editions, is both intentional and meaningful. For this reason, all of Sédir's idiosyncratic spellings of barbarous plant names have been retained. Table 0.1 (p. xxiii) provides variant names for the Chaldean herbs from two of the most important editions of the *Grand Albert,* the first from a Latin incunabulum housed in the Biblioteca Nazionale Centrale in Florence, dated circa 1493, the second from the French edition of 1703, in juxtaposition to Sédir's variants from 1902.

Few save Agrippa have recognized the true antiquity of the chapter on herbs in the *Grand Albert.* When Agrippa says that Pseudo-Albertus "follows Hermes" in assigning particular herbs to the planets, this is not a whimsical inference, as some have claimed.[20] Each of these planetary correspondences has its origin in a late-antique tractate titled *On Plants of the Seven Planets,* which presents itself as an epistolary discourse from Hermes Trismegistus to his disciple Asclepius.* A comparison of the Hermetic and Pseudo-Albertan attributions reveals, however, that some of the plant names in the *Grand Albert* are corruptions. For example, whereas Hermes identifies planetary herb no. 6—the herb of Jupiter—as sugarcane (*Saccharum officinarum* L.), Pseudo-Albertus identifies it as ⋆henbane. Surely this is the result of an error in the manuscript tradition, for the ostensibly Latin plant name *acharonia* is quite obviously a corruption of the Greek plant name *sancharōnion.*† Why exactly a later scribe chose to interpret *acharonia* as henbane remains unclear, but in all probability Pseudo-Albertus inherited this odd interpretation from

*This tractate forms part of a much larger collection of "technical" Hermetica. Whereas the much better known collection of "philosophical" Hermetica, the *Corpus Hermeticum,* elucidates the way (ὁδός) of Hermes, the "technical" Hermetica expound upon the art (τέχνη) of Hermes. The tractate *On Plants of the Seven Planets* exists in two recensions, one short and one long: the short recension, upon which the materials concerning the planetary herbs in the *Grand Albert* are based, consists of a single set of planetary herbs, which Pseudo-Albertus rearranges according to the days of the week, from Saturday to Friday (♄, ☉, ☾, ♂, ☿, ♃, ♀), as in Table 0.2 (p. xvi); the long recension, which is certainly more ancient, contains two sets of planetary herbs in the "horoscopic order" of ancient Greek astrologers (☉, ☾, ♄, ♃, ♂, ♀, ☿).

†Here, for example, one Greek manuscript of the short recension reads ἀλχαράνιος. The change from sugarcane to henbane was possibly incited by the name *saccanaron,* which is said to be the magian name for henbane in some manuscripts of the Apuleian herbal (see Pseudo-Apuleius, *Herbarium* 4).

a Latin translation of the Hermetic tractate, and this would appear to be the case as well with planetary herb no. 3—the herb of the Moon.

TABLE 0.2. THE ARCHETYPAL PLANETARY HERBS

	Planet	Hermes Trismegistus	Pseudo-Albertus (1493)	Papus (1893)
1.	Saturn (♄)	**Asphodel** Greek: ἀσφόδελος	**Asphodel** Latin: *affodilius*	**Black hellebore** French: *ellébore, rose de Noël, offoditius*
2.	Sun (☉)			**a. Heliotrope** French: *héliotrope*
		Knotweed Greek: πολύγονον	**Knotweed** Latin: *poligonia, corrigiola, alchone*	**b. Knotweed** French: *renouée, traînasse, herbe à couchons*
3.	Moon (☾)			**a. Water lily** French: *nénuphar, nénuphar blanc, lis d'eau*
		Whiterose Greek: κυνόσβατος	**Lily** Latin: *chrynostates*	**b. Madonna lily** French: *lis blanc, chinostares, chynostates*
4.	Mars (♂)	**Plantain** Greek: ἀρνόγλωσσον	**Plantain** Latin: *arnoglossus*	**Spurge** French: *euphorbe, ornoglosse (langue d'oiseau)*
5.	Mercury (☿)	**Cinquefoil** Greek: πεντάφυλλον	**Cinquefoil** Latin: *pentaphilon, pentadactilus, calipentalo*	**Cinquefoil** French: *quintefeuille, pe<nta>dactilius, pentafilon*

TABLE 0.2. THE ARCHETYPAL PLANETARY HERBS (*cont.*)

	Planet	Hermes Trismegistus	Pseudo-Albertus (1493)	Papus (1893)
6.	Jupiter (♃)			**a. Alkanet** French: *buglosse*
		Sugarcane Greek: σαγχαρώνιον	**Henbane** Latin: *acharonia, jusquiamus*	**b. Henbane** French: *jusquiame, octharan*
7.	Venus (♀)	**Vervain** Greek: περιστερέων	**Vervain** Latin: *p<er>istereon, hyerobotani, columbaria, verbena*	**Vervain** French: *verveine, p<er>isterion, columbaire*

In 1893, Papus introduced a number of innovations into the Pseudo-Albertan tradition.[21] For example, it was likely for the reason that henbane can hardly be characterized as an archetypal herb of Jupiter that Papus demoted it and gave pride of place to ⋆alkanet. But the same problems of transmission are evident in Papus's disquisition on magical herbs, which relies heavily on French editions of the *Grand Albert.* For example, the plant name *offoditius,* which appears in the Lyon edition of 1800, is undoubtedly a corruption of *affodilius* (⋆asphodel), but Papus identifies planetary herb no. 1—the herb of Saturn—as ⋆black hellebore. Table 0.2 presents a full comparison of the Hermetic, Pseudo-Albertan, and Papusian traditions of archetypal planetary herbs. Sédir almost never diverges from the innovations of Papus, however, and this fact is borne out by his signatures. For example, he places ⋆alkanet under the exclusive rulership of Jupiter and ⋆henbane under the dual rulership of Saturn and Jupiter. Similarly, he places ⋆heliotrope under the exclusive rulership of the Sun and ⋆knotweed under the dual rulership of Jupiter and the Sun.

Sédir's discussion of the occult properties of ⋆spurge, however, is uncharacteristically brief, and this is likely due to the glaring disparity between his primary and secondary sources. Papus understood planetary herb no. 4—the herb of Mars—to be spurge rather than ⋆plantain. The properties Papus ascribes to spurge, however, are virtually identical to those that the *Grand*

Albert ascribes to plantain, but whereas Latin and French editions call the herb *arnoglossa* and *arnoglosse,* both derivatives of ἀρνόγλωσσον, the Greek name for plantain, meaning "lamb's-tongue," Papus calls it *ornoglosse.* This uncommon name would appear to derive from the Greek ὄρνις or ὄρνεον, meaning "bird," and γλῶσσα, meaning "tongue," but in this case the correct compound formations should be ὀρνιθόγλωσσον or ὀρνεόγλωσσον, neither of which is attested in Greek, and *ornithoglosse* or *orneoglosse* in French. In the writings of the old herbalists, however, one frequently encounters *orneoglosson, orneoglossum, ornithoglosson,* and *ornithoglossum* as names for the fruit or seed of the ⋆ash tree. Sédir no doubt recognized Papus's revision to be somewhat problematic, and this is likely the reason why he chose not to rehash the medico-magical properties described in the *Grand Albert* in his entry for spurge. Nonetheless, he places plantain under the exclusive rulership of the Sun and adopts Papus's astral signatures for spurge. These signatures are the same ones Hermes Trismegistus and Pseudo-Albertus assign to plantain according to the doctrine of planetary domiciles, Mars being "in domicile" in Aries and in Scorpio; only Sédir divides these zodiacal signatures among different species of spurge.[22] In any case, Sédir's original placements of materials from the *Grand Albert* have been left as is despite any errors of transmission in this two-millennium-old herbal tradition.

Elemental and Astral Signatures

Sédir compiled zodiacal and planetary signatures from a wide array of sources. A number of his *signa* can be traced back to Pseudo-Albertus, Heinrich Cornelius Agrippa, Nicholas Culpeper, Lazare Lenain, and Papus, but just as many, if not more, are Sédir's own determinations. Readers will likely notice that Sédir's signatures are sometimes markedly different from those supplied by other writers (and English authors in particular). It is important to note, however, as Jean Mavéric points out in *Hermetic Herbalism,* "that the particular nature of a given plant is rarely analogous to that of a single planet and, moreover, that each plant participates in the natures of all of the planets combined, but in varying proportions."[23]* To be sure, one signature can

*Although it may seem at first that one author must be right and another must be wrong, especially when two variant signatures appear to be hopelessly at odds, usually this is the result of two authors focusing on different parts (such as the roots, the stems, the leaves, or the flowers),

be more "correct"—which is to say, more predominant—than another with respect to a given plant as a whole, and for this reason Sédir often includes one signature for a plant or tree and another for its flower, fruit, or root. But more often Sédir includes multiple planetary and zodiacal signatures for a single plant without such specifications. In such cases, the primary signature is always listed as the first *signum* in a series; all subsequent *signa* are secondary—which is to say, less predominant—signatures.

Readers who carefully study Sédir's book will be able to make their own informed determinations of the astral properties of any given plant, and it is this practical aspect of Sédir's book that started a renaissance in the study of occult herbalism in early twentieth-century France. The book's primary purpose, as Sédir states in the introduction, was to incite further study. In this respect, it was a remarkable success. Studies of the type Sédir had hoped for began to emerge less than a decade after its publication. These include Pierre Piobb's *Formulaire de haute magie* (1907), Marc Mario's "La flore mystérieuse," published serially in *La vie mystérieuse* (1910–1911), Jean Mavéric's *La médecine hermétique des plantes* (1911), and François Jollivet-Castelot's *La médecine spagyrique* (1912). Each of these authors adopted and expanded Sédir's system of plant signatures, and their works help flesh out the signatory lacunae in Sédir's dictionary. The French tradition of occult herbalism, in which Sédir stands out as the preeminent figure, is extraordinarily rich. It is at the same time highly innovative and firmly grounded in the Hermetic tradition.

In places where Sédir provides no planetary or zodiacal signatures, signatures have been supplemented with recourse to the writings of his occult colleagues, primarily Piobb, Mario, Mavéric, and Jollivet-Castelot. These supplemental signatures are always followed by parenthetical citations, each conspicuously marked with the Latin preposition *apud* (meaning "at," "by," "among") and followed by the name of one or more authors. *Apud* is a scholarly convention meaning "in the work(s) of" or "according to" and is used to

(*cont.*) different attributes (such as flower color or leaf morphology), or different actions (be they medicinal or magical) of the plant in question. Few authors, however, are as explicit as Sédir in describing the mechanics behind their systems of herbal correspondences, and so it can often be quite difficult to determine the underlying reasoning of some authors' plant signatures. At any rate, contemplating and reconciling variant signatures is just one more way readers may extend ad infinitum Sédir's system of herbal mysticism.

indicate a secondhand reference or source. In places where no signatures are to be found in the writings of Sédir's colleagues, it was necessary to look to his primary sources, such as Agrippa, Culpeper,* Lenain, or Haatan, to fill in the blanks. These supplemental signatures are cited in the same manner as the preceding. Often Sédir's predecessors and contemporaries agree, in which case readers will find citations with a mixture of both. For example, Sédir does not supply a ruling planet for ★lavender in the original French publication, but here it is identified as a Solar plant, following Agrippa, Lenain, and Piobb.

Whereas most encyclopedic works on magical herbalism catalog each plant's corresponding element, Sédir instead enumerates each plant's elemental qualities. The four elemental qualities or modes—hot, wet, cold, dry—are the building blocks of the four elements:

- AIR (🜁) is a commixture of the hot and the wet.
- FIRE (△) is a commixture of the hot and the dry.
- EARTH (🜃) is a commixture of the cold and the dry.
- WATER (▽) is a commixture of the cold and the wet.

To aid readers, I supply the corresponding elemental symbol after each equation of elemental qualities. For example, ★dill is hot and wet (🜁); ★rosemary is hot and dry (△); ★valerian is cold and dry (🜃); and ★chickweed is cold and wet (▽).

From these same equations readers may also determine the plant's binary signature, or "gender"—which is to say, its vibration, polarity, charge, or temperament. In Sédir's system, if the initial component of the equation of ele-

*A word of caution regarding editions of Culpeper's herbal: Nicholas Culpeper published *The English Physitian* first in 1652 and then in a revised and expanded edition in 1653 under the title *The English Physitian Enlarged.* In the latter edition, he rails against the number of misprints and errors in the first edition. A spate of editions have appeared since the early 1800s under the title *Culpeper's Complete Herbal,* by which the work is now commonly known. Those arranged alphabetically starting with amara dulcis (*Solanum dulcamara* L.), such as the augmented and illustrated 1789 edition by Ebenezer Sibly, are more or less trustworthy (these, for the most part, are based on the enlarged edition of 1653), but those arranged alphabetically starting with aconite (genus *Aconitum*), such as the edition published by Richard Evans of London in 1814, do not depend directly on Culpeper's originals. For readers interested in Culpeper's signatures, the so-called aconite editions are best avoided.

mental qualities is *hot,* then the plant's gender is masculine, and if the initial component is *cold,* then the plant's gender is feminine. The old herbalists employ the terms *hot* and *cold,* and a number of them catalog each plant according to its specific degree of hotness or coldness, each of the qualities having been divided into four degrees. Readers, of course, may translate these terms into whichever dualistic jargon they prefer, be it masculine-feminine, positive-negative, electric-magnetic, or, to use the Agrippan terminology, active-passive. As with many other aspects of magical herbalism, the subject of gender is grossly oversimplified in most modern publications.*

TABLE 0.3. ELEMENTAL QUALITIES OF THE SIGNS OF THE ZODIAC

	Zodiacal Sign	Elemental Qualities	Element	Gender (charge)
1.	Aries (♈)	Hot and dry	FIRE (△)	Masculine (+)
2.	Taurus (♉)	Cold and dry	EARTH (🜃)	Feminine (–)
3.	Gemini (♊)	Moderately hot and wet	AIR (🜁)	Masculine (+)
4.	Cancer (♋)	Cold and wet	WATER (▽)	Feminine (–)
5.	Leo (♌)	Hot and dry	FIRE (△)	Masculine (+)
6.	Virgo (♍)	Cold and dry	EARTH (🜃)	Feminine (–)
7.	Libra (♎)	Hot and wet	AIR (🜁)	Masculine (+)
8.	Scorpio (♏)	Cold and wet	WATER (▽)	Feminine (–)
9.	Sagittarius (♐)	Hot and dry	FIRE (△)	Masculine (+)
10.	Capricorn (♑)	Cold and dry	EARTH (🜃)	Feminine (–)
11.	Aquarius (♒)	Moderately hot and wet	AIR (🜁)	Masculine (+)
12.	Pisces (♓)	Cold and wet	WATER (▽)	Feminine (–)

Sédir determines each plant's elemental qualities primarily on the basis of its corresponding zodiacal signature. Table 0.3 provides the elemental

*See further Sédir's discussions of Karl Reichenbach's odic polarities in chapter 2 and Louis Claude de Saint-Martin's theory of binary signatures in chapter 3.

qualities of each of the zodiacal signs along with their equivalent element and gender. For example, ⋆marjoram is hot and dry (△) because it is signed by Aries (♈); ⋆mistletoe is cold and dry (🜃) because it is signed by Taurus (♉); ⋆caraway is hot and wet (🜁) because it is signed by Gemini (♊); ⋆lungwort is cold and wet (▽) because it is signed by Cancer (♋)—and so on. When there are multiple zodiacal signatures, the plant's elemental qualities derive from its primary or predominant zodiacal signature, which, as I have said, is always listed as the first *signum* in each series. There are a couple of exceptions to this rule, however, which Sédir never explains in his book but which must have formed part of his oral teaching at the École hermétique.

⋆Belladonna, for example, bears the astral signatures Saturn (♄) and Scorpio (♏). In this case, the plant's elemental qualities derive from the zodiacal signature ♏. Its corresponding element is WATER (▽)—the sum of the elemental qualities cold and wet—and its gender is feminine (–) because of its elemental quality of coldness. The ⋆spring onion, on the other hand, bears the astral signatures Mars (♂) and Scorpio (♏), but its elemental qualities are hot and dry (△), even though it bears the same zodiacal signature as belladonna. *When the zodiacal signature of a plant is the domicile (or one of the domiciles) of its ruling planet, which is one of the two malefic planets (Saturn or Mars), the plant's elemental qualities derive from its planetary ruler rather than from its zodiacal signature.* Since Scorpio is the domicile of Mars, the elemental qualities of Mars (hot and dry) override the elemental qualities of Scorpio (cold and wet). According to the doctrine of planetary domiciles, each planetary ruler has a much greater sphere of influence when it appears in the sign(s) over which it rules. In astrology, this phenomenon is known as domal dignity. Table 0.4 tabulates the elemental qualities and corresponding domicile(s) of each of the planets.

It is important to note, however, that the two traditional benefic planets (Jupiter and Venus), at least in Sédir's system, do not impact the elemental compositions of plants in the same way. Although their degree of influence may be greater when in domicile, in their role as parents they are less domineering, as it were. Nor does this aspect of Sédir's system impact the elemental compositions of plants governed by the luminaries (the Sun and the Moon) because each luminary has the same elemental qualities as its domicile. Finally, because Mercury is a mixture of all four elemental qualities, the elemental qualities of plants do not change as a result of a ruling Mercury domiciled in Gemini or Virgo. This is all to say that readers need pay special attention only to the elemental qualities of plants governed by Saturn or Mars.

TABLE 0.4. ELEMENTAL QUALITIES AND DOMICILES OF THE PLANETS

Planet	Elemental Qualities	Domicile(s)	
Saturn (♄)	Severely cold and dry (🜃)	10. Capricorn (♑)	11. Aquarius (♒)
Jupiter (♃)	Moderately hot and wet (🜁)	9. Sagittarius (♐)	12. Pisces (♓)
Mars (♂)	Severely hot and dry (🜂)	1. Aries (♈)	8. Scorpio (♏)
Sun (☉)	Hot and dry (🜂)	5. Leo (♌)	
Venus (♀)	Hot and wet (🜁)	2. Taurus (♉)	7. Libra (♎)
Mercury (☿)	Hot, wet, cold, and dry	3. Gemini (♊)	6. Virgo (♍)
Moon (☾)	Cold and wet (🜄)	4. Cancer (♋)	

Sédir hardly ever breaks these rules. The few exceptions concern, once more, plants ruled by Saturn or Mars. Because Saturn is *severely cold,* its excessive coldness can sometimes cancel out the hot quality of a plant's corresponding zodiacal signature, such as the ★horned poppy, for example, which is signed by Saturn (♄) and Gemini (♊) but *cold* and wet (🜄) instead of being hot and wet (🜁). Similarly, because Mars is *severely hot,* its excessive hotness occasionally overpowers the cold quality of a plant's zodiacal signature, such as the "wholly" impotable ★wormwood, for example, which is signed by Mars (♂) and Capricorn (♑) but *hot* and dry (🜂) instead of being cold and dry (🜃).

For the benefit of readers, this edition also includes a concordance of elemental and astral signatures, in which all of the plants in Sédir's dictionary are grouped by elemental composition, ruling planet, and zodiacal signature.

Occult Properties

Sédir draws no real distinction between the medicinal and the magical. Indeed, his dictionary of *magical* plants contains just as much, if not more, information on the *medicinal* properties of plants. For Sédir, as for Paracelsus before him, the medicinal and magical properties of plants are inextricably intertwined. Readers familiar with various ancient and modern definitions of magic will likely not find this conflation problematic. The lexical category occult properties now includes, at a minimum, a list of herbal actions for each plant, whereas in the original French publication Sédir does not always supply this information. These actions have been culled primarily from the botanical sources Sédir recommends in his original bibliography. This edition also includes a glossary for readers who may be unfamiliar with the terminology of herbal actions.

Lastly, some of Sédir's discussions of the occult or medico-magical properties of plants are derivative. In places where it has been possible to track down his sources, the vast majority of which he does not cite, I have used these same sources to flesh out incomplete or cursory entries. Again, all supplements to Sédir's text are clearly identified as such and fully cited in the endnotes.

⊙ IN ♉, ☾ IN ♋,
XWΔ𐌇·X٩Φ ISLAND,
R. BAILEY

R. BAILEY, Ph.D., is editor and translator of Jean Mavéric's *Hermetic Herbalism* (also published by Inner Traditions). His research interests include ancient, medieval, and early modern traditions of herbalism, magic, Hermeticism, and Gnosticism. He is currently working on translations of the Latin herbal of Pseudo-Apuleius and a Greek corpus of "technical" Hermetica.

Introduction

The whole universe is one great work of magic, and the whole plant kingdom is animated by its magical virtue. The reader who takes the title of our book literally might expect it to include a complete exposition of the science of botany. Our ambition, however, is not quite so high, and for good reason.

In botany, as in any discipline, two points of view predominate: an inferior, naturalistic, and analytic point of view and a superior, spiritualistic, and synthetic point of view. Whereas modern science is preoccupied with the former, we have elected to write from the latter because this point of view remains a veritable terra incognita in our modern era and is poorly represented, if represented at all, in modern botanical studies. Surely, someone more qualified will someday come along and present a third point of view, a mean between the extremes of analysis and synthesis that is both comprehensive and holistic.

In short, there are fewer teachings in this work than there are indications for further study. We trust our readers will correct any shortcomings and flesh out any lacunae with supplementary studies.

PART ONE

The Plant Kingdom

To obtain a fair general idea of the nature of the plant kingdom, we must study it first in itself and then in relation to the macrocosmic universe and the microcosmic human being. By following this path, we shall come to understand the first elements of phytogenesis, plant physiology, and plant physiognomy, or plant signatures.

Chapter 1 describes the cosmogonic principles that came to produce the kingdom in question. Chapter 2 contains a study of the vital forces active in plants. Chapter 3 concerns the doctrine of signatures, or science of correspondences, and teaches aspiring occult herbalists how to recognize, by its external features, the qualitative nature of the forces active in any given plant.

1
Phytogenesis

Since we have decided to bring to light only traditional notions on this subject, we shall begin by presenting to the reader only the most authentic ancient teachings.

One of the oldest documents we possess, the Sefer Torah of Moses, preserves the cosmogonic theories of initiates of the secret traditions of the so-called red (Atlantean) and black (Afroasiatic) races.[1] Genesis 1:11 presents the following phytogonic account: "And the Elohim said, 'Let the earth bring forth grass, the herb yielding seed, and the fruit tree yielding fruit after its kind, whose seed is in itself, upon the earth,' and it was so." This act of creation took place on the third day according to the following elemental sequence:

1. FIRE: the first day, the creation of light
2. WATER and AIR: the second day, the fermentation of the waters and their division
3. EARTH: the third day, the formation of Earth and its vegetation
4. FIRE: the fourth day, the formation of the Sun
5. WATER and AIR: the fifth day, the fermentation of the waters and the air, the creation of aquatic and aerial creatures
6. EARTH: the sixth day, the fermentation of Earth, the creation of animals and humans[2]

When the Book of Genesis is considered as a whole and from the cosmogonic point of view, it is the figure of Isaac who, so the initiates teach, is emblematic of the plant kingdom. His nearly consummated sacrifice, his filiation, the names of his parents and sons, and the symbolic acts of his life all provide ample proof of this undeniable analogy. We shall not belabor the matter here, however, so as not to overburden our readers with such arduous symbolism, but its grasp lies within the reach of any conscientious student.*

HERMETIC THEORY

The Hermetic philosophers envisaged at the primordial origin of things a chaos in which all the forms of the universe were prefigured, a cosmic matrix or matter, and a generative, spermatic fire whose reciprocal action constituted the monad—the philosopher's stone (*lapis philosophorum*), or philosophical egg (*ovum philosophicum*)—the means and the end of all forces.

The generative fire is hot and dry, masculine and pure. It is the spirit of God that moved upon the waters, the head of the dragon (*caput draconis*), the Sulfur of the alchemists. The chaos is a spermatic water, hot and wet, feminine and impure, the Mercury of the alchemists. The action of these two principles in heaven constitutes the principle of the good, of light, heat, and generation or life. The action of these two principles on Earth constitutes the principle of evil, of darkness, the cold, and putrefaction or death.

On Earth this pure fire became a formless and void primal matter, the *tohu wa-bohu* of Genesis 1:2, a humid earth, vain and confused, a moon, a mercurial water. Paracelsus refers to this primal matter by the names *limus* (or *limbus*) *terrae, iliaster,* and *Mysterium Magnum.* The pure and celestial water, conversely, became a terrestrial matrix, cold, dry, and passive, the Salt of the alchemists.

Thus, all things in nature pass through the following three ages:

Age 1. In the beginning there is a confrontation between these creative principles. Their interaction produces light, then darkness, and then a confused and mixed matter: this is fermentation, the first age.

*[According to Sédir, Abraham represents mineral life; Isaac, vegetal life; and Jacob, animal life. See further Sédir, *L'enfance du Christ,* 60, n. 1. —*Trans.*]

Age 2. Fermentation leads to a general decomposition, or putrefaction, after which the molecules of the matter in question begin to coordinate according to their subtlety: this is sublimation, or the life of the thing, the second age.

Age 3. Finally, there comes a moment in time when sublimation ceases. A separation is established between the subtle and the gross. The former ascends to the heaven, the latter descends to the earth, and the rest is dispersed into the aerial regions. This is death, the third age.

We have already outlined the succession of the four modalities of the universal substance known as the four elements. Here the elements FIRE, AIR, WATER, and EARTH are all easily recognizable. The student is encouraged to organize all of these notions in a table of correspondences and use the Pythagorean triangle as a hermeneutical key.[3] The same pattern is equally discernable in the Sankhya system of Hindu philosophy and in the sephirotic and tarotic system of Jewish mysticism.

These principles, moreover, act in each of the three worlds—the cosmic, the planetary, and the sublunar in Hermetic terminology—in the following ways:

1. In the first world, the uncreated fire, or spirit of God, fertilizes the subtle, chaotic water, which is the created light or the soul of bodies.
2. In the second world, the chaotic water, which is igneous and contains the Sulfur of Life, fertilizes the median water, which is a viscous, moist, and unctuous vapor or the spirit of bodies.
3. In the third world, the spirit, or elemental FIRE, fertilizes the igneous ether, which is, again, a thick water, a subtle or androgynous earth, a silt or first solid, a fertile mixed body.[4]

Thus, each earthly creature is a by-product of the actions of three grandiose series of forces in the empyrean or intelligible world, in the ethereal world comprised of the zone of the fixed stars and the planets, and in the hylic, or material, world. The cosmic or empyrean world produces the *Anima Mundi,* the *Spiritus Mundi,* and the *Materia Mundi* as well as the viscous vapor and the uncreated universal seed. The planetary or ethereal world engenders the

Sulfur of Life, the intellectual ether or Mercury of Life, and the watery principle or Salt of Life, in addition to the created seed and the second matter of bodies. The sublunar or hylic world generates the elemental FIRE, the elemental AIR, and the elemental WATER—which is the vehicle of life—as well as the receptacle of seeds and the innate seed of bodies.

MANIFESTATIONS OF THE PLANT KINGDOM

For the plant kingdom to manifest, a planet must be sufficiently evolved to crystallize its atoms into solid earth and produce waters and an atmosphere, as the creation account of Genesis clearly illustrates. Under such conditions, a wave of new life descends and serves as the vehicle of first animation on the planet. The plant kingdom as a whole is governed by Venus, and it is therefore a symbol of beauty.* Its representative geometric pattern is the spiral, and this is why occult herbalists use phyllotaxis to measure the degree of vital force within each plant.

Vegetal life results from the reciprocal action of the Solar light and the greed of the inner Sulfur. No plant can grow without the force of the Sun, which each plant attracts by its essential principle. Thomas H. Burgoyne explains the process that led to the evolution from mineral to vegetable as follows: "The atoms of oxygen and hydrogen by a certain combination produce water. In this union both become polarized and form a substance which is the polar opposite of their original inflammable states. From this change of polarity we have clouds, oceans, and rivers."[5]

The heat of the Sun decomposes an infinitely small fraction of these waters into the gaseous state, and the atoms of the water molecules assume a different angle of motion. Before they rotated in a circle, but now they ascend in a spiral. During their ascension, they attract or are attracted by the atoms of carbonic acid, giving rise to a third type of motion—namely, precipitated rotation. They combine, and it is in these new combinations that the germ of physical life is born. Under the impetus of a central atom of fire,

*The lush greenery of vegetation corresponds to the green sea of the foam-born goddess of love (namely, Aphrodite, the Greek equivalent of the Roman Venus), which is fixed upon the surface of Earth.

the predominant forces being oxygen and carbon, this union produces yet another change in polarity, and the atoms become attracted once more to the earth. The waters receive them, and thus form the first vegetative *sphagnum mosses. When these first vegetative forms decay, their atoms resume their ascending spiraliform path and are attracted once more by the atoms of the air. The same process of polarization is then repeated, forming lichens and successively more and more perfect plant species.[6]

As the spirituous essence of the Sun penetrates to the center of Earth by the attraction of each mixed body, it coagulates into an aqueous fire and ascends, wishing to return to its source. But it is retained in the matrices of the various species as it ascends, and because these matrices each have a particular virtue in their species, "in one thing it becomes one thing, and in another it becomes another, always engendering their like. And as the spirituous essence becomes even more subtle, it passes through Earth's surface and makes seeds grow, each according to their kind."[7]

The same theory is presented in a more concise manner in the kabbalistic system of classification known as the Fifty Gates of Understanding. The enumeration of the gates composing the second class (gates 11–20), or "decad of mixed bodies," begins as follows:

Gate 11. Manifestation of minerals by the disjunction of Earth
Gate 12. Organization of flowers and saps for the generation of metals
Gate 13. Secretion of seas, lakes, and flowers through the alveoli of Earth
Gate 14. Formation of herbs and trees or vegetative nature
Gate 15. Evolution of the forces and seeds of each plant species[8]

Lastly, to conclude this brief exposition, we shall summarize Jacob Boehme's theory of the creation of the vegetable kingdom. Its relationship to the two preceding theories will be immediately apparent.

God created plants on the third day by the *fiat of Mars,* which is bitterness and the source of motion. Vegetal life emerged from the flash of fire within this bitterness, when God separated the universal matrix and its igneous form, and because he wished to manifest himself in this exterior and sensible world, the *fiat* that issued forth according to his will called

into vigorous action the aqueous property of the Sulfur of the first matter.*
We already know that the element WATER serves as an attractive matrix and may therefore link Boehme's vegetal cosmogony with the previous theories.

Before the Fall, all plants were united to the inner paradisiacal element. After the Fall, sanctity fled from their roots, which remained buried in the terrestrial elements. Flowers alone, as we shall later demonstrate, are the most perfect vestiges of paradise.

THE STATIC CONSTITUTION OF THE PLANT

Before endeavoring to sketch the rudiments of plant physiology, it would be prudent to consider the principles and forces active in the plant kingdom to better understand their functioning. When we examine the plant from the constitutive point of view, we recognize five distinct principles in action:

1. A matter formed of *vegetative* WATER.
2. A soul formed of *subtle* AIR.
3. A form comprised of *concupiscent* FIRE.
4. A matrix composed of *intellective* EARTH.
5. A universal and primitive QUINTESSENCE, or *indelible mixed body,* formed of the four elements, which determines the four phases of motion: fermentation, putrefaction, formation, and growth.

When we examine the plant from the generative point of view, we discover seven distinct forces in action:

*[*Fiat* is the third person singular present subjunctive of the Latin verb *fieri,* which functions as the passive of *facere,* "to do" or "to make." It is the first word uttered by the Elohim in the Vulgate translation of the first creation account: *Fiat lux,* or "Let there be (made) light" (Genesis 1:3). According to Sédir's dictionary of Boehmic argot, *fiat* refers to "the creative Word that separates forms and acts according to desire and the essential light" (Sédir, *Le bienheureux Jacob Boehme: Le cordonnier-philosophe,* 29). See further Sédir's translation of and commentary on Boehme's *De signatura rerum,* published under the title *De la signature des choses ou l'engendrement et de la définition de tous les êtres,* 25, 67, 82–83, 104, 146, 201; cf. Hartmann, *The Life and Doctrines of Jacob Boehme,* 128–30. —*Trans.*]

1. A matter or patient, formed of light and darkness, or a chaotic and vegetative water: it is here that the Paracelsian *derses* resides, the occult exhalation of the earth by which plants are enabled to grow.[9]
2. A form, an active agent or fire.
3. A link between the two preceding forces.
4. A motion, the result of the action of the agent on the patient. This motion is propagated by the four elements and manifests in the four phases of motion determined by the indelible mixed body.

These preliminary occult actions give rise to three visible outgrowths:

5. The soul of the plant or the corporified seed, which Paracelsus calls *clissus,* the occult power and vital force that passes from the roots to the stem, leaves, flowers, and seeds and causes the latter to produce new vegetal organisms.[10]
6. The organized spirit or mixed body, which Paracelsus calls *leffas*—that is, the primordial juice or astral body of the plant.[11]
7. The corporeal, or physical, body of the plant.

To gain a more comprehensive picture of these two classifications, we encourage students to search for analogies in Greek mythology, which is very expressive and lends considerable scope to meditations on the occult properties of plants.[12]

2
Plant Physiology

Nothing is so simple as the structure of the plant. Its anatomical parts are reducible to three, and it is these three parts that, through successive processes of individualization, form each of the plant's organs.

PLANT ANATOMY

Organ 1. The general mass of the plant is formed by the *cellular tissue,* which may be regarded as the digestive organ of the plant. The *root* is formed by the individualization of the cellular tissue and is therefore a repetition of the cellular tissue in a higher degree. The root functions as the plant's stomach. In the anatomy of the flower, which represents the highest degree of plant metamorphosis, the mature ovule or seed (the embryo or fetus) is a repetition of the root and ultimately of the cells.

Organ 2. The intervals between the ordinarily hexagonal cells form tubes that extend throughout the entire plant and conduct the sap by which the plant is nourished. These *intercellular ducts,* or passages, are to plants what the blood vessels and veins are to animals. The *stem* is formed by the individualization of the veins and is therefore not a sui generis formation but a repetition of the intercellular ducts in a higher degree. The stem functions as the plant's circulatory system. In the flower, the capsule (or female organ) is a repetition of the stem and ultimately of the intercellular ducts.

Organ 3. In the cellular tissue of most plants there are additional tubes

> formed of fibers twisted in a spiral that conduct air* throughout the entire plant. These tubes or *spiral vessels* are to plants what tracheae are to animals. For this reason, Marcello Malpighi, the father of microscopic anatomy, gave them the name *tracheae.* The *leaves* are formed by the individualization of these spiral vessels and are therefore a repetition of the tracheae in a higher degree. The leaves function as the plant's lungs. In the flower, the corolla is a repetition of the leaves and ultimately of the spiral vessels.[1]

From this preliminary sketch we can begin to appreciate the functional relationships among these organs. Papus describes the essential features of these interrelationships as follows:

1. The root, which functions as the plant's *stomach,* plunges into EARTH. It digs in search of alimentary matter, the materials needed to stimulate its growth.
2. The leaves, which function as the plant's *lungs,* reach out into AIR, either the open air or air dissolved in WATER. They seek out the light and gases needed to renew the force expended on the alimentary matter in the intimate recesses of the cellular tissues. This force finds its expression in the "green blood" called chlorophyll, which mediates the primary act of photosynthesis.
3. The stem, which functions as the plant's *circulatory system,* contains the vessels that raise the result of the digestive process in the form of a lacteal juice. This is the *ascending sap,* which is analogous to chyle in humans—a milky bodily fluid consisting of lymph and emulsified fats. Some vessels in the stem lower the result of the action of the lungs, the air absorbed by the leaves, while others lower the action of this air on the nourishing (or ascending) sap. This is the *descending sap.*
4. The flower, which is the seat of the plant's reproductive organs, is the result of a superfluity of force.

*[What early botanists deduced to be "air" is actually water and mineral salts. —*Trans.*]

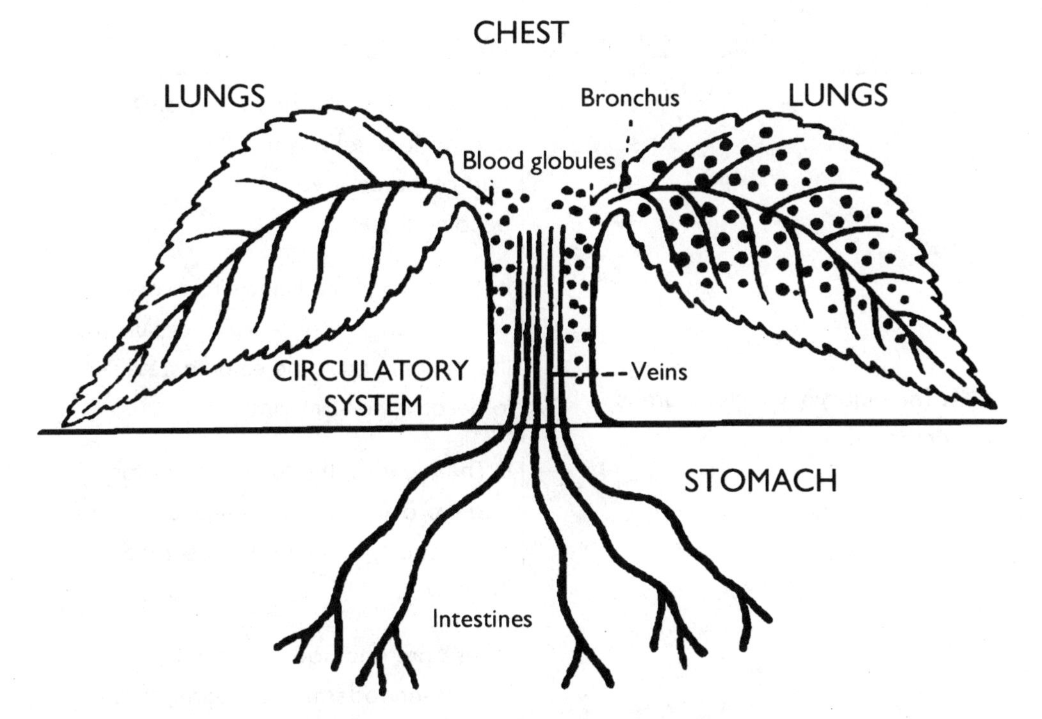

Fig. 2.1. Anatomy of plant organs. From Papus, *Traité méthodique de science occulte*, 262.

In sum, the plant may be understood synthetically as a being who possesses a matter-generating stomach (the root), force-generating lungs (the leaves), and a rudimentary circulatory system (the stem), all of which constitute what may be construed as a chest. The plant's green pigment, chlorophyll, behaves like red blood cells coursing through the veins (intercellular ducts) of the circulatory system. Add to this the reproductive organs in flowering plants (angiosperms), and you now have a complete picture of the plant from the physiological perspective.[2]

We shall return to these organic relationships and study them in greater detail below. In fact, as we shall demonstrate, the art of Hermetic herbalism relies heavily on the knowledge of these functions.

PLANT EMBRYOGENESIS

The seed is that element that gives birth to the new plant. It is composed of two main parts, the embryo and the cotyledons.

TABLE 2.1. ANATOMY OF THE SEED

1. The *embryo,* which in turn is composed of	**a. The *radicle,*** the future digestive and abdominal organs, the lower part of the embryo destined to become the ROOT; **b. The *epicotyl,*** the future circulatory organs and general center of evolution, the median part of the embryo destined to become the STEM; and **c. The *plumule,*** the future respiratory and thoracic organs, the upper part of the embryo destined to become the LEAVES. (These are analogous to the three layers—endoderm, mesoderm, and ectoderm—of the human embryo.)
2. The *cotyledons,* which constitute the material destined to nourish the embryo as it develops.	

The embryological development of the plant consists of three stages, which Papus enumerates as follows:

STAGE 1. The localization of the seed in a suitable matrix (wet earth): Certain conditions must be met before the seed can begin to develop. The seed must be placed in an environment where it will not endure any action harmful to its development. We may call this locus the *vegetable matrix,* the plant's maternal center par excellence. This matrix must contain two essential elements: EARTH and WATER. When the vegetable matrix fulfills these requirements, the nutritive materials in the elements increase in volume until they are assimilated by the seed. The seed awakens and begins to produce heat. The interaction of these elements and qualities in the vegetable matrix incites the formation of the three main parts of the seed-plant embryo (the radicle, the epicotyl, and the plumule).

Stage 2. The development of the embryo at the expense of the cotyledons: The three main parts of the embryo begin to vegetate as they feed on its own reserve materials or cotyledons, which nourish it during the most critical period of its growth. The lower part of the embryo, the radicle, borrows the nutritive elements necessary for its development from the cotyledons and begins to push down toward the earth. The intermediate part of the embryo, the epicotyl, borrows nutritive materials from the cotyledons and pushes up toward the sky, carrying at its summit the plumule, which will evolve into the leaves. The cotyledons gradually move to opposite sides of the new stem, or caulicle. At this stage, the seedling absorbs no nutritive elements from its external environment and lives exclusively on its own reserve materials.

Stage 3. The individualization of the digestive and respiratory systems: The root develops on one side and begins to absorb and digest the nutritive substances contained in the earth. The plumule develops on the other side and begins to absorb and breathe in the products of the sky (light and air). The plant is individualized by its digestive and respiratory functions, and the reserve materials, or cotyledons, now useless, wither and fall off. In other words, the plant is born and is now capable of sustaining itself.[3]

Every seed containing a potential tree encases the Great Mystery, or *Mysterium Magnum*. The development of the seed is a mirror image of the creation of the cosmos. The ★oak tree, for example, begins to manifest as soon as the acorn finds a place in its natural matrix, wet earth. But wet earth by itself is merely a passive matrix and cannot ignite the vital spark or awaken the seminal *ens.** For this to occur, the light and heat of the Sun is a desideratum because it is Solar light and Solar heat that kindle the subterranean "cold fire" of putrefaction. Only then can the seed be stimulated into development and proceed through each of the subsequent stages of its evolution. We shall discuss later, in the chapter on occult horticulture, what happens when the earth matrix is unsuited to the needs of the seed entrusted to it.

*[*Ens* is the present participle of the Latin verb *esse,* "to be," meaning "being" or "essence." It is often defined as "that which is" or "that which has the act of being." —*Trans.*]

PLANT DEVELOPMENT AND MORPHOGENESIS

In the above account of plant embryogenesis, we see the mutual interaction of three *enses,* or dynamisms: the terrestrial *ens,* the seminal *ens,* and the Solar *ens.* Each of these three *enses* consists of a triad of constituent principles: Salt, Sulfur, and Mercury. The magnetic attraction of the first and last *enses* encourages the seed to develop in opposite directions: the terrestrial *ens* incites growth in the descending axis, or root; the Solar *ens* incites growth in the ascending axis, or stem. These two organs, as we already know, will fulfill two contrary but complementary roles over the life span of the plant. The harmony of the three *enses* depends on the healthy state of the root, which should be either thick and multifurcated—divided and forked in multiple directions—or lean and dry, and the healthy state of the stem, which should be either smooth and green or knotty and black.

From the constitutive point of view, the life and magnetic sensibility of the root reside in Mercury. The subterranean Mercury of minerals (*Mercurius mineralium*) is almost always toxic and laden with impurities.[4] It is quite literally in hell—which is to say, it finds no food other than itself and no object beyond itself. But as soon as the Solar vibration reaches it, it devours the Salt in its body and the Sulfur in its mother, and the two become intimately united with its essence. Then Earth opens her womb, the root's atoms obtain a relative freedom, and the homogenous elements of the plastic body—the Salt that languished in a Saturnian torpor—become susceptible to the attraction of the seminal *ens.*

Usually, the base of the growing stem is white, the middle brown, and the top green. The white color at the base is indicative of the stem's sudden tendency to expand as soon as it is freed from the constrictive powers of the root. The brown color of the median part of the stem is indicative of a Saturnian impression resulting from the divine curse.[5] The bark is that part of the plant that remains in limbo. Indeed, although the Great Mystery may be epitomized in trees, the entire plant kingdom was affected, like the whole of creation, by the Fall of Adam and Eve. Nevertheless, more so than in any other living creatures, we can perceive the splendors of paradise in the beauty of their flowers and in the sweetness of their fruits.* Trees are certainly the most perfect specimens of all vegetal beings. They receive influences from the stars, the elements, the *Spiritus Mundi,* and the *Mysterium Magnum,* which itself is both fire and light, anger and love, the pronounced Word of the eternal Father.† Lastly, the green color of the upper part of the stem is indicative of the Mercurial Life of the root, which winds its way upward into the Jupiterian and Venusian components of the foliage.

Plant growth occurs by the mutual emulation of two *enses,* the outer Sun (☉) and the inner Sun, toward the achievement of their purpose, which is to produce a fresh water that will furnish the plant's flowers with the elements of their elegant form and beautiful color. Thus, seven forms of outer nature act upon the plant: Jupiter (♃), Venus (♀), and the Moon (☾) naturally cooperate with the expansive action of its inner Sun. Mars (♂), which is none other than the igneous spirit of Sulfur, accelerates this expansive action, while the Mercurial Life (☿) looms frighteningly before it and Saturn (♄) congeals and corporifies its fear. It is in this manner that the stem produces nodes.

Branches are the result of the struggle between these natural forces as they try to maintain communication with the outer Sun. They are, if you

*[Sédir's conception is wholly Boehmic. "Furthermore," Boehme writes, "we see the *Mysterium Magnum,* the Great Mystery, in trees. Though indeed they are different and mixed, we may still discern their *paradisiacal* form, for they bear their fruit upon branches, and the fruit is something different from the tree: the tree is bitter, but the fruit is sweet. And we argue that the trees and fruits we have nowadays would be *paradisiacal* if the curse did not continue to enter into them. Paradise fled from them, and *now* all fruit is *but such* as the apple was from which Eve had eaten of death. Furthermore, you must realize that the kingdom of anger also *pressed* into the Garden of Eden, which brought forth a tree that bore such fruit as all the trees we feed upon nowadays produce. . . . But when the earth was cursed, the *curse* entered into all fruits" (Boehme, *The High and Deep Searching Out of the Three-fold Life of Man,* chap. 9.14–16, 130 [translation slightly revised]). *—Trans.*]

†[Boehme's conception of the *Mysterium Magnum* is intimately connected to his notion of *fiat:* "But we must understand that in the *Verbum fiat,* in the Word *fiat,* the *Mysterium Magnum* is compacted or conceived into a *substance,* namely, out of the inward spiritual substance into a palpable one, and in the palpability lies the science or root of life." Thus these two, Boehme later elaborates—namely, the Word and the *Mysterium Magnum,* the Great Mystery—are in one another as soul and body, "for the *Mysterium Magnum* is the substantiality of the Word, wherein and wherewith the invisible God in his Trinity is manifested and becomes manifested from eternity in eternity, for what the *Word* is in power and sound, of that the *Mysterium Magnum* is a *substance,* the eternal substantial Word of God" (Boehme, *Concerning the Election of Grace,* chaps. 4.27 and 8.61, 31, 90 [translation slightly revised]). *—Trans.*]

will, like the gesticulations of a plant that feels oppressed and desires to climax by its own free will. Just as the vital force in humans brings out inner venoms in the form of boils, so the vital heat of plants pushes out buds, and this occurs especially in the spring when the call of the outer *ens* is at its most urgent. In other words, the Salt encircled by Saturn—which is to say, the corporified fear of the Mercurial Life—heats up in a desperate struggle and eventually transforms into a Sulfur. This Sulfur breathes new life into its son Mercury, which radiates until Venus supplies it with the plastic substance of buds and twigs.

Inflorescence occurs when the Sun gradually overcomes the excesses of Mars and the bitterness of the plant diminishes. Jupiter and Venus exhaust their activity and dissolve in the matrix of the Moon, and the two *enses* unite so that the inner Sun, the vital force of the plant, recovers its principle, passes to a state of Sulfur, and returns to a regime of divine freedom.* In paradise, the seven forms of this same regime were inversed, inwardly and upwardly, and collaborated in a symphony of complete harmony: the image of eternity was formed in time, the Sulfur of the plant returned to the latent state, the Salt transmuted itself, and the reign of the son was inaugurated by a paradisiacal joy, which the Edenic flowers exhaled as a perfume.† This paradisiacal perfume is what Paracelsus calls tincture. After the Fall, as we have said, the vegetal paradise came to an end and entered into the darkness of the seed, where the two Suns now come to hide.

The occult spirit of the elements governs the process of infructescence. Fruits possess both good and bad qualities corresponding to the kingdoms of anger and love. They are not entirely under the kingdom of anger because the unique Word, which is everywhere immortal and imputrescible, even in the subterranean putrefaction of the seed, still reverberates within them: the Word grasps the earth, but the earth has not grasped the Word. And so we are left with the triumph of the kingdom of love—which is to say, with the

*There are two modes of inflorescence: *indeterminate* or *racemose inflorescence,* where the growth commences from the center (as exemplified by the ★lily and the ★rose), which symbolizes spiritual development, and *determinate* or *cymose inflorescence,* where the growth commences from the circumference, which symbolizes material development.

†It is in this very same manner that the bodies of saints emit the exquisite fragrance known as the odor of sanctity.

process of the generative organs of the plant. When infructescence or fructification occurs, the *ens* is transported to the fruit buds, where it amasses a great number of plastic elements, or "moons," which the heat of the outer Sun transforms into Venusian components. In this manner the pulp of the fruit develops around a center, the son of the inner Sun. Thus, all seven planets manifest within fruits in varying proportions and determine their distinct flavors, until Saturn returns them to the earth from whence they came.

We normally give fruits the qualification mature (French: *mûr*) to describe the moment of their perfection, but the period when their juice is sweetest is poorly designated by this term, which indicates, on the contrary, their state of agony. English *ripe,* German *reif,* and Dutch *ryp,* the latter being the metathesis of French *pur,* are much more appropriate and expressive.* Maturation is the result of a kind of vertigo that the Sun or the *ens* causes in the paternal principle of the Sulfur, which precipitates it from eternal into temporal life. Later on, we shall give some indication of the significations of the various fruit flavors.

We deliberately made this rapid sketch using Jacob Boehme's esoteric nomenclature. With recourse to naturalistic or Ionian Buddhist theory, however, we can explain vegetal creation much more simply as the result of the interactions of three primary forces:

1. The *force of expansion:* light or sweetness (typified by Abel)
2. The *force of contraction:* darkness or roughness (typified by Cain)
3. The *force of rotation:* anguish or bitterness (typified by Seth)

Each of these forces operates in the process of plant development and morphogenesis. When the seed is situated in the earth, its sweetness flees from the darkness and anguish that pursue it: this incites the growth of the plant. Once the seed is subjected to the heat of the Sun, the struggle between these three forces intensifies. The forces of contraction and rotation are exalted and overwhelm the force of expansion: this causes the production

*We should add that even though the word *pure* (French: *pur*) designates something healthy or wholesome (French: *sain*), it also designates the purulent effusion (French: *sanie*) of compost (French: *purin*). [Cf. Lenglet-Mortier and Vandamme, *Nouvelles et véritables étymologies médicales tirées du gaulois,* 181–84. —*Trans.*]

of bark and nodes. But the force of expansion, when left the briefest respite by its adversarial forces, spreads out in all directions and pushes out buds and branches, inscribing itself in the color green, and gives itself over to the vivifying forces of the Sun, which carry it toward perfection into the flowers. The force of contraction unites the various organs of the plant, which anguish had divided into parts, into a homogenous whole. They begin to work in tandem because, coming as they do from below, they must obey the Solar forces that come from above. Thus, the fruit forms and continues to develop until the force of expansion is depleted, at which point the fruit falls off and gives birth to a new vital *circulus,* or cycle.

PLANT PSYCHOLOGY

Karl Reichenbach's discovery proves that everything in nature emits a sort of exhalation that is invisible to the human eye under ordinary conditions but clearly visible to sensitives. This radiation, which Reichenbach calls the *od,* or odic force, varies in color, intensity, and quality.[6]

As a general rule, a positive *od* predominates in the ascending axis, or stem, and a negative *od* predominates in the descending axis, or root. This general rule remains true no matter which part of the plant is presented to sensitives for examination. Hence, fruits and flowers are odically positive, whereas tubers are odically negative. With respect to flowers and fruits, however, the rachis, the major axis holding the products of inflorescence and infructescence, is odically positive, whereas the peduncle, the stem holding the whole inflorescence and infructescence, is odically negative.[7] Reichenbach's odic polarities are still in vogue among the successors of Count Cesare Mattei, the father of electrohomeopathy, but we personally do not believe that Mattei's "vegetable energies" are especially meaningful.[8]

In any case, that plants possess a soul is indisputable. Without referencing Brahmanic, Buddhist, Taoist, Egyptian, Platonic, or Pythagorean doctrines, all of which are more or less endowed with the initiatic spirit, let us recall that philosophers like Democritus, Anaxagoras, and Empedocles also supported this thesis. Pseudo-Aristotle says in his treatise *On Plants* that these pre-Socratics believed plants "moved by desire" (ἐπιθυμίᾳ . . . κινεῖσθαι), "experienced sensations and felt pleasure and pain" (αἰσθάνεσθαί τε καὶ λυπεῖσθαι ἥδεσθαι διαβεβαιοῦνται), and "possessed both mind and cognition" (καὶ νοῦν καὶ γνῶσιν . . . ἔχειν), and according to Plutarch, they even characterized plants as "animals stuck in the earth" (ζῷον . . . ἔγγειον τὸ φυτὸν εἶναι).[9]

In a wonderful book titled *The Soul of Plants,* Arnold Boscowitz documents the growing number of naturalists and botanists who similarly came to attribute individuality and personhood to plants.[10] For example, to cite just a few of the many proponents of this view, Erasmus Darwin describes plants as animated beings who possess powers of volition.[11] John Percival, too, claims that the movements of plant roots are voluntary. Gerardus Vrolik, Johann Hedwig, Charles Bonnet, Friedrich Ludwig, and James Edward Smith all maintain that plants experience sensations—even sensations of happiness and sadness—and feel both pleasure and pain. Karl Friedrich Philipp von Martius independently draws the very same conclusions in several of his works.[12] Finally, Gustav Theodor Fechner devotes an entire monograph to the subject of the "soul-life" (*Sellenleben*) of these "earth-stuck animals."[13]

Plants exhibit several characteristics indicative of beings endowed with personality:

1. The act of "breathing" is carried out by the contraction and expansion of the spiral-wound cellular tubes that Malpighi dubbed tracheae.
2. Air is essential to their vitality, and its effect on the sap is analogous to its effect on human blood (so Denis Papin, Jean-Louis Calandrini, Henri-Louis Duhamel du Monceau, and Pierre Bertholon).
3. The epidermis of the lower surface of their leaves is studded with stomata, which serve as the plant's organs of respiration (so Jan Ingenhousz, Stephen Hales, Nicolas-Théodore de Saussure, Hugo von Mohl, and Lazare Garreau).
4. They keep oxygen in the air and exhale carbonic acid (so Lazare Garreau, Hugo von Mohl, and Julius von Sachs).
5. They feed on carbon, which they extract from carbonic acid, and consequently exhale large quantities of oxygen during the daytime.
6. Roots serve as their stomachs, just as leaves serve as their lungs, and their sap is analogous to chyle in humans.
7. The nutritive function of plants is so active that an ⋆oak, for example,

absorbs approximately 600,000 pounds of food over the course of one hundred years (so Richard Bradley).[14]

8. Almost all vegetal excretions are vivifying substances for humans, much like the excretions of animals are for plants.
9. Although some take exception to the analogy between the circulation of sap in plants and the circulation of blood in humans, we know for a fact that plants transpire very strongly.[15]

How else can we explain the movements of plants, their spiraliform nutations, their quest in search of the Sun, their zest for light and sustenance, their lust for propitious earth? How else can we explain their amorous power or the emissions of heat and energy they give off at the moment of fertilization? From where come the magical resuscitative properties of the resurrection lily (*Lycoris squamigera* Maxim.) and the ⋆rose of Jericho? Modern initiates who observe all these phenomena cannot but admire the ingenious wisdom and penetrating intuition of their ancient ancestors, who assigned each tree its hamadryad, each flower its fairy, and each blade of grass its genius. But what are we to make of all these scientific observations? Do they not evince the overt movements of elemental souls striving toward consciousness?

The ingenious Genevan naturalist Charles Bonnet devotes a large part of his masterpiece *The Contemplation of Nature* to such parallels between plants and animals. From these numerous parallels he draws the following conclusion:

> Nature descends by degrees from the human being to the polyp, from the polyp to the sensitive plant, and from the sensitive plant to the truffle. The superior species always adhere by some character to the inferior species, and the latter to species more inferior still. . . . *Organized* matter has received an almost infinite number of variations, and all are nuanced like the colors of the prism. We make dots on the image, trace some lines over it, and call this making genera and classes. But we perceive only the dominant hues, and the delicate nuances escape us. Plants and animals are mere variations of organized matter. They all participate in the same essence, yet the distinctive attribute is unknown to us.[16]

TABLE 2.2. THE FOUR KINGDOMS

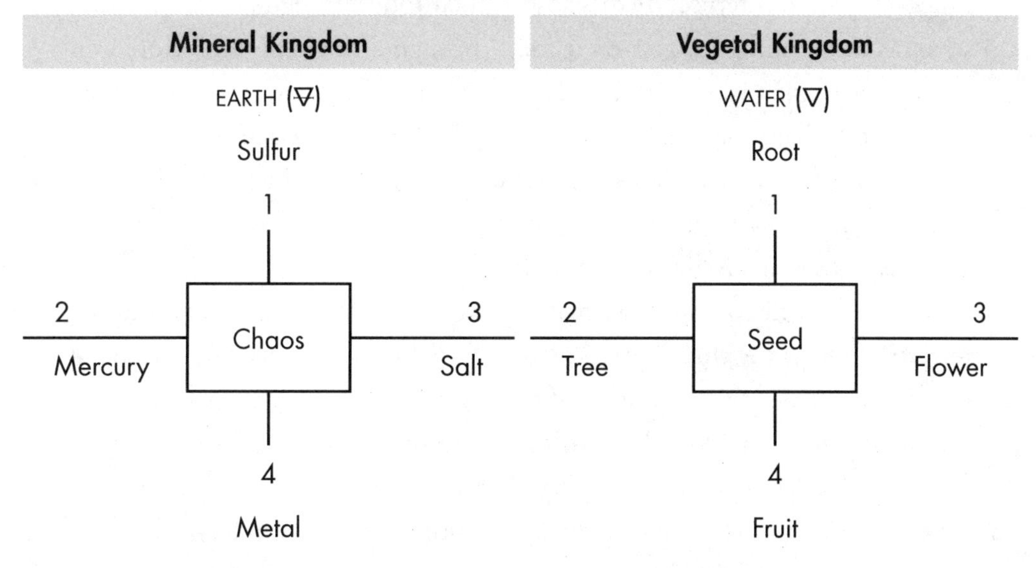

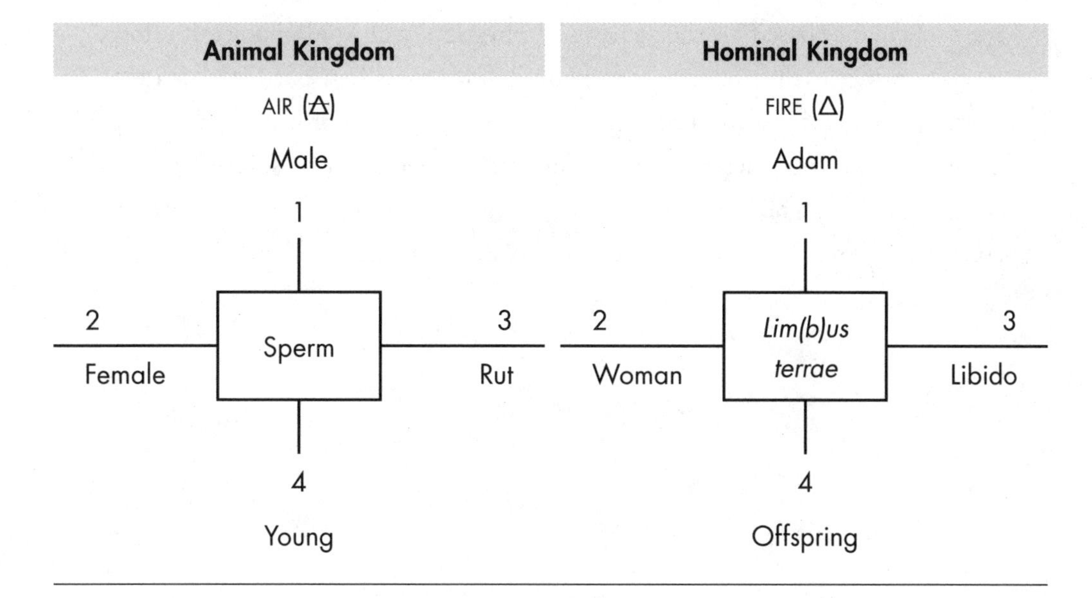

Plants, like animals, vegetate, feed, grow, and multiply, but there are some notable differences between denizens of the plant and animal kingdoms. For example, the majority of plants are hermaphroditic and produce cosexual flowers, whereas most animal species are unisexual. Plants produce seeds in much greater quantities than animals produce eggs or fertilized ova, with the exception, of course, of some inferior species. Similarly, an organism in the former kingdom produces many more buds than an organism in the latter kingdom produces fetuses. Plants absorb their food through porous surfaces, whereas most animals ingest their food through a single mouth. Moreover, plant alimentation is ongoing and executed by external roots, whereas animal alimentation happens at intervals and is executed by internal roots (or chyliferous vessels). Finally, plants are motionless except for the sunward (heliotropic) or moonward (selenotropic) movements of some flowers and leaves, whereas most animals are freely mobile.

From this brief overview we may conclude that the general progression of terrestrial life in the three lower kingdoms—from the infinity of the mineral kingdom, to the individualization of the vegetal kingdom, to the spontaneous motion of the animal kingdom—is indicative of a colossal effort on the part of an *organized power*—namely, physical nature—to attain *free will,* a defining characteristic of the hominal kingdom, the kingdom of humanity. Table 2.2 on p. 23 contains a graphic representation of this grandiose progression in four schematics based on the alchemical philosophy of Madathanus. It allows us to consider each kingdom as a medium for atoms in a particular state of motion: a state of rest (mineral kingdom), a state of equilibrium (vegetal kingdom), a state of turbulence (animal kingdom), and a state of resolution (hominal kingdom).[17] The fifth, sixth, and seventh kingdoms of nature represent spiritual states greater than the current evolutionary state of humanity.

3

Plant Physiognomy

Every plant is a terrestrial star. The plant's celestial properties are inscribed in the color of its petals; its terrestrial properties, in the shape of its leaves. No form of magic is extrinsic to the plant kingdom because, as we have already demonstrated, the plant kingdom encompasses the entire spectrum of astral powers and influences. There are three different keys to the external properties of plants that occult herbalists use to unlock their inner virtues. These are the *binary* key, the *quaternary* (elemental or zodiacal) key, and the *septenary* (or planetary) key.

BINARY SIGNATURES

Louis Claude de Saint-Martin, the self-styled *Philosophe Inconnu,* or "Unknown Philosopher," elucidates the theory of binary signatures and provides two examples of practical application in his work *On the Spirit of Things.*

In all things, whether material or immaterial, there is an impulsive force, and this force is the principle from which everything receives its existence. But the universal impulsive force we observe in nature would not exist if an opposing compulsive force did not constrict it and augment its intensity. It is this opposing force that, by the action of resistance, simultaneously effectuates the development and appearance of all properties and forms engendered by the impetus of the impulsive force.

Nowhere are these two laws more distinctly discernible than in the plant kingdom. In the stone or pit of a fruit, for instance, resistance outweighs force, and hence the stone remains inactive. Once planted, however, it becomes active and begins to vegetate as force counters resistance and sets itself on an equal footing. As the fruit ripens, force overcomes resistance and succeeds in subverting all of its obstacles. Even so, the fruit itself is a by-product of the union of force and resistance, insofar as it is composed of substantialized properties and an envelope that encases them, bands them together, and preserves and supports them according to universal law. This picture allows us to imagine what calamities primitive and eternal nature—which we wrongly consider to be the sole prerogative of humankind—must have endured.[1]

The object of vegetation, the Unknown Philosopher continues, is to transmit to us those rays of beauty, color, and perfection that have their source in the higher worlds and that otherwise tend merely to flit into our inferior world unseen. Thus, every grain of seed is a "little chaos."

Everything in nature is composed of two actions: a divisive action (force) and a divisible action (resistance). The latter when deprived of the former produces WATER, but when it does not suffer such privation produces FIRE. Just as the union of FIRE and WATER is manifest in the green color of the plant's leaves, putrefaction is localized in its roots and sublimation in the vivid colors of its flowers and fruits. As the prisons of higher powers, seeds analogically retrace the story of the Fall and the myth of Saturn devouring his own children.[2] Thus, generation is a brutish struggle whose phases are expressed in signatures, and there is nothing extant that does not retrace the history of its birth by its external form.

For example, the harsh and bitter taste of the acorn kernel indicates that the *oak tree is subjected to the action of an especially violent resistance, an action that aims at nothing less than its annihilation. If we consider in the same light the leaf of the *grapevine, the pip of the grape, and the properties of wine, we shall come to realize that

1. WATER is extremely concentrated in the pip by the action of resistance and this is what makes it spread so profusely throughout the vines;
2. the form of the leaf bears the sign of this expansion of WATER, which is so plentiful as to be altogether separate from its FIRE, and its determinatives are binary, as in an infinity of other plants;

3. consequently, its FIRE is equally divorced from its WATER, which is apparent particularly in the branches of the vinestock, where the leaves and the peduncles of the grape clusters alternate on opposite sides near the base of the shoot;
4. according to its law, the FIRE invariably rises higher than the WATER and makes itself known to the peduncle of the grape cluster, which always sits just above its corresponding leaf petiole;
5. this FIRE is so close to primitive life as to be virtually indistinguishable from it, for which reason the grape takes on such a regular, spherical shape—as though the entire band of astral potentialities, whose number encompasses the entire circumference, had been sapped by its stamens and pistil—as it establishes an equilibrium between force and resistance;
6. for this reason, the grape is so healthy and salutary when consumed in moderation but, because of the divided or binary source from which it derives, can be extremely detrimental when consumed in excess quantities; and
7. such excesses are quite remarkable insofar as they can lead to quarreling, absence of reason, combativeness, or even homicide, insofar as they can incite lust, a characteristic that is so blatantly inscribed in the shape of the pip, and insofar as drunkenness, despite the fact that it incites lust, is nonetheless far from being fatal to generation.[3]

QUATERNARY SIGNATURES

The four elements and the QUINTESSENCE each correspond to one of our five senses—which is to say that each of these five forms of motion reveals the qualities of objects through the vibration of one of our sensory nerve centers:

1. EARTH corresponds to our sense of smell (to the odor of the plant).
2. WATER corresponds to our sense of taste (to the flavor of the plant).
3. FIRE corresponds to our sense of sight (to the shape of the plant).
4. AIR corresponds to our sense of touch (to the body of the plant).
5. QUINTESSENCE corresponds to our sense of hearing (to the spirit of the plant).

TABLE 3.1. ELEMENTAL SIGNATURES

Botanical Element	Odor of Flowers	Flavor of Fruits	General Form (Color)	Plant Size
EARTH	Fatty	Sweet	Stocky (yellow)	Small
WATER	Scentless	Sour	Climbing (greenish)	Small stem, large leaves and fruits
FIRE	Penetrating	Spicy	Twisted (red)	Medium size, radiating branches
AIR	Stinking	Bitter	Slender (bluish)	Very large

Table 3.1 provides correspondences between plant physiognomies and the four elements. Occult herbalists will notice, however, that this table includes only simple types, which are purely theoretical. In reality, plants have much more complex elemental compositions consisting of combinations of elements. Table 3.2 presents the various combinations and their corresponding zodiacal signatures, which specify their general character.

TABLE 3.2. ZODIACAL SIGNATURES

Elements	FIRE	EARTH	AIR	WATER
FIRE	△ *FIRE*	♉ 2. Taurus	♊ 3. Gemini	♋ 4. Cancer
EARTH	♈ 1. Aries	🜃 *EARTH*	♎ 7. Libra	♏ 8. Scorpio
AIR	♌ 5. Leo	♍ 6. Virgo	🜁 *AIR*	♓ 12. Pisces
WATER	♐ 9. Sagittarius	♑ 10. Capricorn	♒ 11. Aquarius	▽ *WATER*

If, then, we wish to know a priori the vegetal qualities of a particular zodiacal signature—for example, those of Aries—we may first consult

Table 3.2 and determine that Aries is a FIRE (vertical column) of EARTH (horizontal row) sign. The Arien plant, then, according to the sensory and physiognomic features outlined in Table 3.1, will have a penetrating and fatty odor and a spicy but not unpleasant flavor, its flowers will be orange-red in color, and it will be small but robust. This example should suffice to demonstrate the ingenuity of this method. In addition, we provide below a more detailed and descriptive list of the vegetal characteristics of each of the zodiacal signatures, which we have compiled from a great number of authors. With the aid of this catalog, occult herbalists will be able to perfect the practice of identifying the quaternary signatures of plants on the basis of their external features.

1. *Aries* (♈): FIRE of EARTH. Plants with Arien signatures are hot and dry. The element FIRE predominates in them. Their physical structures will more or less resemble the human head or its anatomical subdivisions such as the eyes, the nose, the mouth, the teeth, or the tongue. Their medicinal actions cure diseases of the head and the face, and they have therapeutic effects for individuals born under the sign of Aries. They have yellow flowers, an acrid flavor, thin stems, and diphyllous or bipetalous leaves. The characteristic perfume of Aries is ⋆myrrh.

2. *Taurus* (♉): EARTH of FIRE. Plants with Taurean signatures are cold and dry. The element EARTH predominates in them. They are sour to the taste and sweet scented. They are tall, give off aromatic fragrances, freeze easily, and produce large quantities of fruit. Some Taurean plants assume the shape of the throat or the esophagus. Their medicinal actions cure diseases of the neck and the throat, and they have therapeutic effects for individuals born under the sign of Taurus. They often bear androgynous flowers. The characteristic perfume of Taurus is costus or kuth, the essential oil extracted from the root of *Saussurea costus* (Falc.) Lipsch., a species of thistle native to India.

3. *Gemini* (♊): AIR of FIRE. Plants with Geminian signatures are hot and moderately wet. The element AIR predominates in them. Geminian herbs are very green, bear white or pale-colored flowers, and are sweet to the taste. They are often lactescent, and with respect to anatomical structure, some resemble the shoulders, the arms, the hands, or the breasts. Their medicinal actions cure diseases of the shoulders, the arms, and the bronchi, and they have therapeutic effects for individuals born under the sign of Gemini. Many

are heptaphyllous, or seven leaved. The characteristic perfume of Gemini is mastic, the resin extracted from the mastic tree (*Pistacia lentiscus* L.).

4. *Cancer* (♋): WATER of FIRE. Plants with Cancerian signatures are cold and wet. The element WATER predominates in them. They are insipid, peaty, and have white or ash-colored flowers. They commonly grow along the banks of water courses. Their leaves often take the shape of the lungs, the liver, or the spleen. Their medicinal actions cure diseases of the chest, especially the lungs, and they have therapeutic effects for individuals born under the sign of Cancer. They are usually spotted, swollen, and pentaphyllous, or five leaved. The characteristic perfume of Cancer is ⋆camphor.

5. *Leo* (♌): FIRE of AIR. Plants with Leonian signatures are hot and dry. The element FIRE predominates in them. They have red flowers, a penetrating, bitter, or fiery and peppery flavor. Their fruits can take the shape of the stomach or the heart. Their medicinal actions cure diseases of the stomach and the heart, as well as the liver and the spleen, and they have therapeutic effects for individuals born under the sign of Leo. They are often cruciferous. The characteristic perfume of Leo is ⋆frankincense.

6. *Virgo* (♍): EARTH of AIR. Plants with Virgoan signatures are cold and dry. The element EARTH predominates in them. They are usually creeping plants with hard and brittle tissues. Their leaves and roots can have the appearance of the abdomen or the intestines. Their medicinal actions cure diseases of the abdomen and the intestines, especially digestive disorders, and they have therapeutic effects for individuals born under the sign of Virgo. The vast majority of Virgoan flowers are pentapetalous, or five petalled. The characteristic perfume of Virgo is Indian sandalwood, the essential oil of ⋆white sandalwood.

7. *Libra* (♎): AIR of EARTH. Plants with Libran signatures are hot and wet. The element AIR predominates in them. Their flowers are tawny, and their stems are tall, soft, and flexible. The shape of their fruits or leaves is often reminiscent of the kidneys, the umbilicus, or the bladder. Their medicinal actions cure diseases of the kidneys and the bladder, and they have therapeutic effects for individuals born under the sign of Libra. They are sweet to the taste and prefer to grow in rocky terrain. The characteristic perfume of Libra is galbanum, the gum resin extracted from the fern species *Ferula galbaniflua* Boiss. et Buhse.

8. *Scorpio* (♏): WATER of EARTH. Plants with Scorpion signatures are cold and wet. The element WATER predominates in them. They can be insipid, aqueous, viscous, lacteal, or malodorous and are often phallic in form. Their medicinal actions cure diseases of the genitals and the anus, and they have therapeutic effects for individuals born under the sign of Scorpio. The characteristic perfume of Scorpio is red coral (*Corallium rubrum* L.).*

9. *Sagittarius* (♐): FIRE of WATER. Plants with Sagittarian signatures are hot and dry. The element FIRE predominates in them. They are bitter and can assume the shape of the gluteal region. Their medicinal actions cure diseases of the buttocks and the thighs, and they have therapeutic effects for individuals born under the sign of Sagittarius. The characteristic perfume of Sagittarius is ★aloeswood.†

10. *Capricorn* (♑): EARTH of WATER. Plants with Capricornian signatures are cold and dry. The element EARTH predominates in them. Their flowers are often greenish in color. Their juices coagulate and can become toxic. Their medicinal actions cure diseases of the knees and the knee joints, and they have therapeutic effects for individuals born under the sign of Capricorn. The characteristic perfume of Capricorn is spikenard, the

*[The majority of Sédir's zodiacal perfumes derive from Heinrich Cornelius Agrippa's *De occulta philosophia libri tres,* but here Sédir diverges from Agrippan tradition. Agrippa identifies the characteristic perfume of Scorpio as opopanax (from Greek ὀποπάναξ), the gum resin extracted from Hercules's allheal (*Opopanax chironium* W.D.J.Koch). Because of its branch-like shape, red coral was long held to be a plant species instead of an animal species, although the spagyrists more frequently classify it among precious stones and minerals (see, for example, Mavéric, *La médecine hermétique des plantes,* 182–83; English edition: *Hermetic Herbalism,* 161–62). Agrippa places coral (*corallus*) under the dominion of Venus and includes red coral (*corallum rubeum*) as an ingredient in his magical perfume of Venus (see Agrippa, *De occulta philosophia libri tres,* liber I, capita XXVIII et XLIIII, xxxiiii, li; English edition: *Three Books of Occult Philosophy or Magic,* bk. 1, chaps. 28 and 44, 102, 136). This same recipe is reproduced in the eighteenth-century grimoire known as the *Petit Albert* (Pseudo-Albertus, *Les solide tresor des merveilleux secrets de la magie naturelle et cabalistique du Petit Albert,* 66). It is interesting to note, however, that François Jollivet-Castelot, who adopts most of Sédir's innovations (for example, spikenard as the characteristic perfume of Capricorn), lists no corresponding perfume for the Martian sign of Scorpio (see Jollivet-Castelot, *La médecine spagyrique,* 15). —*Trans.*]

†[Here Sédir mistakenly writes *aloès* (aloe) instead of *bois d'aloès* (aloeswood). See *Les plantes magiques,* 49. On Sédir's confusion of the plant names *aloès* and *bois d'aloès,* see the entry ★aloe in Sédir's dictionary in part 3 and the corresponding footnote. —*Trans.*]

essential oil of the species *Nardostachys jatamansi* (D.Don) DC, otherwise known as nard, nardin, or muskroot.*

11. *Aquarius* (♒): AIR of WATER. Plants with Aquarian signatures are moderately hot and wet. The element AIR predominates in them. They are often very aromatic, and in shape they can resemble the legs. Their medicinal actions cure diseases of the tibias and the calves, and they have therapeutic effects for individuals born under the sign of Aquarius. The characteristic perfume of Aquarius is euphorbium, the milky resin, or latex, of ★spurge.

12. *Pisces* (♓): WATER of AIR. Plants with Piscean signatures are cold and wet. The element WATER predominates in them. Their flavor is typically bland, and in form they can resemble the toes or the feet. Their medicinal actions cure diseases of the ankles and the feet, and they have therapeutic effects for individuals born under the sign of Pisces. They often grow in cool, dark places along the banks of water courses. The characteristic perfume of Pisces is red storax, the oleoresin obtained from the bark of the storax tree (*Styrax officinalis* L.).[4]

SEPTENARY SIGNATURES

Here, in summary form, are the most basic correspondences of the septenary or planetary system of classification:

1. Saturn (♄): astringency and concentration
2. Jupiter (♃): radiation and majesty
3. Mars (♂): anger and spinosity
4. Sun (☉): beauty, nobility, and harmony
5. Venus (♀): sweetness

*[Here again, Sédir diverges from Agrippan tradition. Agrippa identifies the characteristic perfume of Capricorn as *asa.* The Latin term *asa* (literally "altar") was applied to two distinct aromatic species: *asa dulcis,* the "sweet-smelling" variety, and *asa foetida,* the "stinking" variety. The former is another name for benzoin, the balsamic resin obtained from the gum benjamin tree (*Styrax benzoin* Dryand.). The latter is now known as asafetida, the gum oleoresin of *Ferula assa-foetida* L., a species of giant fennel. English translators invariably take Agrippa's *asa* to be the sweet-scented, or *dulcis,* variety, but it is important to note that Capricorn is a Saturnian sign and that Sédir characterizes Saturnian plants as having *pungent* aromas. See Agrippa, *De occulta philosophia libri tres,* liber I, caput XLIIII, lii; English edition: *Three Books of Occult Philosophy,* bk. 1, chap. 44, 136. —*Trans.*]

6. Mercury (☿): indeterminacy
7. Moon (☾): strangeness*

When these general characteristics are developed further, we obtain the various correspondences provided in Table 3.3 (p. 34).

The Salt of the earth in which the plant grows gives it its flavor. The flavor of the plant indicates its ideal type and signposts the particular path occult herbalists must follow to extract its balm. The leaves and the stem reveal the plant's ruling planet. In general, the root embodies the influence of Saturn; the seed and the bark, the influence of Mercury; the hardwood, the influence of Mars; the leaves, the influence of the Moon; the flowers, the influence of Venus; and the fruit, the influence of Jupiter.

Saturn (♄): Plants with Saturnian signatures are heavy, glutinous, astringent, bitter, pungent, or acetous. Saturn governs the roots, plants that produce fruits without evident flowers (known as the *vanaspati* in Ayurvedic medicine), plants that reproduce without seeds, asporous nonflowering plants, and plants that produce black berries. Their odor is penetrating, their form often frightening, and their shade ominous. Saturnian plants are resinous; have narcotic, psychotropic, and soporific properties; and tend to grow rather slowly.

Jupiter (♃): Plants with Jupiterian signatures have sweet, smooth, subtle, styptic, and even sour flavors. Jupiterian plants are balsamic and majestic. Jupiter governs all fruit-bearing plants, even those without evident flowers (the *vanaspati*), but especially those that produce an abundance of fruit and are lush in appearance.

Mars (♂): Plants with Martian signatures are acidic, amaroidal or bitter, acrid, and spicy. They are often bitter and poisonous due to an excess of the elemental quality of hotness. Martian plants are spinose and bristly, can irritate the skin when touched, and often possess a sap that causes an intense burning sensation when it comes into contact with the eyes.

Sun (☉): Plants with Solar signatures are aromatic and have an acidulous flavor. They guard against lightning and serve as excellent counterpoisons.

*[Sédir lists the planets and luminaries according to the so-called Chaldean order—that is, from the slowest moving celestial body in the night sky (♄) to the fastest moving celestial body in the night sky (☾). The planetary hours are tabulated according to this same Chaldean schema (see Table 7.1, p. 68). —*Trans.*]

Many Solar plants remain green all year-round. They are especially well suited for botanomancy and for protection from hexes and evil spirits. They typically turn toward the Sun or bear its image in the shape of their leaves, flowers, or fruits.

Venus (♀): Plants with Venusian signatures have a sweet, agreeable, and unctuous flavor. They produce colorful flowers without bearing fruits. Venusian plants produce large quantities of seeds and often serve as excellent aphrodisiacs. Their aroma is invariably sweet.

Mercury (☿): Plants with Mercurial signatures have mixed flavors. They produce flowers and leaves without fruits and often grow very quickly. Their leaves are small and their colors varied.

Moon (☾): Plants with Lunar signatures are insipid and thrive in or next to water. Lunar plants are cold, lacteal, narcotic, and anaphrodisiac. Their leaves are often abnormally large.

TABLE 3.3. PLANETARY SIGNATURES

Planet	Plant Size (Character)	Flower Color (Size)	Odor of Flowers	Flavor of Fruits
♄ **Saturn**	Large (melancholic)	Black and gray	Stinking	Bitter (poisonous)
♃ **Jupiter**	Large (dense foliage)	Blue or white	Scentless	Sweet, sour
♂ **Mars**	Small (thorny)	Red (small)	Spicy and unpleasant	Hot, peppery (poisonous)
☉ **Sun**	Medium size	Yellow	Aromatic	Sour, pleasant
♀ **Venus**	Small (flowery)	Multicolored roses (large, beautiful)	Exquisite, strong	No fruits or sugars
☿ **Mercury**	Medium size (sinuous)	Multicolored (small)	Penetrating or malodorous	Mixed
☾ **Moon**	Various sizes (strange looking)	White	Scentless or bland	Insipid, saccharine

On the basis of the above zodiacal and planetary signatures, occult herbalists may also determine the beneficent or maleficent relationship between one plant and another for the purpose of polyculture, or companion planting. The following are the most important beneficent allelopathic relationships:

1. Taurean plants are favorable to Cancerian and Sagittarian plants.
2. Geminian plants are favorable to Libran and Aquarian plants.
3. Cancerian plants are favorable to Libran plants.
4. Virgoan plants are favorable to Taurean plants.
5. Scorpion plants are favorable to Cancerian plants.

The following are the most important maleficent allelopathic relationships. Plants that bear such zodiacal relationships should never be planted in close proximity to each other.

1. Taurean plants are unfavorable to Libran and Scorpion plants.
2. Geminian plants are unfavorable to Capricornian plants.
3. Cancerian plants are unfavorable to Sagittarian plants.
4. Virgoan plants are unfavorable to Arien and Leonian plants.

The same beneficent and maleficent relationships also apply to planetary signatures. For example, Saturn, Mars, and the Sun are enemy planets, whereas Venus is friendly to all other planets and to Mars especially, and Mercury is likewise friendly to all other planets and to Jupiter in particular.

Naturally, it is often the case that plants have more than one ruling planet. We provide a few examples below to help students understand the characteristics and properties of plants that fall under the dominion of two or more governing astral bodies.

When Saturn is the dominant planet, for example, it gives the plant a black or charcoal-gray color; a hard, rough stem; a sour, bitter, and salty flavor; and a tall and slender form with inky black or ash-gray flowers. Saturn typically calls upon Mars, and then the plant becomes bumpy, gnarled, many branched, wild, and twisted. The combined influences of Saturn and Venus produce large trees that are hard and strong because the Venusian sweetness supplies the tree with an abundance of matter, which grows in accordance with the Saturnian Sulfur.

If Jupiter combines with Venus, the plant will be full of strength and virtue. If Mercury comes to influence the plant together with Jupiter and Venus, it will produce an even more perfect and beautiful specimen that is average in size and has white or blue flowers. If the Sun comes to join in with the preceding planets, the flowers will turn a yellow or xanthous color, and if Mars does not oppose them, the plant will be able to resist all maleficent influences and will render the choicest of herbal remedies. Such a combination is quite rare, however, as it verges on the paradisiacal form of plants. If the influences of Mars and Saturn oppose each other in combination with the influences of Mercury, Venus, and Jupiter, the tree will become poisonous, have reddish or whitish-red flowers (white because of the Venusian influence), be rough to the touch, and have an execrable flavor. But if Jupiter and Venus are more powerful than Mercury, even though Mars and Saturn oppose each other, the plant will be predominantly hot in quality and have healing properties; its stem will be delicate, somewhat rough, and spinose; and its flowers will be whitish in color. Finally, if Venus is close to Saturn, if the Moon is not thwarted by Mars, and if Jupiter is left unencumbered, the plant will be beautiful, tender, delicate, white flowered, and innocuous but not very useful.

PART TWO

Plants and Humans

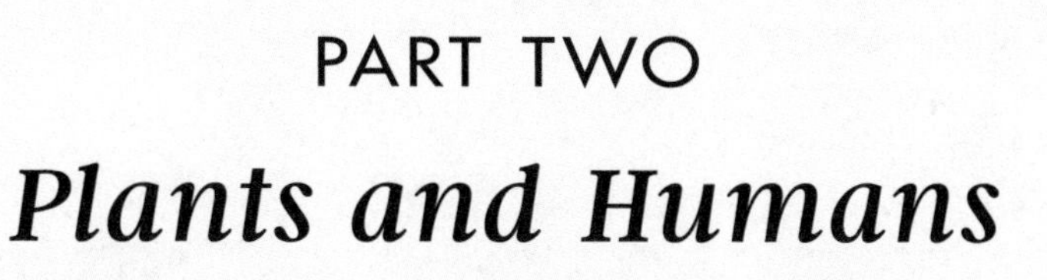

Since the plant kingdom as a whole falls under the dominion of Venus, it has one primary function vis-à-vis humans: to provide them succor and sustenance.

Plants can nourish humans in three ways:

1. They can restore organic deficiencies in the physical body through their consumption.
2. They can restore deficiencies in the electromagnetic body through their use in herbal therapeutics.
3. They can restore deficiencies in the astral body through their integration into somnambulistic, ecstatic, and divinatory practices and into rituals of ceremonial magic.

But the plant-human relationship should by no means be a one-sided alimentary relationship. Humans in turn can nourish plants in three ways:

1. They can cultivate them using the methods of occult horticulture.
2. They can restore them using the methods of vegetation magic.
3. They can resurrect them using the methods of plant palingenesis.

We shall study each of these articles in separate chapters.

4
Alimentation

We have no intention to rehearse the many arguments in favor of vegetarianism. Other more learned authors have already covered this subject far more thoroughly and authoritatively than we ever could. Instead, we offer here only a few practical guidelines for vegetarian beginners and some general advice on the preparation and consumption of meals. Suffice it to say that the vegetarian regimen, which quickly eliminates all corporeal resistance to the will, is indispensable to students of practical magic. Students who do not wish to adopt an exclusively vegetarian diet can greatly enhance their astral vision and psychic abilities by abstaining from animal foods for periods ranging from seven to forty days.[1]

PRACTICAL GUIDELINES FOR VEGETARIAN BEGINNERS

1. Don't quit meat cold turkey. Transition slowly from the carnivorous to the vegetarian diet, but do not substitute water or milk for fermented beverages until you have successfully made this transition—to facilitate this change, increase your consumption of fleshy fruits.
2. Make this dietary transition in the countryside.
3. Remain vegetarian when visiting large cities, and Paris especially, unless you are dining out at a restaurant and there are no vegetarian options or you are suffering from general weakness and fatigue.
4. Don't be afraid to consume plant foods in much larger quantities than the animal foods you used to eat.

5. Keep fish on your menu longer than any other meat, and do not abstain from eggs, milk, or butter except during periods of rigorous ascesis or spiritual exercise.
6. Finally, teach yourself during this period of transition to control your physical organism and to gain mastery by means of the will over the minor functional irregularities you are likely to experience.

HOW TO TAKE YOUR MEALS

As a general rule, the more force one expends to accomplish an action, the more profitable the action becomes for us. Taken to its logical extreme, this means that we should cultivate, harvest, and prepare all of our plant foods ourselves and use tools and utensils reserved exclusively for these purposes. In naturalistic and pantheistic initiations, students develop from below to above and from the outer to the inner by purifying and perfecting their astral body first and their intelligence last. For this reason, Brahmins and Hindu ascetics are directed to prepare their own foods and never to allow any other person to touch their copper pots, with the exception—for Brahmins—of their spouses.

The same reasoning gave rise to various prescriptions concerning the positioning of the body during mealtime. There is clearly a relationship between the electromagnetic currents of a planet and the individuals who live on its surface, but space does not allow for a full exposition of this theory here. Suffice it to say that for inhabitants of our region of the world, it is best to eat while facing north.

Other prescriptions consist of ritual ablutions. For example, before taking meals, Hindu priests ritually wash their hands, feet, mouth, nose, eyes, and ears while reciting a sacred invocation. For Western esotericists like us, this practice corresponds to the recitation of the *benedicite* formula, which, when uttered with the correct magical intonation, has a truly powerful and energizing effect.*

*[The French Roman Catholic form of the *bénédicité,* or "blessing," runs as follows: *V. Benedicite. R. Dominus. Nos et ea quae sumus sumpturi benedicat dextera Christi in nomine Patris et Filii et Spiritus Sancti. Amen.* (The abbreviations *V.* and *R.* stand for the liturgical directives *versicula,* or "versicle," and *responsa,* or "response.") Anglophones usually recite this prayer in the abridged form "Bless us, O Lord, and these thy gifts, which we are about to receive from thy

Another prescription worth mentioning here is the strict observance of silence during meals as practiced by religious devotees the world over. By remaining silent and focusing your attention on the action of eating, you can reduce the volume of food needed to feel satiated to smaller and more sensible portions. Thus, the process of digestion will put less strain on the solar plexus and, as a result, produce a surplus of nervous force, which may be put to greater use in exercises of contemplation. But for those who have to live in and too much with the world on a daily basis, and especially in the heavy atmosphere of a big city, conviviality is not only the best digestif to stimulate a lazy stomach but also more effective than all the world's spirits.

(*cont. from p. 39*) bounty through Christ, our Lord. Amen." Most likely Sédir, who had been consecrated bishop of Concorezzo in the Église gnostique de France (Gnostic Church of France) under the episcopal name Tau Paul, would have recited some variation of this Latin formula. Jules Doinel, who founded the Église gnostique circa 1890, revised a number of Latin prayers and liturgical formulae and incorporated them into the rituals of the Gnostic Church. He invariably changed the masculine form *Dominus,* "Lord," to the feminine form *Domina,* "Lady," the *Domina* in question being the Celestial Sophia (or her personification as Helen-Ennoia, the consort of Simon Magus). See further, for example, Doinel, "Rituel du Consolamentum"; and the Gnostic Rite of Ordination in Kostka, *Lucifer démasqué,* 149–51. —*Trans.*]

5
Phytotherapy

The healing virtues and curative properties of plants have always been famous. The ancient Druids showed remarkable intuition in bestowing upon their god of medicine—Asklepios in Greek, Aesculapius in Latin—the name Aesc-heyl-hopa, which means "the hope of healing (or salvation) is in the wood."[1] According to Porphyry, Asklepios, as the god who heals solutions of continuity in the tissues, exemplifies the Solar faculty of corporeal regeneration.*

Plants can be used as medicines in three different states: living, dead, or resurrected. The living plant, especially when it is aromatic, functions as a medium of transformation. Plant aromas can cure all inflammations of the respiratory mucous membranes. Thus, those suffering from tuberculosis would do well to inhale the scents of ★pine, ★lavender, ★rosemary, ★basil, and ★mint. This is an exoteric method of healing by means of living plants. Paracelsus designates their esoteric or occult use under the rubric "the transplantation of diseases."

THE TRANSPLANTATION OF DISEASES

Any disease may be transplanted from a sick person to another living being. To do this, you must first extract the *mumia*—that is, the vital principle or

*[Here, Sédir shows equally remarkable intuition. Porphyry is actually reported to have claimed that medicine came from Athena, "because Asklepios is Lunar Intellect, just as Apollo is Solar Intellect" (διότι καὶ ὁ Ἀσκληπιὸς νοῦς ἐστι Σεληνιακός, ὥσπερ ὁ Ἀπόλλων Ἡλιακὸς νοῦς). Iamblichus rightly criticizes the veracity of this statement and reaffirms the Solar affinities of Asklepios (according to Proclus, *Commentary on Plato's Timaeus* 1.24b–c). Moreover, in fragment 8 of his work *On Images,* Porphyry states that "Asklepios is the symbol of the Sun's healing power" (according to Eusebius, *Preparation for the Gospel* 3.11). —*Trans.*]

vehicle of life—from the patient (from the blood or the sperm, for example) and use it to water some earth in a pot in which you have planted the seed of a plant possessing the same astral signature as the disease. Once the plant has grown, it should be thrown into running water if the disease is predominantly hot in quality, like most fevers or inflammations, but if the disease is predominantly wet in quality, it should be burned and reduced to smoke. This method of occult medicine accords with the ancient aphorism of sympathy—namely, *similia similibus curantur,* or "likes are cured by likes."[2]

Here are a few recipes for such transplantational procedures. For ulcers and wounds, use ⋆knotweed, especially the species known as spotted lady's thumb (*Persicaria maculosa* Gray), ⋆comfrey, or ⋆olive.* The plant should be brought into contact with the ulcer before being burned. For toothaches, rub the gums until they bleed with the root of the groundsel (*Senecio vulgaris* L.), then replant it. For uterine menorrhea, the *mumia* should be extracted from the patient's groin and then planted with ⋆knotweed. For dysmenorrhea, however, the *mumia* should be planted with ⋆pennyroyal. For pulmonary tuberculosis, plant the *mumia* with an ⋆orchid in the vicinity of an ⋆oak or ⋆cherry tree.[3]

In our modern era, we have made giant steps toward understanding the inner workings of the transplantation of diseases through various experiments on the remote action, or "action at a distance," of medicinal substances on hypnotic subjects. On this topic, interested readers may consult the works of Henri Bourru and Ferdinand Burot, professors of the Rochefort School of Medicine, Jules Bernard Luys, Hector Durville, and the magnetizers of the nineteenth century.[4] We offer only a few isolated examples here because students will be able to multiply them at their leisure according to the doctrine of signatures.

*[Instead of ⋆olive (*Olea europaea* L.), Sédir reproduces the regrettably imprecise Latin name used by Paracelsus, *Botanus europaeus,* literally "European plant." If readers search Sédir's dictionary in part 3, however, they will notice that he gives identical planetary signatures—namely ♃ and the ☉—to both ⋆knotweed and ⋆olive. As for ⋆comfrey, Sédir assigns it to the exclusive rulership of ♃, but in this entry he reaffirms its medicinal action against wounds and ulcers. The ⋆olive tree, moreover, has a long history of use in the treatment of ulcers. See especially Culpeper, *Culpeper's English Physician and Complete Herbal,* 277–78. —*Trans.*]

MEDICAL HERBALISM (EXOTERIC PHYTOTHERAPY)

Plants may be used exoterically as medicines in the following forms:

1. Juices
2. Powders
3. Decoctions
4. Infusions (made in boiling water), which are more active than decoctions
5. Magisteries
6. Tinctures (in alcohol)
7. Essences

Plant medicines are always more active when they have been prepared by a robust person animated by a desire to heal. This is one of the great secrets of the success of homeopathic globules and dilutions. We knew an old health officer in the Saint-Georges quarter in Paris who could cure the most obstinate dyspepsia by means of bread dumplings. He would spend only two or three hours each morning kneading the dough himself in his pharmaceutical laboratory.

What follows are some practical examples of exoteric pharmacopeia from the dusty tomes of the old herbalists. Students can find in these forgotten books a great number of herbal preparations—which are often strikingly similar to many of our so-called modern medicines—and make them themselves without too much difficulty.[5] We shall focus here only on the examples of ⋆black hellebore, ⋆pine tar, and ⋆hemlock.

Black Hellebore

Popular misconceptions about hellebore have prevailed to such an extent that most people believe its incorporation in medicines will drive the patient to madness, despite the fact that it is known not only to cure and prevent a number of maladies but also to preserve the vitality and prolong life. Its efficacy and virtue reside in its rejuvenative nature: it not only corrects the blood but also purges impurities arising from excess, inanition, and suppression, which are the causes of so many health problems and imbalances. Physicians in antiquity were more than willing to make use of it, but modern physicians have derogated from ancient herbalism to such an extent that the virtue of hellebore must be restored to its original dignity.

By far the most virtuous species is the ⋆black hellebore (ἐλλέβορος μέλας) described by Theophrastus,[6] an opinion that countless professors of medicine have reiterated over the centuries, because this species is known to have softer and more favorable effects than other hellebores, such as ⋆green hellebore, otherwise known as bastard hellebore, and the ἐλλέβορος λευκός of Dioscorides,[7] a probable reference to our ⋆white hellebore or false helleborine.

Take the root of ⋆black hellebore, cut it into small pieces, and stuff them into an apple. Let it air overnight, and in the morning cook the apple slowly. Remove the root and grind it into a powder. The dose is roughly the weight of a French half *écu* coin,* taken three hours before eating and only three to four times a year, ideally in the autumn or spring. This preventive medicine has powerful purgative properties that will cleanse the body of unwanted humors and cure even the most vexing indispositions. If necessary, the dosage may be increased.

To make a corrective, cook the leaves along with the root in rye bread and grind it into a fine powder. The dose is from thirty to forty grains, or more, if need be, for more robust patients, taken either as a pill or in a wafer, baked inside an apple, or in a broth. The whole plant can also be administered in the form of a powder, at a dose of one lot, without any preparation, as they used to do in Rome. Another method is to cut the root and cook it with the flesh of a fruit into a broth, coulis, jelly, or tincture, which serves as a milder purgative, or laxative. To this recipe you may add any ingredient you wish, according to your taste.

To better access the rejuvenative virtue of ⋆black hellebore and purify the blood, some accustom themselves gradually and imperceptibly to the virtue of its leaves, which should be harvested in good season,† dried in the shade,

*[The French silver half *écu* coin weighs approximately 14.33 grams (about a half ounce or one tablespoon). This is roughly equal to the old half ounce, or lot (see Table 5.1). —*Trans.*]

†[As a general rule, leaves should be harvested before flowering, when vegetative growth is at its most energetic. Harvesting at this time not only increases leaf production but also maximizes flavor—many leaves will have less flavor or become bitter if they are harvested after flowering. Sédir's colleague Jean Mavéric similarly recommends harvesting leaves "just before the flower opens or blooms" (Mavéric, *Hermetic Herbalism,* 37). Black hellebore, which is toxic if consumed in large quantities, usually flowers from midwinter to early spring—in

and mixed with an equal portion of sugar. Whoever takes this medicine will live to a ripe old age and be free from those illnesses and ailments, both internal and external, that come with the onset of old age. Initially, it should be taken periodically at a dose of ten to fifteen grains. Once the body becomes accustomed to this dosage, it should then be increased to twenty grains, and then to thirty grains per day. And once the body becomes accustomed to this dosage, the remedy should be taken at a dose of one dram, but no more than once every six days. In this manner, black hellebore becomes ordinary and familiar. After it loses its purgative power, it will take on rejuvenative and corrective properties.

TABLE 5.1. OLD WEIGHTS AND MEASURES

Old Weights	Equivalences	Weights in Grams
Medicinal pound	12 ounces	367 grams
Merchant pound	16 ounces	489 grams
Half pound	6 (or 8) ounces	186 (or 245) grams
Ounce	8 drams	30.59 grams
Half ounce, or lot	4 drams	15.30 grams
Dram	3 scruples	3.82 grams
Half dram	36 grains	1.91 grams
Scruple	24 grains	1.27 grams
Half scruple	12 grains	0.64 grams
Grain		0.05 grams

From Mavéric, *Hermetic Herbalism*, 107.

[NB: All recipes in this book that include ingredients in pounds call for the old medicinal pound (12 ounces, or 367 grams). —*Trans.*]

(*cont.*) December in warmer winter climates, and hence the common name Christmas rose (French: *rose de Noël*), but in February or March in colder winter climates. Sédir clearly states that the remedies provided in this section are forms of exoteric herbalism. In the practice of esoteric herbalism, of course, additional factors must be considered when harvesting plants or plant parts. According to Sédir, black hellebore is best harvested while Saturn is in his domicile of Capricorn, but after December 17, 2020, Saturn will not return to Capricorn until 2047. See further the entry *black hellebore in Sédir's dictionary in part 3. —*Trans.*]

When black hellebore is reduced to a balm by the industry of a pharmaceutical craftsman, the dose of its balsamic virtue is about ten grains. However, it is possible to extract an excellent herbal essence from ⋆black hellebore that surpasses all of the preceding preparations in both artifice and beneficence. This extract should be taken at a dose of five to six drops in a suitable liquor, such as Carmelite water (an alcoholic extract of ⋆lemon balm and other herbs) or an essence extracted from fruits.

From the entire plant, after washing it thoroughly and moistening it with *acetum scylliticum,* or vinegar of squill (*Drimia maritima* L.), you can make a syrup that will purge the body of all black and earthy humors, separating the pure from the impure and pestilential. This syrup, which is both more dependable and more benign than any other purgative, will uproot virtually any malady it encounters. We prefer the syrup to the extract, but both have exclusively purgative properties. Neither are powerful enough to rectify the blood and maintain the health in a firm and stable state.

The protracted use of this simple, and primarily its root, has the marvelous virtue of separating and removing the most obstinate diseases from the body, in addition to the remarkable faculties of rejuvenating the body, purifying the blood, and purging decay, which is most often the cause of declining health and death. In a certain sense, ⋆black hellebore could justly be described as a "lesser universal medicine."[8]

Tar Water

To make a good tar water for internal use, pour four pints of cold water over one pint of ⋆pine tar, then stir the mixture with a wooden spoon or mixing stick. Carefully seal the vessel and let it rest for at least twenty-four hours so that the tar has time to precipitate. Decant the clear liquid, having previously skimmed the surface of the supernate (the liquid above a precipitate or sediment) with great care so as not to disturb the precipitate. Store the water for later use in tightly sealed glass bottles. The residual tar will be devoid of virtue but may be used for other mundane purposes.[9] Good tar water should have a color somewhere between the color of a white wine from France and the color of a white wine from Spain, and it should be just as translucent.

To make a good tar water for external use, pour two quarts of boiling water over one quart of tar, then stir the mixture well with a mixing stick or ladle for about fifteen minutes. Let it rest for ten hours, decant the clear liquid as before, and store it in tightly stopped glass bottles. The water may be made weaker or stronger as needed.[10] This lotion may be prescribed for kidney stones, scabies, ulcers, scrofula, and even leprosy. It is also effective against smallpox, hemorrhages, ulcerative colitis, inflammations, gangrene, scurvy, erysipelas, asthma, indigestion, bladder stones, edema, and hysteria.

The best tar comes from the pitch pine (*Pinus rigida* Mill.) and the silver fir (*Abies alba* Mill.), but the latter species, if selected, should be situated on high and dry ground and exposed to the north wind.

Hemlock

Take some fresh ⋆hemlock (stems and leaves), as much as you like, and express the juice. Evaporate it over a very slow fire (*ignis lentus*) in an earthenware vessel, stirring it from time to time to keep it from burning. Let it cook until it reduces to the consistency of a thick extract, then add a sufficient amount of ⋆hemlock powder so that it forms into a mass from which you can make pills weighing about two grains each. If fresh ⋆hemlock is not available, an extract can be made from a decoction of dried ⋆hemlock, but this preparation will possess less virtue.[11] This medicant should be taken in very small doses in the beginning and increased gradually to one and a half lots. Immediately after ingesting a pill, it is advisable to drink some tea, veal broth, or an infusion of ⋆elderberry flowers.

Dried and chopped ⋆hemlock leaves may also be put in a sachet, soaked for a few minutes in some boiling water, and administered in the form of a poultice. The sachet should be laid gently on the skin while it is still warm.

Each of these preparations have emollient, resolvent, and calmative properties. Although the plant is now known by the Latin binomial *Conium maculatum* L., over the years botanists have called it by various names, such as *Cicuta officinalis* Crantz, *Cicuta major* Bauh., *Cicuta vulgaris major* Park., *Cicuta vulgaris* Clus., *Cicuta vera* Gesner, and *Conium seminibus striatis* L. Theophrastus rightly states that the best ⋆hemlock (κώνειον in Greek) grows in the shade in colder regions.[12] Thus, hemlocks in Vienna and around Soissons are more active than those that grow in Paris or Italy. Hippocrates, Galen, Girolamo Mercuriale (otherwise known as Hieronymus Mercurialis), Jean Astruc, and many other physicians from antiquity to the Middle Ages and the Renaissance used preparations of hemlock for internal use to resolve tumors and cure colic and pelvic inflammatory disease.[13] Our ancestors also prescribed an herbal essence made from ★greater celandine, ★lemon balm, ★valerian, ★betony, ★saffron, and ★aloe as a general tonic.

HERMETIC HERBALISM (ESOTERIC PHYTOTHERAPY)

Hermetic herbalism, which begins with the harvesting of plants under the most favorable astral conditions, is markedly different from the more mundane medical variety.[14] Its purpose is not to prepare the physical qualities or juices of the plant in the most profitable manner but rather to extract the plant's living, or vital, force—that is, the quintessence or soul of the plant, which the ancient Hermeticists call the *balsamus,* or "balsam."

The balsam is the essential oil of the plant. This is no vulgar plant oil, however, nor is it the salt, the earth, or the water of the plant. It is, rather, something infinitely more subtle—namely, the plant's *astral body.* The essential oil must be obtained, as Herman Boerhaave asserts, by fire rather than by fermentation.[15] This is what Paracelsus calls the *arcanum,* a fixed, immortal, and, in a certain sense, incorporeal substance that alters, restores, and preserves the body. To obtain this force, which is enveloped in a paradisiacal tincture, the spagyrist must reduce the vegetal body from its second matter (*materia secunda*) to its first, or primal, matter (*materia prima*) or, to paraphrase Paracelsus, from the *cagastric* body to the *iliastric* body.[16]

Properly speaking, the healing power of the plant resides in its spirit. In its natural state, however, the activity of the spirit is encumbered, its light being obscured by the garment of matter. It is therefore necessary to destroy these envelopes, or at least to transmute them into something pure and fixed. To achieve this transmutation, the plant matter must undergo a period of digestion, and a substance capable of absorbing its impurities must be added to the plant matter while it digests. The choice of solvent or menstruum should be dictated by the plant's flavor, which signposts the plant's ideal type. The occult herbalist must first discover which planet governs the plant's particular flavor, and then begin the digestion with a mineral salt having the same "planetarism"—that is, the same configuration of planetary influence. Successful digestions will render three components: a salt, a first matter, and a Mercury, or "fixed water." "First we extract the water from plants," Pseudo-Aquinas says in his *Tractatus de lapide philosophico,* "then we burn them in the furnace of calcination and convert the lime into water, then we distill and coagulate it, transforming it into a stone endowed with the virtues of whichever plants were used."[17]

There are three Salts, or vegetal powers, that are particularly useful in occult therapeutics:

1. The Salt of Jupiter, which has a pleasant odor and taste, is produced internally by the force of divine expansion and externally by the Sun and Venus. It is not strong enough, however, to cure diseases by itself. It is the enemy of the venomous life of igneous substances and incites harmony or a conveyance toward gentleness.
2. The Salt of Mars, which is bitter, igneous, and astringent.
3. The Salt of Mercury, a great stimulator that triggers beneficent reactions.*

The first matter extracted from plants almost always appears in the form of an oil. This oil is highly nutritive and will not only restore humoral equilibrium but also strengthen the temperament of the patient. The Mercury of Life (*Mercurius vitae*), on the other hand, is regenerative and revivifying. This Mercury can only be extracted from almost perfect, sweet-flavored plants signed by the Sun, Venus, and Jupiter. Harsh-tasting plants do not tap in to the root of this Mercury. This is why they can only act upon the four elements, whereas the Mercury of Life can act upon the astral body.†

*[Salt of Jupiter, *sal* or *vitriolum Jovis,* is usually identified as stannous acetate or tin(II) acetate—chemical formula: $Sn(C_2H_3O_2)_2$. Salt of Mars, *sal* or *vitriolum Martis,* is usually identified as ferrous sulfate or iron(II) sulfate—chemical formula: $FeSO_4$. Salt of Mercury, *sal* or *vitriolum Mercurii,* is usually identified as mercuric nitrate or mercury(II) nitrate—chemical formula: $Hg(NO_3)_2$. —*Trans.*]

†[The true identity of Paracelsus's *Mercurius vitae* remains a great alchemical mystery. Sédir is clearly not of the opinion that it is antimony oxychloride, as many have claimed, since he says it can only be extracted from certain plants. Paracelsus describes its regenerative properties in some detail, asserting that "this arcanum casts off the nails from the hands and feet of a man, the hair, skin, and everything that belong to him, makes them grow afresh, and renews the whole body" (Paracelsus, *The Hermetic and Alchemical Writings,* 2:39). These properties are strikingly similar to those Nicaise Le Fèvre, a disciple of Paracelsus, ascribes to the *primum ens,* or "first being," and specifically to the *primum ens melissae,* the "first being of ★lemon

In general, it is better to use the Salts of Mars and Mercury first, since they are more active, and then use the Salts of Jupiter and Venus, their respective antidotes, to unite them and extinguish the fire of their ire. Once this action has been accomplished, the cure comes into being—which is to say, the plant's original harmony has been restored—and then all that remains is to "give it some Sun" to set everything back into motion.*

Occult herbalists should always be aware that good plants can be spoiled by bad aspects, especially from Saturn and Mars, and that poisonous plants can be made benevolent by good aspects from the Sun, Venus, or Jupiter. Furthermore, occult herbalists must cure likes by likes (*similia similibus curantur*). Do not simply prescribe a plant for a disease. Instead, administer a plant in which you have spagyrically enhanced the ire of Mars through the *amor* of Jupiter and Venus. The more its anger has been transmuted into love, the more virtuous the plant medicine will be, for if the plant venom should become the sole property of Mercury, it can have disastrous effects on the body and even result in death.†

To conclude this chapter, we offer a recipe for what is perhaps the most celebrated remedy of spagyric medicine, Paracelsus's *primum ens melissae.*[18]

Primum Ens Melissae

Take a half liter of pure potassium carbonate and expose it to the air until it dissolves. Filter the liquid and add as many ⋆lemon balm leaves as possible, such that all of them are immersed in the liquid. Let the leaves soak

(*cont. from p. 49*) balm." See further Mavéric, *La médecine hermétique des plantes,* 147–49 (English edition: *Hermetic Herbalism,* 129–30). To make matters still more mysterious, no plant in Sédir's dictionary—not even ⋆lemon balm—is signed simultaneously by the ☉, ♀, and ♃. On the rarity of this trinitarian "planetarism" in the plant kingdom, see Sédir's comments in chapter 3. —*Trans.*]

*[Salt of Venus, *sal* or *vitriolum Veneris,* otherwise known as blue vitriol or Roman vitriol, is usually identified as copper sulfate or copper(II) sulfate—chemical formula: $CuSO_4(H_2O)_x$. As for Sédir's final directive, "give it some Sun," this most likely refers to the use of the Sun as a heat source in spagyric distillation. See further Mavéric, *La médecine hermétique des plantes,* 157–58 (English edition: *Hermetic Herbalism,* 135–38). —*Trans.*]

†One of the most active counterpoisons against plant venoms is the following composition: Heat tartar with alcohol at a moderate but constant temperature until a red oil endowed with its particular properties distills into the retort. Set this red oil to digest, and then cohobate three additional times and it will render a magnificent antidote.

in a closed vessel at a gentle heat for twenty-four hours. Decant and then pour over the pure liquid a layer of one or two inches of alcohol. Let it rest for two days, or until the alcohol turns a beautiful green color. Remove the alcohol and preserve it for other uses, then replace it with fresh alcohol until all of the coloring matter is completely absorbed. Next, distill the alcohol and evaporate it to the consistency of a thick syrup. The alcohol and the alkali should be highly concentrated.

6
Plant Magic

Magic, being first and foremost a practical art, studies living beings from the standpoint of individuality and personhood. The magician is therefore concerned primarily with the *spirits* of plants—that is to say, with dryads, hamadryads, sylvans, and fauns. These plant spirits, or elementals, are known by many names: they are the *dusii* of Augustine,[1] the fairies of various peoples of the Middle Ages, the *doire-òigh* of the Welsh, and the grove maidens of the Irish. Paracelsus similarly draws a distinction between the spirits of forest plants, which he calls sylvesters, and those of aquatic plants, which he calls nymphs.[2]

PLANT MYSTICISM

The various mystical traditions teach an array of methods for exploiting the occult forces of plants. Any plant may be used in its entirety (that is, in its entire individuality) or in one of its parts (namely, the root, the stem, the leaves, or the flowers).

The primary method of plant magic is through the evocation of plant spirits, or elementals. These beings are inhabitants of the astral plane who aspire to incarnate in human form. They are endowed with a certain instinctive intelligence and adapt their form according to whichever vegetal being they attach themselves. It was through the evocation of these elementals that the old Rosicrucians were able to perform their miraculous cures, for the herbaceous elementals are *familiar spirits* and quite naturally obey the commands of *spiritual* human beings. They exert a considerable influence over the material plane because they dwell on the borders of this plane and the astral. The elementals of the vegetable kingdom effectuate healings and induce astonishing visions in much the same way that the elementals of the mineral kingdom, when directed accordingly, produce all manner of alchemical phenomena, and the elementals of the animal kingdom produce the vast majority of spiritualistic manifestations.[3]

Here is a very simple procedure for evoking the spirits of forest plants to visible appearance on the material plane: After the usual purifications, go into the woods early in the morning and look for a place where the tree cover is so thick it blots out the sky completely. You should be able to find such a place without too much difficulty, especially among ⋆pines or other more leafy trees. Find a place to sit down comfortably and be perfectly still. Hold your eyelids half closed and fix your gaze while mentally evoking the sylvans or sylvesters, or whichever name you prefer to give them. Do not stare exactly but rather look intently among the trees while holding your eyes still. You may have to hold your breath at first to keep your eyes from moving, but a little practice will enable you to do this as though it were second nature. The spirits of the wood will surely appear before you, especially if you offer them some water or ⋆wheat and continue the evocation for several days in a row. The spirits inhabiting the wood are as vigorously and personally alive as you are, and your contact with them will be as real as your contact with any human individual.[4]

Another particularly interesting aspect of plant magic is a kind of ritual pact by which the fate of a newborn child is linked to the fate of a particular tree. A close relationship develops between the two co-created beings: the child benefits from the vigor of the tree and vice versa, but if the tree were to be cut down, the child would suffer and die. This plant-human pact is still very much in vogue among natives of Central America, New Guinea, New Zealand, India, and some parts of Germany.

In the same vein, there is not a single village in the East Indies that does not have a tree inhabited by a guardian spirit, to which members of the lower classes render worship. This is very similar to those Hellenic traditions that maintain that every forest has its spirit and each tree its hamadryad, or nymph. It is, moreover, not uncommon to see in the Nilgiris District of Tamil Nadu large trees with figures traced in vermillion and with three red-painted stones situated at their base. To these "haunted trees" the lower classes elect and dedicate one child from a family in the community for periods of seven years or more.* Such trees serve as the locus of acts of religious devotion, sacrifices, and worship (*pūjās*), as well as prayers for the sick or possessed. Natives call these guardian tree spirits *munispurams.* They are, more often than not, beneficent spirits, but their power is restricted to a single object. Such trees belong almost exclusively to the genus *Ilex* (★holly), although the genera *Eugenia* (in the ★myrtle family) and *Cinnamomun* (★cinnamon) occasionally serve as abodes for these guardian tree spirits.[5]

The mystical symbolism of the tree is so highly developed in the sacred texts of ancient religions that we must limit ourselves here to recalling only the Tree of the Knowledge of Good and Evil and the Tree of Life in the Garden of Eden, which are emblematic of the two paths Adam could follow to accomplish his mission; the sephirotic tree of the kabbalah; the *aśvattha,* or sacred fig tree (*Ficus religiosa* L.), of Hindu mythology; the Zoroastrian *haoma,* which Zoroaster employed as a symbol for the circulatory and nervous systems of the microcosmic human being and the macrocosmic universe; the *zampun* of Tibetan mysticism;† the Norse Yggdrasil; the sacred oak of the Druids; and the winged oak of the pre-Socratic philosopher Pherecydes.[6] Each of these arboreal symbols has multivalent meanings. We merely reference them here, however, so as not to distance ourselves too much from the subject at hand, which pertains to our mental development. Virtually all religious traditions concerning high adepts and initiates present them acquiring enlightenment beneath such a tree. Only the Christ, who among other things represents knowledge itself (γνῶσις),

*Children suffering from the matted hair disease known as *plica polonica,* or Polish plait, are considered to be especially favored by the tree gods and are often selected for dedication.

†[Sédir's source here is Helena Petrovna Blavatsky, who describes the Tibetan *zampun* as follows: "The first of its three roots extends to heaven, to the top of the highest mountains; the second passes down to the lower region; the third remains midway, and reaches the east" (Blavatsky, *Isis Unveiled,* 1:152). Several theosophical writers repeat this information but without ever clarifying the origin or spelling of the name *zampun,* which appears to be Blavatsky's own idiosyncratic transliteration, or perhaps a dialectical variation, of the Tibetan "tree" known as the three roots (ཙ་བ་གསུམ་ in Tibetan)—that is, *tsawa sum,* or *rtsa ba gsum* in the Wylie transliteration schema. The three roots, traditionally identified as *lama* (spiritual teacher), *yidam* (meditation deity), and *khandroma* (spiritual muse), are sometimes characterized as a metaphorical tree of life or tree of health in Tibetan medical texts. —*Trans.*]

is never portrayed with this particular mythological motif. The reason for this, which is rather esoteric, concerns the very definition of his being or, if you prefer, the dual use his being could make of his free will. Thus, a complete religious symbolism should comprise two trees, but only the kabbalistic or Egyptian tradition identifies two distinct trees, for which reason it was destined to be crowned by the descent of the Son of God. The other traditions, as the inheritances of cultural disintegrations, preserve only outward memories of the Tree of Knowledge.

In terms of naturalistic initiation, the Tree of Knowledge is nothing other than the image of the inner microcosmic human being: its trunk is the spinal cord; its branches are the 72,000 nerves described by the Hindu yogis; its seven flowers are the seven centers, or chakras, of the astral body; its leaves are the dual respiratory apparatus of the lungs; its roots are the genital center and the legs; and its sap is the cosmic electricity that courses through the nerves as it descends from the cerebral ether to the spermatic earth.

The words *yoga* and *religion* are synonyms. The word *yoga* derives from the Sanskrit noun योग, which means "union" and refers to the act of yoking, joining, uniting, or harnessing. The word *religion* derives from the Latin verb *religare,* meaning "to reconnect" or "to bind together." Both terms refer to the bond that unites the microcosmic human being with the macrocosmic universe and God. This process of unification is the same as the process by which the seed borrows energy from the dark earth and uses it to guide its molecules in their development and transform into a fragrant flower. Serious practitioners of yoga know that yogic exercises transmute the impure molecules of the physical body into fixed and unalterable molecules, the base passions into energized enthusiasm, and intellectual ignorance into the light of truth. This is the real reason why yogis are so often depicted sitting beneath a sacred tree.

MAGICAL PHILTERS

A philter is any potion whose composition contains substances that have been magically prepared for the occult attainment of a specific end. The three kingdoms of nature furnish countless materials for such preparations, but we concern ourselves here only with materials that come from the plant kingdom.

Generally speaking, ointments, electuaries (powdered herbs mixed with a sweetener, such as honey), unguents, collyriums (eye salves), and magical potions all fall under the domain of "black magic." Their number is great, and this number can be multiplied indefinitely by the ingenious mind. The Chinese Taoist priests, for example, employ only thirteen vegetal, animal, and mineral substances in all of their branches of medicine, psychology, and magic, but they know how to extract a prodigious number of permutations from these thirteen substances.

Such preparations can be used on oneself or on others. They act on the astral body—that is, on one of its three great centers:

1. The instinctive center (the solar plexus)
2. The passionate center (the cardiac plexus)
3. The mental center (the cervical plexus)

In the first center, they produce health, illness, and all possible physiological phenomena. In the second center, they induce love, hatred, and other passions. In the third center, they cause phenomena such as somnambulism, clairvoyance, clairaudience, or divinatory phenomena of a higher and more developed order.

The folkloric accounts of the witches' sabbat and the reports of remote poisonings and assassinations at a distance of people or animals all describe the actions of magical plant substances on the instinctive center. Similarly, the numerous tales of love potions and magical philters describe their occult actions on the passionate center. The mystical use of psychotropic plants to provoke psychic phenomena via the mental center, however, is somewhat less well known. This art is still practiced in the East in many Buddhist convents, as well as among the Chinese Taoists, Tibetan lamas, Bhutan tantrikas, Turkestanian shamans, and some orders of Muslim dervishes, not to mention the less sophisticated, mechanical use of such plants to achieve similar ends by tribes in various remote parts of the world.

Hashish and opium are two of the most well-known psychotropic plant substances, but no one in the West, unless they were initiated into their mysteries in the East, seems to know the proper mystical and occult manner of utilizing these substances. The stories of Thomas de Quincey and Charles Baudelaire, regardless of their literary value or artistic merit and sincerity, do not offer a window into the magical possibilities of these mystical adjuvants.[7] All we are permitted to say is that the use of such drugs will lead to intellectual ecstasy only if subjects have previously been able, without the use of a stimulant and by sheer force of the will alone, to master their mental powers and control the association of ideas, which is no mean feat. Otherwise, hashisheens who have not previously fixed their minds are setting out to sea in a sieve, to a sea far more terrible than the cyclone-prone Indian Ocean, from which they shall return, if they return at all, with madness as their companion.[8]

In *Occult Masonry and Hermetic Initiation,* Jean-Marie Ragon, the preeminent interpreter of Freemasonry, tabulates the results of certain magnetic experiments made with colored disks and psychotropic plants. He took disks of the different colors of the spectrum, coated them with the juices of various plants, and presented them to magnetists to hold in their gaze and contemplate. We reproduce the results of his experiments in Table 6.1 on the following page.[9]

We do not recommend that anyone repeat these experiments since they clearly disrupted the nervous systems of Ragon's unfortunate subjects under the fallacious pretext of immediate scientific utility. We also reject the practices of natural and psychic magic except for therapeutic purposes. The egoistic satisfaction of personal desire, which is the foundation of all erotic and aggressive magic, and the futile acquisition of intellectual knowledge are not ends important enough to contravene the free will and hamper the laws of the universe. Only one thing is necessary, and that is to love God and to love one's neighbor.* Everything else is vain and ephemeral.

*[Here Sédir paraphrases Jesus's answer to the question concerning which is the greatest commandment in the law: "Thou shalt love the Lord thy God with all thy heart, and with all thy soul, and with all thy mind. This is the first and greatest commandment. And the second is like unto it, Thou shalt love thy neighbor as thyself" (Matthew 22:37–39). —*Trans.*]

TABLE 6.1. PSYCHOTROPIC PLANTS

Disk	Plants	Effects
1. Violet	★**Belladonna** (*Atropa belladonna* L.) ★**Henbane** (*Hyoscyamus niger* L.) ★**Indian hemp** (*Cannabis indica* Lam.) ★**Jimsonweed** (*Datura stramonium* L.) **Snakewood** (*Strychnos colubrina* L.)	Continuous movement of the arms and legs; a desire to touch or walk over ordinary objects; screeching, barking, or imitating other animal sounds; an urge to bite or fight; total inebriation; an appearance of happiness; illusions of possessing anything one desires; a clear memory of all actions and visions.
2. Indigo	★**Black pepper** (*Piper nigrum* L.) **Sabadilla** (*Veratrum sabadilla* Retz.)	Febrile excitement; weakness in the abdomen and legs; subject kneels down as though to pray but cannot recall a single word; loss of vision but not of mobility, hence subject frequently bumps into walls; tremors of the eyelids, occlusion of the eyes, deep sleep; subjects awaken only when water is poured over their face.
3. Blue	**Asafetida** (*Ferula assa-foetida* L.) ★**Camphor** (*Cinnamomum camphora* L.) **Cubeb pepper** (*Piper cubeba* L.f.) ★**Hemlock** (*Conium maculatum* L.)	General excitement; convulsive movements; a desire to fall asleep, drowsiness; loss of the ability to reason; despondency.
4. Green	**False angostura bark** (the bark of *Strychnos nux-vomica* L.) ★**Mandrake** (*Mandragora officinarum* L.) **Wild lettuce** (*Lactuca virosa* L.)	Copious tears; subject fidgets and plays with his hands; a desire to run (subject claims he could run faster than a horse); a thrilling sensation in all the muscles; subject wishes to say his final farewells, as though he is about to die; a general numbness; lethargy.

TABLE 6.1. PSYCHOTROPIC PLANTS (*cont.*)

Disk	Plants	Effects
5. Yellow	***Asparagus** (*Asparagus officinalis* L.) ***Lettuce** (*Lactuca sativa* L.) ***Opium poppy** (*Papaver somniferum* L.) **Saint Ignatius's bean** (seeds of the fruit of *Strychnos ignatii* P.J.Bergius) **Strychnine** (*Strychnos nux-vomica* L.) ***White hellebore** (*Veratrum album* L.)	Swaying of the head back and forth; general numbness; drowsiness (when subject opens his eyes, the color of the disk puts him in a furor); voluptuous dreams, chills, and extreme pallor; severe despondency; renewed somnolence; subjects are in a zoomagnetic or hypnotic state but remain mobile and responsive; no memory of what transpired.
6. Orange	**Essential salt of opium** (*sal essentiale opii* or morphine) **Jalap** (*Ipomoea purga* [Wender.] Hayne) ***Tobacco** (*Nicotiana tabacum* L.) ***Valerian** (*Valeriana officinalis* L.)	Subjects experience intense joys, numbness of the limbs, and drowsiness; when subject opens his eyes, the colored disk makes him want to laugh, but he is interrupted by an inexplicable moral suffering; tears; a general tendency toward great lucidity.
7. Red	***Foxglove** (*Digitalis purpurea* L.) **French lavender** (*Lavandula stoechas* L.) **Heal-all** (*Prunella vulgaris* L.) **True lavender** (*Lavandula angustifolia* Mill.)	Subject is paranoid and fears that people are lurking in wait and mean to do him harm; acute and intermittent screams; this state lasted from two and a half hours in some subjects to four to five hours in others; each required a lengthy period of time to recover.

[Ragon's description of the mechanics of these experiments is extremely vague. He adds only that the colored disks were actually nine in number: "disk numbers 1 through 7 represent the primitive colors, while disk number 8 is white and disk number 9 is black, signifying the beginning and end of the color spectrum. Each colored disk equally impacts the imagination of the subject, but each one does this in its own unique way" (Ragon, *Maçonnerie occulte,* 83). —*Trans.*]

WITCHES' OINTMENTS

One of our primary sources of information on the curious effects of witches' ointments is the French physician and demonologist Jean de Nynauld's *Lycanthropy, Transformation, and Ecstasy,* a rare book, which we had the opportunity to consult in the library of dearly departed master Stanislas de Guaïta.[10]

"Of all the simples that the devil uses to disturb the senses of his servants," Nynauld writes, "some have the virtue of causing a deep sleep, while others are mildly soporific or hallucinogenic and tempt the senses with bizarre figures and phantasmagoria, as much in the waking state as in the sleep state."[11] According to Nynauld, the most powerful of these so-called diabolical herbs are the following:

1. ★Belladonna root (French: *la racine de belladonna*)
2. ★Belladonna (French: *morelle furieuse*)[*]
3. Bat's blood (French: *sang de chauve-sourris*)[†]
4. Hoopoe's blood (French: *<sang> d'huppe*)[‡]
5. ★Wolfsbane (French: *l'aconit*)
6. Water parsnip (*Sium suave* Walter) (French: *la berle*)
7. ★Belladonna (French: *la morelle endormâte*)[§]

*[*Morelle furieuse* (deadly nightshade), diabolical herb no. 2, is a French common name for belladonna. Although it is possible that Sédir understood this to refer to jimsonweed (see the entry ★jimsonweed in Sédir's dictionary and the corresponding footnote), the latter is already listed as diabolical herb no. 15. Possibly, since Nynauld identifies diabolical herb no. 1 with belladonna *root,* the name *morelle furieuse* here signifies belladonna *leaves.* —*Trans.*]

†‡[It is probable that most if not all of the stomach-churning ingredients of the ancient and medieval grimoires functioned as a sort of witches' argot to keep the magical recipes secret. This appears to have been the opinion of both Papus and Sédir (see especially the entries ★coltsfoot and ★cyclamen in Sédir's dictionary and their corresponding footnotes). The true identities of diabolical herbs nos. 3 and 4 remain unclear, but these very same ingredients are at least as old as the Greek and Demotic magical papyri. See, for example, *PGM* II. 1–64, XXXVI. 231–55; and *PDM* XIV. 1–92, 295–308, 1070–77, 1110–29. —*Trans.*]

§[Diabolical herb no. 7, *morelle endormante* (sleeping nightshade), which is the French equivalent of the Latin name used by the old herbalists, *solanum somniferum* (cf. the footnote to diabolical herb no. 18), is invariably understood as a variant for *morelle furieuse.* The name could also conceivably refer to jimsonweed, another soporific member of the nightshade family, which was known by the French common names *endormeuse, endormie,* and *pomme endormante,* although it is equally possible, given Nynauld's use of *morelle furieuse* (diabolical herb no. 2), that *morelle endormâte* here signifies the belladonna fruit. —*Trans.*]

8. ★Celery (French: *l'ache*)
9. Soot (French: *suye*)*
10. ★Cinquefoil (French: *le pentaphyllon*)
11. ★Sweet flag (French: *l'acorum vulgaire*)
12. ★Parsley (French: *le persil*)
13. ★Poplar leaves (French: *feuilles du peuplier*)
14. Juice of the ★opium poppy (French: *l'opium*)
15. ★Jimsonweed (French: *hyosciame*)
16. ★Hemlock (French: *cyguë*)
17. Other species of poppy (French: *les especes de pauot*)
18. Poison darnel (*Lolium temulentum* L.) (French: *l'huyroye*)†

The aforementioned simples behave in the same manner as the *synochitides,* or "holding stones," which are capable of bringing the qliphoth, demons, and other infernal shadows to visible appearance, as opposed to the *anancitides,* or "stones of necessity," which reveal images of holy angels.‡

*[Nynauld characterizes these ingredients as simples—that is, plants with medicinal properties. This has led some to speculate that the Latin *fuligo,* "soot," of which Old French *suye* is merely the translation (cf. the footnote to diabolical herb no. 18), might be an error for Latin *siligo,* "wheat" or "wheat flour." This is unconvincing not only from a paleographical standpoint but also for the reason that the *fuligo* in question is no doubt the *fuligo ligni,* or "wood soot," of the old herbalists, which, like several other black or ash-colored Saturnian herbs, was held to have spasmolytic properties. —*Trans.*]

†[The herb Nynauld calls *hyuroye* is most likely the plant known as darnel, poison darnel, or darnel ryegrass (*Lolium temulentum* L.) or *ivraie* in modern French, which derives from the Latin *ebriacus,* meaning "drunk" or "inebriated" (much like the word *temulentum* in the modern Latin binomial, which translates to "drunken ryegrass"). According to Robert Estienne's *Dictionarium seu latinae linquae thesaurus,* 17 s.v. *aera,* the herb *hiuroie* is otherwise known as *lolium,* or ryegrass. Several of the ingredients Nynauld mentions derive from the recipes for witches' ointments (*lamiarum ungenta*) in Giambattista della Porta's *Magiae naturalis.* In the first Latin edition of this work, Porta describes these ointments as mixtures of boiled fat with *eleoselinum* (★celery), *aconitum* (★wolfsbane), *frondes populneas* (poplar-like leaves), and *fuligo* (soot) or, alternatively, with *sium* (from Greek σίον, water parsnip), *acorus vulgaris* (★sweet flag), *pentaphyllon* (from Greek πεντάφυλλον, ★cinquefoil), *vespertilionis sanguis* (bat's blood), *solanum somniferum* (★belladonna), and *oleum* (oil). It is to be noted as well that Nynauld relies on François Arnoullet's 1615 French translation of Porta's work and that these recipes were censored from subsequent editions of *Magiae naturalis* and, as a result, are nowhere to be found in the English translation of 1658. —*Trans.*]

‡[According to Pliny the Elder, the *synochitides,* or συνοχίτιδες (from συνοχή, meaning "holding together"), were used to summon and detain "shades from below" for nefarious magical ends, whereas the *anancitides,* or ἀναγκίτιδες (from ἀναγκή, meaning "necessity"), were

Nynauld also recognizes three main types of ointments in the diabolical pharmacopeia.[12] The first type, which induces dream visions, consists of fat, ★celery juice, ★wolfsbane, ★cinquefoil, ★belladonna (French: *morelle endormante*), and wood soot. By virtue of the second type, witches who have rubbed their bodies with the unguent acquire the ability to fly through the air—according to folklore, by means of a broomstick situated between their legs—and ride off to their sabbats at an incredible speed. According to Nynauld, "no narcotic simples factor into the composition of this unguent, but only those which have the virtue of disturbing and alienating the senses, such as an immoderate portion of wine, cat's brains,* ★belladonna, and other ingredients concerning which I shall remain silent for fear of giving the wicked the power to do evil."[13]† The third type of ointment was allegedly imparted to witches by the devil himself and has the power to transform them into beasts. Its composition, concerning which Nynauld is extremely vague, consists of various parts and organs of the toad, the snake, the hedgehog, and the fox, as well as human blood and certain herbs and roots that have the virtue of troubling and disturbing the senses and affecting the imagination.[14]

In the same vein, the German mystic Karl von Eckartshausen, who published a series of important works at the end of the eighteenth century, gives the following formula for a magical suffumigant that has effects similar to the first of Nynauld's witches' ointments but is designed specifically to manifest spirits and apparitions on the physical plane: burn incense pellets made of dried ★hemlock, ★henbane, ★saffron, ★aloe, opium, ★mandrake, ★opium poppy, asafetida (*Ferula assa-foetida* L.), and

(*cont. from p. 61*) employed in hydromantic rituals to summon the gods to visible appearance. See Pliny, *Natural History* 37.73. Qliphoth (literally "peels," "shells," or "husks") are, in Jewish mysticism, evil or impure spiritual forces. —*Trans.*]

*[The ingredient cat's brains (French: *cervelle de chat*) is commonly listed in old books of poisons and venoms among toxic, narcotic, soporific, and psychotropic plants; see, for example, Ardoynis, *Opus de venenis,* liber IV, caput XXI, 250–51. According to Pietro Andrea Mattioli, *cervelle de chat* is what the ancient Greek's called αἴλουρος, meaning "cat," which was the ancient Greek common name for the scarlet pimpernel (*Anagallis arvensis* L.). See Mattioli, *Les commentaires de M. P. André Matthiole,* 569; cf. Sédir's entry ★pimpernel in part 3. —*Trans.*]

†We shall imitate the prudent reserve of Dr. Nynauld and refrain from mentioning either the ingredients of this concoction or the manner of its preparation.

★parsley.* The following magical suffumigant, on the other hand, from the magical treatise titled *Pneumatologia occulta et vera,* is exorcistic in action and drives away evil spirits and apparitions: burn incense pellets made of sulfur, asafetida, castoreum, ★Saint John's wort, and vinegar.†

Nynauld also provides a number of recipes for magical perfumes or suffumigants. He attributes these compositions to an anonymous doctor, but the "certain doctor" to whom he refers is actually none other than Heinrich Cornelius Agrippa von Nettesheim. Of Nynauld's collection of magical perfumes, the following are the most practical:

1. To achieve supernatural visions, burn incense made from ★heather root, ★hemlock juice, ★henbane juice, and ★opium poppy seed.‡

*[Karl von Eckartshausen's original formula includes the following ingredients: ★hemlock, ★henbane, ★saffron, ★aloe, opium, ★mandrake, ★belladonna, ★opium poppy, ★parsley juice, ★southernwood, and bulrush root (*Typha latifolia* L.). See Eckartshausen, *Aufschlüsse zur Magie aus geprüften Erfahrungen,* 2:101–2. Sédir's version omits *Nachtschatten* (belladonna) and *Moosrohr-Wurzel* (bulrush root) and has *assaoetida* (*sic*), a typographical error for asafetida (*Ferula assa-foetida* L.), instead of *Ferula (Gertenkraut)* (★southernwood). Sédir's source appears to be Karl Kiesewetter's article on magical suffumigants, which also has *Nachtschatten* (belladonna) and includes *Asa foetida* (asafetida), but instead of bulrush root has *Sumpfporst,* or marsh laborador tea (*Rhododendron tomentosum* Harmaja); see Kiesewetter, "Magishe Räucherungen." Moreover, it is clear from Eckartshausen's citation of the Austrian physician Joseph Jakob Plenck that *Nachtschatten* (literally "nightshade") refers to belladonna or deadly nightshade rather than black nightshade (*Solanum nigrum* L.) —*Trans.*]

†[Again, this formula appears to be Sédir's own variation. The original formula from the *Pneumatologia occulta et vera* includes only four ingredients: sulfur, asafetida, castoreum, and rue (*Ruta graveolens* L.) in equal portions. See Horst, *Zauber-Bibliothek,* 2:92–93; cf. Kiesewetter, "Magishe Räucherungen," 221. —*Trans.*]

‡[Sédir here passes over Nynauld's supplementary clause "along with some other ingredients" (*De la lycanthropie,* 73). According to Agrippa's original Latin, this magical suffumigant should include the following ingredients: root of *canna ferula,* juice of ★hemlock (*cicuta*), juice of ★henbane (*hyosciamus*), juice of ★mullein (*tassus barbassus*), ★red sandalwood (*sandalum rubeum*), and ★opium poppy (*papaver nigrum*). Agrippa further goes on to say that if ★celery (*apium*) is added to this composition, the suffumigant will become exorcistic in action. As for Agrippa's *canna ferula,* which Nynauld identifies as ★heather (French: *bruyere*), English translators apparently equate it with the *ferula medica* of the old herbalists, namely, sagapenum (from Greek σαγάπηνον), the oleo-gum resin of the Persian fern (*Ferula persica* L.), but Georg Franck von Franckenau equates it with *narthex,* which he identifies as *Ferula galbanifera,* that is, the galbanum plant (*Ferula galbaniflua* Boiss. et Buhse). See Franck von Frankenau, *Flora Francica aucta,* 130 s.v. *canna ferula,* 253 s.v. *ferula,* 430 s.v. *narthex.* Latin *narthex* derives from νάρθηξ, the ancient Greek name for the giant fennel (*Ferula communis* L.). —*Trans.*]

2. To have premonitions of future events, burn incense made from the seeds of ★flax and psyllium (*Plantago psyllium* L.) and the roots of ★violet and ★celery.
3. To drive away evil spirits and harmful ghosts, burn incense made from lesser calamint (*Clinopodium nepeta* [L.] Kuntze), ★peony, ★mint, and ★castor bean.
4. To make your house appear full of water or blood or to make the ground tremble, burn incense made from cuttlefish ink,[*] red storax,[†] ★rose, and ★aloeswood, then douse it with either water, blood, or earth.[15]

*[Nynauld's *fiel de seche* is an error for *fiel de sieche,* literally "cuttlefish gall." Although Agrippa's *fel sepiae* similarly refers to the "gall" or "bile" of the cuttlefish (Latin *sepia,* Greek σηπία, from which the modern genus name *Sepia* derives), surely this must be a reference to its dark-colored ink. In fact, the ancient Greek word for gall or bile, χολή, can also mean "cuttlefish ink." —*Trans.*]

†[Here Sédir has *thymiame,* Nynauld *thymiamas,* and Agrippa *thimyama.* Latin *thymiama* derives from Greek θυμίαμα, both of which were used in a general sense to mean "incense." Among the old herbalists, however, *thymiama* could refer to red storax, otherwise known as *thus Judaeorum,* or Jewish frankincense, the oleoresin obtained from the wounded bark of the storax tree (*Styrax officinalis* L.), or to ammoniacum, the gum-resin exuded from the stem of the gum ammoniac tree (*Dorema ammoniacum* D.Don.). Sédir uses the same term, but spelled *thymiane,* to identify the characteristic perfume of Pisces in chapter 3. See further Franck von Frankenau, *Flora Francica aucta,* 47–48 s.v. *ammoniacum,* 627 s.v. *storax,* 651 s.v. *thymiama.* —*Trans.*]

7
Occult Horticulture

The precepts and modus operandi of occult horticulture are poorly known in modern times. The fundamental basis of this art consists in sowing the seed in the precise earth matrix that complements it. Just as persons who find their soul mate, according to various mystical traditions, become more powerful in both word and deed, seeds sown in their "soul matrix" come to attain their generic perfection.

CULTIVATING MAGICAL PLANTS

Occult herbalists sow under the auspices of Saturn, the god of generation and agriculture. According to the well-known folk etymology, Saturnus, the Latin name of the Roman god Saturn, derives from *satus,* the past participle of the verb *serere,* meaning "to sow." The Gauls, moreover, called the seed *sat* and the sower *satur.*[1] To sow is to plant something in darkness, in the depths, and in isolation. The darkness provokes the light, and the formless mass of putrefied cotyledons evokes the resplendent flower and the majestic tree.

Unfortunately, the vast majority of farmers sow in earth matrices that do not complement in any way the seeds entrusted to them.[2] We have seen that the subterranean development of the seed takes place at the expense of the Salt, Sulfur, and Mercury of the earth. The Sun is the universal dispenser of life, but its invisible rays of vitality are assimilable to the seed only if they appear in complementary synergy. If, then, the earth in which the seed resides does not satisfy these conditions, the seminal *ens* will extend its radicles and exhaust its forces as it searches for nutritive elements in its external environment. In such cases, the root and the stem will grow dry and gnarled: the Salt, Sulfur, and Mercury are squandered as they fruitlessly consume the Solar vitality, which comes to them in an environment ill suited for its assimilation.[3]

The art of occult horticulture remedies this fundamental deficiency in one of two ways: if the seed has yet to be sown, through the careful election of the earth matrix best suited to the fertilization of the seed, or if the seed has already germinated, through the administration of a vital stimulant. In the first case, it is essential to know the exact proportions of Salt, Sulfur, and Mercury in the compositions of the earth matrix and the seed; in other words, one must know the exact chemical composition of both earth and seed. In the second case, the occult herbalist produces—over the course of preparing the philosopher's stone and especially by means of the *via sicca,* or "dry way"*—various liquors that serve as suitable medicines for sickly plants. We shall discuss these vital liquors in greater detail in the chapter on vegetation magic.

In addition to knowing the relationship between the plant and the soil that physically nourishes it, the occult herbalist must also be cognizant of plant sympathies and antipathies—that is, the allelopathic relationships the plant has with other plant species—and carefully choose its polycultural society. Some plants thrive when their companions please them but languish when their neighbors displease them. For example, the ⋆olive tree is the friend of the ⋆grapevine, but the ⋆cabbage is the enemy of the ⋆grapevine. The ⋆buttercup is the friend of the ⋆water lily, and the rue (*Ruta graveolens* L.) is the friend of the ⋆fig tree. Students can easily determine the sympathetic and antipathetic relationships among plants by analyzing the relationships between their planetary and zodiacal signatures.[4]

Lastly, there can be no question that external agents, and light especially, also have a great influence on plant life. Red light abnormally accelerates the growth of plants, whereas green light forces plant growth to a lesser extent. White light is normative and therefore has a natural effect on plant growth,

*[The alchemists developed two primary methods for creating the philosopher's stone: the *via humida,* or "wet way," and the *via sicca,* or "dry way." Whereas the *via humida* used natural, potential, "philosophical," or "cold" fires and took lengthy periods of time, the *via sicca,* often considered to be the more dangerous of the two paths, employed the intense heats of the actual or artificial alchemical fires and hence could be completed in shorter periods of time. —*Trans.*]

but blue light is harmful and significantly retards plant growth. Camille Flammarion has conducted conclusive scientific experiments on the effects of these colored lights on the development of plants.[5]

HARVESTING MAGICAL PLANTS

According to ancient medicine, the astrological conditions under which the plant is harvested greatly influence the virtue of the simple. For this reason, we indicate the ideal astrological conditions for harvesting each of the plants listed in our dictionary in part 3 of this work. In addition, according to the doctrines of Hermetic herbalism, it is essential that the herbal remedies prepared from these plants be administered under the appropriate astrological conditions, in other words, during conjunctions of the benefic planets and when the malefics are in exile. Since these astrological conditions ultimately depend on the particulars of the patient's natal chart, we cannot be as thorough on the subject of astrological or Hermetic herbal medicine, but our dictionary will nonetheless offer some guidance in the different cases practitioners are likely to encounter.[6]

Occult herbalists may additionally enhance the medicinal or magical potency of the herbs they gather by harvesting them on the respective day and hour of their ruling planet. All herbs and plants under the dominion of the Sun are best gathered on Sundays; those under the dominion of the Moon on Mondays; those under the dominion of Mars on Tuesdays; those under the dominion of Mercury on Wednesdays; those under the dominion of Jupiter on Thursdays; those under the dominion of Venus on Fridays; and those under the dominion of Saturn on Saturdays. The ruling planet of the day of the week governs *the first and the eighth hour of the day,* each day being divided into twelve equal parts with the first "hour" beginning at sunrise, and *the third and the tenth hour of the night,* each night being divided as well into twelve equal parts, but with the first "hour" beginning at sunset. The planetary hours of the day tabulated in Table 7.1 on the following page are nearly twice as long at the height of summer (roughly eighty minutes) as they are in the midst of winter (roughly forty minutes).[7]

The practice of harvesting medicinal simples under favorable astrological conditions has long been condemned by the church. Penitential canons such

TABLE 7.1. PLANETARY DAYS AND HOURS

HOURS								
Day	Night	Sunday	Monday	Tuesday	Wednesday	Thursday	Friday	Saturday
1	3	☉	☾	♂	☿	♃	♀	♄
2	4	♀	♄	☉	☾	♂	☿	♃
3	5	☿	♃	♀	♄	☉	☾	♂
4	6	☾	♂	☿	♃	♀	♄	☉
5	7	♄	☉	☾	♂	☿	♃	♀
6	8	♃	♀	♄	☉	☾	♂	☿
7	9	♂	☿	♃	♀	♄	☉	☾
8	10	☉	☾	♂	☿	♃	♀	♄
9	11	♀	♄	☉	☾	♂	☿	♃
10	12	☿	♃	♀	♄	☉	☾	♂
11		☾	♂	☿	♃	♀	♄	☉
12		♄	☉	☾	♂	☿	♃	♀
	1	♃	♀	♄	☉	☾	♂	☿
	2	♂	☿	♃	♀	♄	☉	☾

From Bélus, *Traité des recherches*, 50.

as Archbishop Theodore of Canterbury's *Paenitentiale Theodori,* the Anglo-Saxon monk Bede's *Paenitentiale Bedae,* Archbishop Rabanus Maurus of Mainz's *Liber paenitentium,* and Bishop Halitgar of Cambrai's *Paenitentiale Halitgari,* as well as collections like the French Benedictine Luc d'Achery's *Collectio Dacheriana,* Bishop Isaac of Langres's *Canones Isaac episcopi lingonensis,* and Archbishop Ecgbert of York's *Paenitentiale Ecgberti,* and even works on canon law such as Burchard of Worms's *Decretum Burchardi* and Bishop Ivo of Chartres's *Decretum Ivonis* are all unanimous in condemning the doctrine of plant signatures. According to these canons, collections, and decrees, whoever observes harvests according to the "superstitious signs" attributed to plants and trees must do penance for two years on "lawful feast days"—which is to say, on Sundays and holy days of obligations—and whoever gathers medicinal herbs while uttering invocations, magical words of power (*voces magicae*), or barbarous names (*nomina barbara*) must do penance for twenty days. The decrees of Jean-François Bonhomme, bishop of Vercelli and apostolic envoy of the Italian provinces of Novara and Como under Pope Gregory XIII, forbids the harvesting of fern, fernseed, and any other herb or plant on a specific day or night under the belief that the simple in question would be rendered ineffective if it were to be gathered at any other time. "If anyone is found guilty of such superstitions or others of a similar nature," the decree proclaims, "let them be punished severely according to the severity of their crime and as it pleases the local ordinary."[8]

We should add that the outlandish proscriptions of the church against herbal rites and rituals do have their reason for being, but this secret is known only to a select few. Suffice it to say that from a truly mystical point of view, in the divine plan every magical act is an act of revolt. For this reason, occult herbalists must exercise a great deal of judgment and act with equal caution. In any case, the feast day of Saint John the Baptist (June 24), or the days before and after, is an ideal time for gathering all sorts of herbs, but especially ★chicory, ★elderberry, ★elecampane, ★male fern, ★Saint John's wort, and ★wheat. In addition, each plant has a few days each year when its force is supremely exalted, and the planetary hours of the night are usually more favorable in this respect. Lastly, occult herbalists should harvest plants, after consecrating them with the signs, invocations, and words of power germane to their signature, with a special knife used solely for this purpose.

8
Vegetation Magic

Simon Magus alleged he could make trees shoot up suddenly and produce sprouts instantaneously. "In short," he says, "once when my mother Rachel ordered me to go to the field to reap, I saw a sickle and commanded it to go and reap, and it reaped ten times more than the other workers." The learned reader will no doubt recognize in the conspicuous Saturnian symbol of the sickle or scythe (♄) an ancient reference to occult horticulture. "Lately," Simon continues, "I have produced many new sprouts from the earth and made them bear leaves and produce fruit in an instant."[1] There are numerous eyewitness accounts of performances of vegetation magic, by which magicians, physicians, fakirs, and spiritualists alike have utilized occult forces to stimulate and accelerate plant growth. We mention here only a few of the most interesting cases cited by Dr. Carl du Prel in his fascinating work on the subject of *Forciertes Pflanzenwachstum,* or "forced plant growth."[2]

FORCED PLANT GROWTH

Christoph Langhans witnessed a fakir's public performance of vegetation magic during his travels in India in the early 1700s. The fakir asked the crowd for an orange, which Langhans repeatedly calls *Apfelsine,* or "Chinese apple." As soon as someone in the crowd handed the fakir an orange, he immediately cut it open, pulled out a single seed, and buried it in the earth, which he had moistened a little beforehand. He covered this spot with a small basket and placed a handful of tobacco in his mouth. For some time he worked a wax thread under his lip, coating the thread from end to end with the sticky substance, each time adding a thicker coat of tobacco on the wax thread. Suddenly he lifted the basket and revealed to the crowd that a plant had grown out of the ground within a half hour. He covered the plant once more and performed a few more antics similar to the trick with the wax thread. When he removed the basket for the second time, the plant was as tall as the basket itself and bore fragrant flowers. He covered the small tree with the basket and performed a few more contortions. Upon removing the basket for the third time, the crowd saw that the tree had borne fruit. To make the fruit mature and ripen, he began once more to coat the wax thread with tobacco, and after a quarter of an hour the fakir presented five beautiful ripe oranges to the crowd. Langhans tasted one and remarked that its flavor was almost indistinguishable from any naturally grown orange. The fakir then uprooted the ★orange tree and threw it into the river.[3]

James Hingston reports witnessing a similar magical feat on the verandah of his hotel on Popham's Broadway in Madras. He saw a group of four fakirs come and seat themselves on the stone flooring of the verandah. Their clothing consisted of the usual bit of rag about their loins, so as to testify that nothing could be concealed up a sleeve or in the folds of their garments. "Yet these men were the best of their kind that I ever saw," Hingston affirms. "One of these jugglers then laid a nut on the stone floor of the verandah and covered it with two pieces of toweling. He raises these now and again to show the process that is going on. The nut is sprouting, and the sprout grows more each time it is covered, until it is, in ten minutes, a veritable little tree, the roots of which are shooting out of the other side of the nut."[4]

Louis Jacolliot reproduces ritual prescriptions for yet another variation of vegetation magic from the so-called *Agrouchada-Parikchai.** This plant ritual

*[The authenticity of the text Jacolliot calls *Agrouchada-Parikchai* is often called into question. Many believe the work to be a pastiche of sacred Hindu scriptures of Jacolliot's own making. The first element of the book title is a compound of Sanskrit *agra* (अग्र), meaning "foremost" or "best," and a dialectical variant of Sanskrit *auṣadha* (अउषध), meaning "herb," "drug," "herbs" (collectively), or "herbal medicine." Surely these are what are commonly classed as *agrya aushadha* or *aushadhi,* "best herbs" or "drugs of choice," in Ayurvedic medicine (in the Ayurvedas such and such an herb is often characterized as the "best herb" or "drug of choice" for such and such a disease). The second element is a dialectical variant of Sanskrit *parīkṣā* (परीक्षा), meaning "investigation," "examination," or even "trial by ordeal." The book title *Agrauṣadhaparīkṣā* literally means *Examination of the Best Medicinal Herbs,*

was performed during the Upanayana ceremony. On the second day of the celebration, all the married women in the village would gather in the forest to search for a nest of white ants. Once they located one, they would fill ten earthenware pots with the disturbed earth of the ant's nest and plant a different kind of seed in each pot. The Brahmin would then cover each of the pots with a fine cloth and recite the following invocation: *Agnim pā pātra paryāya parōxa.* This magical formula, which was to be recited eighty-one times, contains the following Sanskrit words:

1. *Agni* (अग्नि): sacred fire
2. *Pā* (पा): holy water
3. *Pātra* (पात्र): purified vessel
4. *Paryāya* (पर्याय): magical vegetation (literally, "creation" or "artificial production")
5. *Parokṣa* (परोक्ष): invisible

Like many ancient invocations, the magical formula अग्निं पा पात्र पर्याय परोक्ष is ungrammatical but highly alliterative. Over the course of its repeated recitation and intonation, attendants would see the cloths covering the pots slowly rise. The seeds would not only germinate but grow into plants and shrubs with beautiful flowers and flavorful fruits.[5]

No one, however, excited so much attention as Dr. Georg Andreas Agricola, who by means of a secret vegetable *mumia* could generate from a single tree as many more of the same species as the tree had boughs, branches, and buds. Each of the new trees would have roots and shoot forth branches and leaves within an hour and bear fruit within a year. He conducted the following experiments in Regensburg before Count Wratislaw in 1715:

(*cont. from p. 71*) but the meaning of this title seems to have been completely lost on Jacolliot, who frequently calls it either the Book of Spirits or the Book of the Pitṛs, a datum that argues strongly against the forgery hypothesis, so, too, on Helena Petrovna Blavatsky, who takes it to mean "Rules of Initiation" (*The Secret Doctrine,* 1:299), and on all other commentators as well, whose transliterations of the title vary widely. To judge from the portions of the treatise Jacolliot reproduces in translation, however, the text goes far beyond your garden-variety Ayurvedic commentary, such that one wonders whether the manuscript Jacolliot had before him was not actually a *florilegium* of herbal, traditional, and magical texts. On the ancestral spirits called *pitṛs,* see the entry *sesame in Sédir's dictionary in part 3. —*Trans.*]

1. Out of twelve small ★lemon trees he made as many large ones, each having roots, branches, and leaves proportionate to its great size.
2. He performed the next experiment on six different species of tree, such as the ★apple tree, the ★peach tree, and the apricot tree (*Prunus armeniaca* L.). The trees had grown only four to five feet high, but he immediately converted each one into a large tree, and each of the newly enlarged trees was perfectly furnished with roots, branches, and leaves and bore fruit within a year.
3. He made a third experiment on fifteen slips of carnation, or clove pink (*Dianthus caryophyllus* L.), from which he produced an equal number of large plants in the space of an hour, to the pleasure and amazement of the audience.[6]

We conclude our brief survey of accelerated plant growth with a still more remarkable account in which a spirit produces the same phenomenon. A medium known as Madame d'Espérance, who led an English spiritist circle, materialized on several occasions a spirit who called herself Yolande. During one of Yolande's materializations, she issued from the spirit cabinet and signaled for a vase, some water, and some sand. Yolande squatted on the floor in the sight of all who were present and called upon one of the sitters to put some of the water and sand in the vase. She planted the seed of a Chinese ixora (*Ixora chinensis* Lam.) in the vase and placed it near the center of the room. After a few circumambulations, she covered the vase with a muslin coverlet of white spirit drapery and then retired beside the spirit cabinet, which was about three feet away from the vase. Immediately, the sitters saw something project upward and expand until it reached a height of about fourteen inches. Yolande then rose up and removed the white coverlet, and the sitters saw that a plant with roots, a stem, and several green leaves had grown in the vase. In a matter of minutes, the plant produced a large flower and grew an additional six inches (see Fig. 8.1 on p. 74). On another occasion Yolande performed the same feat with the seed of a flamingo flower (*Anthurium scherzerianum* Schott), which produced a profusion of leaves but no flower.[7]

Similar accounts may be found in the writings of Jean-Baptiste Tavernier, Roger Gougenot des Mousseaux, and Johann Joseph von Görres.[8]

Fig. 8.1. Chinese ixora produced for William Oxley of Manchester at a séance held August 4, 1880. From d'Ésperance, *Shadow Land, or Light from the Other Side*, 264

Louis Jacolliot's well-known account of a fakir's spontaneous vegetation of a papaya, or pawpaw, seed (*Carica papaya* L.) provides further confirmation of these eyewitness accounts.[9]

THE VITAL FORCE

Progressive philosophers have attempted to explain these accounts of mediumistic *Pflanzenwachstum.* "We know," writes the German philosopher Eduard von Hartmann, "that the physiological functions of vegetal life can be powerfully excited, as much by super-refrangible rays of light as by electricity and chemical stimulants, that even in the hominal kingdom an exceptional four-year-old boy can attain the development of a thirty-year-old man, and that the growth of certain fast-growing vegetable germs can be artificially accelerated."[10] Accordingly, Hartmann asserts, it is quite conceivable that the mediumistic force functions in the same manner as a vegetable stimulant.

Hartmann's assertion incited Carl du Prel to construct the intriguing theory that the organic and intellectual life of humans offers a parallel to the accelerating action of vegetation magic described in the aforementioned accounts. The author refers to our subjective experience of time in certain dream phenomena, during which images pass before our eyes in processions that seem to last for hours but in reality span only a few brief moments.[11] In the maternal womb, moreover, the human being traverses a biological process that in external nature spans millions of years.[12] Why should it be impossible for a practiced will to construct an invisible material around a plant or animal, or even a mineral, that would provide its *ens* with a much more dynamic sustenance—which is to say, a much more spiritual alimentation? This, according to Papus, is precisely what the fakir does: "The will of the fakir acts upon the dormant life within the seed. He not only sets his own vital force in motion but also provides the seed with elements that are more active than those which nature generally provides it."[13] In other words, the fakir's vitality develops the seed upon which he lays his hands, while his soul is concentrated in that center of his astral body known as the sacral, or *svādhiṣṭhāna,* chakra, which is the second primary chakra located at the root of the sexual organs, just below the navel. These are the vital forces that produce the phenomena of vegetation magic.

THE VITAL LIQUOR

Whereas the fakir takes the invisible vital force of his own human organism and transfers it to the seed or plant, the alchemist creates from nature a visible *archaeus*—which is Paracelsian argot for the vital force, otherwise known as the *liquor vitae,* the "vital liquor" or "liquor of life"—and transfers it to the seed or plant.* In fact, Dr. Agricola is said to have artificially reproduced trees by first steeping their seeds in a secret liquor and then planting them in the ground.[14]

We provide below only two recipes from among the panoply of alchemical formulae, the first for the liquor of life (*liquor vitae*), and the second for the famous potable gold (*aurum potabile*).

*[Franz Hartmann defines *archaeus,* which derives from the Greek word ἀρχαῖος, meaning "ancient" or "original," as "the formative power of nature, which divides the elements and forms them into organic parts. It is the principle of life; the power which contains the essence of life and character of everything" (Hartmann, *The Life and the Doctrines of Philippus Theophrastus,* 31 s.v. *archaeus*). See further appendix 2. —*Trans.*]

Liquor Vitae

Take one ounce of Mars (♂) and one ounce of naked Venus (♀) and set them to digest at 75 degrees in a thick glass balloon until the mixture leaves a green or red *caput mortuum* and renders a greenish solvent liquor.* Distill it to dryness and cohobate five or six times until nothing remains in the retort, then evaporate and it will produce a red fixed salt. If you infuse some seeds in water to which only a small pinch of this salt has been added, the seeds will grow at an accelerated rate, the plant's leaves will obtain a luminous golden luster, and its fruits will have more flavor.

Aurum Potabile

Heat at 400 degrees some magistery of sulfur† until gelatinous. Melt the mass again and distill until it leaves a residue. Collect this residue and mix it thoroughly with a salt capable of fixing it. Distill the mixture over a strong fire (*ignis fortis*), then sift the *caput mortuum* through a sieve and repeat until the distillation renders nothing more than an insipid water. When treated with pure alcohol, you will obtain both an oil and a water, which must be decanted. The water dissolves the Salt of Gold.‡ When saturated with metals, it is good for watering sick vines and stunted fruit trees.

*[In alchemy, *caput mortuum* (literally "dead head") signifies the "worthless remains," "dead earth," or "feces" left over from a spagyric chemical operation. Often, as here, the digested or distilled liquor would be cohobated—that is, poured back over the *caput mortuum* and redistilled multiple times. In alchemical tractates, each planetary symbol corresponds to a particular metal and to a particular color. For example, in his description of the separation of the elements from metals, Paracelsus notes that the color of the oil depends on the practitioner's choice of metal: gold (☉) will turn the oil a light and brilliant red; silver (☾), a light azure; iron (♂), a dark red or garnet; copper (♀), a bright green; mercury (☿), a bright white; lead (♄), a livid gray; and tin (♃), a luteous or xanthous color. See Paracelsus, *The Hermetic and Alchemical Writings,* 2:15. The reader will likely notice that these color attributes are strikingly similar to the attributes of flower color in Sédir's tabulation of planetary signatures (see Table 3.3 on p. 34). As for the expression "naked Venus," Sédir is here making use of an alchemical metaphor—*sahe ich Fraw Venerem ganz bloß*—from the Fifth Day of the *Chymical Wedding of Christian Rosenkreutz.* See Andreä, *Chymische Hochzeit Christiani Rosencreutz,* 97. —*Trans.*]

†[The *magisterium sulphuris* (magistery of sulfur) of the alchemists is presumably one and the same as the *sulphur praecipitatum* (precipitated sulfur) of the old pharmacists. Precipitated sulfur is a fine yellow powder prepared by boiling sulfur and quicklime in water and then precipitating with hydrochloric acid. —*Trans.*]

‡[The Salt of Gold to which Sédir refers is the *sal auri philosophorum* of the alchemists, which is usually identified as potassium bisulfate or acid potassium sulfate, a white, inorganic compound that is water soluble and frequently used in fertilizers. —*Trans.*]

9
The Vegetable Phoenix

We now turn to the mysteries of biology in the three lower kingdoms of nature. The most intuitive of our contemporaries feel there is something secret lurking behind all ordinary chemistry, botany, and zoology. The great initiates of all ages have always known the identity of this secret something, and they have even allowed a few clues concerning its identity to slip into the mundane world. But whereas alchemy or occult chemistry is famous in the history of the scientific development of the West, occult botany is much less known, and occult zoology is hardly known at all. These three disciplines are the successive developments of a single notion: *terrestrial life.* We could attempt to reconstruct the secret art and science of each of the three kingdoms of terrestrial life as practiced in the ancient temples of wisdom, but this is not the place for seductive hypotheses, and so instead, we shall extract from these occult syntheses only those elements that pertain to the subject at hand.

THE PLANT SOUL

There is an intermediate world between the material and spiritual worlds. This is the astral world. The astral repeats itself across the three kingdoms of nature. In the plant kingdom, it is called the *leffas,* to use the Paracelsian terminology. The *leffas,* or astral body, of the plant, when combined with the vital force of the plant, constitutes the *primum ens,* or "first being," and it is this first being that is the object of plant palingenesis.

Plant palingenesis is a tripartite art consisting of three principal operations:

1. The revitalization of the soul of the plant or, if you prefer, the ghost or phantom of the plant
2. The resurrection of the body and soul of the plant
3. The creation of the plant in body, soul, and spirit using materials derived from the mineral kingdom

Here we provide the operative procedures only for the first method of plant palingenesis. Nothing, in fact, has ever been published concerning the bodily resurrection of plants or their creation from minerals.

The celebrated alchemist and spagyrist Joseph Duchesne, alias Josephus Quercetanus, reports meeting a certain Polish adept in Krakow who knew how to enclose the phantoms of plants in vials.[1] At any moment, opportune or otherwise, he could make a plant gradually appear within a seemingly empty vial. At the bottom of the glass vessel was a small portion of earth having the appearance of ashes, and the vessel was sealed with the seal of Hermes.* Whenever he wished to expose such a vial to public view, he would gently warm the bottom of the vessel, and the heat as it penetrated the center of the cinerous matter would cause first a stem to appear, then branches, leaves, and flowers, each according to the nature of the plant whose soul he had enclosed in the vessel. The plant phantom would be visible to viewers for as long as the soft, stimulating heat lasted.[2]

The vegetal soul invariably materializes in an ephemeral objectification in the morphic pattern of the original plant—that is, in its *sidereal,* or potential body, a substratum of visible matter (reduced to its *caput mortuum*). The plant soul presides in a vegetative mode over the molecular grouping of the "fledgling ice."†

*[That is to say, the vessel was hermetically sealed. For one method of forming a "seal of Hermes," using clay instead of crushed glass and borax, see Mavéric, *La médecine hermétique des plantes,* 160 (English edition: *Hermetic Herbalism,* 141). —*Trans.*]

†[The phrase *fledgling ice* is Guaïta's (*Le serpent de la Genèse,* 2:694). In Sédir's usage, this merely refers to the ghost-like formation and appearance of the vegetable phoenix. But there was also a variant palingenetic procedure, sometimes called icy palingenesis, which involved mixing the plant's ashes with water and setting it out to freeze during the winter. After freezing, the image of the plant would appear temporarily in the ice in the same ghostly spiderweb-like appearance. See further Kiesewetter's essay in the following chapter. —*Trans.*]

SECRET OPERATION OF PLANT PALINGENESIS

The anonymous work titled *The Apocalypse of Hermes: Philosophical Revelations from the Great Book of Nature,* published in 1790 by a Rosicrucian chapter, provides step-by-step instructions for a secret operation to obtain a vegetable phoenix. The author, who is identified solely by the initial D, uses this collocation as a metaphor not just for the ash-like substance at the bottom of the vessel but also for the prepared vessel of the palingenetic experiment. As for the essential manipulations of palingenesis, it is with some reservation that we reproduce these step-by-step instructions here in all their meticulous detail, but Athanasius Kircher already let the cat of the bag, as it were.*

Step 1. Collect four pounds of ripe seeds of whichever plant you wish to resurrect from its ashes or whose soul you wish to disengage. Grind the seeds in a marble mortar and put them in a glass vessel that is very clean and the same size as the plant from which you collected the seed. Stop the vessel and keep it in a temperate place.

Step 2. Choose an evening when the heavens are very calm and serene. Transfer the plant matter into a large basin and expose it to the nocturnal humidity so that it is imbued with the vivifying virtue of the dew.

Step 3. Tie a large clean cloth to four stakes in a meadow to collect eight pints of the same dew; wring out the cloth, and pour the water into a clean glass vessel.

*[This operation, inspired by Paracelsus's *De natura rerum: IX Bücher,* 36r–43r (Liber VI. *De resuscitatione rerum naturalium*), was first published in Latin in 1665 by Athanasius Kircher in *Mundus subterraneus in XII libros digestus,* 1:414–15 (Experimentum I. παλιγγενεσία, *seu regeneratio plantarum ex cuiuscunque plantae semine*), then in a French translation in 1790 by Touzay Duchanteau—the anonymous author D.—in *La grand livre de la nature, ou l'apocalypse philosophique et hermétique,* 16–18, a work that alleges to be published *par une Société de Ph[ilosophes] Inc[onnus]*—that is, "by a Society of Unknown Philosophers." Guaïta and Sédir both claim that the work was published by a chapter of the Rose + Croix, but Oswald Wirth demonstrates in his introduction to the 1910 edition of this work that the group in question was otherwise known as Les Philalèthes—that is, the Philaletheans, or "Lovers of Truth" (from Greek φιλαλήθης)—which formed within the Masonic lodge Les Amis Réunis in Paris in 1773. Guaïta presents a slightly different French version in *Le serpent de la Genèse,* 2:695–96, and it is Guaïta's version on which Sédir primarily relies. Presented here is a somewhat fuller, synoptic version of this secret operation. —*Trans.*]

Step 4. Transfer the dew-soaked plant matter back into the vessel. Do this before the Sun rises, otherwise the dew will evaporate. Set the vessel as before in a temperate place.

Step 5. If you have collected enough dew water, first filter it and then redistill it to remove all impurities. Calcine the remaining *caput mortuum,* or feces, and extract the salt, which has curious properties and is very pleasant to behold.

Step 6. Pour the distilled dew water imbued with its salt over the plant matter, then hermetically seal the vessel with crushed glass and borax. Set the vessel in this state in horse manure for one month.*

Step 7. Remove the vessel at the end of the month and you will see that the plant matter has congealed at the bottom of the vessel. The spirit will look like a "coat of many colors" floating over the plant matter. Between the coat and the silty substance there will be a kind of greenish dew signifying a good harvest.

Step 8. Expose this well-sealed vessel during the summer to the Sun by day and to the Moon by night. If the weather becomes rainy, the vessel must be kept warm until the good weather returns—in this case, move the vessel to the same dry and temperate place as before. Sometimes this operative procedure can be completed in one to two months, but other times it could take several months or even a year. The sure signs of the operation's success are when you see the silty substance swell up and rise, the spirit or small coat diminish each day, and the plant matter begin to thicken. When you see subtle exhalations rising in the vessel and the formations of faint clouds, these are the first rudiments of vegetal resurrection.

Step 9. Finally, the plant matter will transform into a bluish powder. Whenever the vessel is exposed to a gentle heat, the stem, branches, leaves, and flowers of the original plant will rise up from the blue ashes. The moment this gentle heat subsides, however, the spectacle will cease, and the matter will precipitate back to the bottom of the vessel. And this is how you create a vegetable phoenix.

*[One month in alchemical time, otherwise known as a philosophical month, is usually equal to forty days or to six weeks. —*Trans.*]

Plant palingenesis would merely be an object of amusement if this operation did not suggest greater and more useful applications. The spagyric art of alchemy or occult chemistry is capable of reviving other bodies. It destroys them with fire and brings them back to their first form. The alchemical transmutation of metals by means of the philosopher's stone is a form of palingenesis in the mineral kingdom. Furthermore, what can be done to vegetal beings can also be done to animal beings, but the strength of our commitments is such that we must refrain from openly discussing the subject of *palingenesia animalium*.* Perhaps the most marvelous aspect of palingenesis is the art of resuscitating the remains of animals and humans. How enchanting it is to entertain the resurrected shadow of a loved one or family member who is no longer with us. Artemisia II of Caria is said to have swallowed the cremated ashes of her deceased husband (and brother), Mausolus.† She did not know, alas, this occult secret of cheating grief.[3]

Have you grasped the import of all this? The homogeneity of universal nature authorizes adepts to infer by analogy. If they reason justly, their experiments will always confirm their inductions. What is possible in the plant kingdom must by analogy also be possible in each of the other kingdoms of nature: as in the lower kingdoms of nature, so in the higher. The transmutation of metals in the mineral kingdom is equivalent to the posthumous revivification of departed forms in the vegetal, animal, and hominal kingdoms.[4]

*The author D. (Touzay Duchanteau) claims that the study of palingenesis is the sole province of the Unknown Philosophers. "It is from them," the anonymous author asserts, "that I have received the truths recorded in this work" (Duchanteau, *La grand livre de la nature,* 22).

†[According to Pliny, the plant name *Artemisia*—that is, ⋆mugwort (*Artemisia vulgaris* L.)—derives either from the abovementioned Artemisia II of Caria (d. 350 BCE) or from Artemis Eileithyia, Eileithyia being the Greek goddess of childbirth (Pliny, *Natural History* 25.36). Each of these onomastic origins is grounded in the widespread belief that the herb ⋆mugwort is a protector of women. —*Trans.*]

10

Plant Palingenesis in History and Practice

BY KARL KIESEWETTER

Since the subjects of vegetation magic and the vegetable phoenix have already been introduced, we trust that a more expansive overview of plant palingenesis in history and practice will not be without interest to readers.[1] Such an overview will allow researchers to recognize, if they have not already, the true value of this extremely obscure subject matter. Having spent years amassing all available resources pertaining to palingenetic experimentation, many of which are now quite rare and some of which remain unpublished, we have come to distinguish two main types of *palingenesia:*

1. *Shadow palingenesis,* which has for its object the manufacture of the astral bodies of vegetal and animal beings
2. *Corporeal palingenesis,* which involves the acceleration of plant vegetation (or forced plant growth) and aims to reconstitute organized bodies that have been destroyed

The latter, in its final reaches, enters into the domain of the homunculus, that chemical evocation of the human being in which the extremes of mysticism and materialism collide. Readers should note, however, that we exclude from this study all experiments that early practitioners mistook for genuine palingenesis, for example, the phenomena of *generatio aequivoca,* or spontaneous generation from decaying matter; the arborescent metallic precipitates and their crystallization; and all that concerns the "palingenesis" of the ★nettle in the congealed lye of its salt, the latter being the experiment of Joseph Duchesne, physician-in-ordinary to Henri IV.[2]*

Ovid speaks of forced plant growth in no uncertain terms when he describes Medea's magical rejuvenation of Aeson. The famous sorceress is said to have brewed a magical potion consisting of acerbic juices, roots, flowers, and seeds from plants native to the valleys of Thessaly, mixed in some elixirs, sands, white hoarfrost, several mysterious animal ingredients, and "a thousand others impossible to name" and stirred it with a dead olive branch. "And while she stirred the withered olive branch in the hot mixture," Ovid writes, "it began to change from brown to green, presently put forth new leaves, and soon was heavy with a wealth of luscious olives. As the ever-rising fire threw bubbling froth beyond the cauldron's rim, the ground was covered with fresh verdure, flowers, luxuriant grasses, and green plants."[3]

The alchemists repeatedly performed palingenetic experiments. Abū Bakr Muhammad ibn Zakariyyā al-Rāzī, better known by his latinized name, Rhazes, and Albertus Magnus both took a special interest in palingenesis.[4] The latter is alleged to have fashioned a homunculus or artificial human head that was capable of speech (which, legend has it, Thomas Aquinas eventually destroyed), and even Johann Isaac Hollandus alludes to plant palingenesis in his influential work *Opera vegetabilia.*[5] It is only in the writings of Paracelsus, however, that we find more detailed descriptions of the two types of palingenesis.

On the subject of shadow palingenesis, Paracelsus writes the following:

---

*One could even go so far as to see in the ice crystals of windowpanes covered with hoarfrost the palingenesis of plants that have been burned to extract potash from them (cf. Eckartshausen, *Aufschlüsse zur Magie,* 2:399). In any case, whoever wishes to get a better idea of the vast array of experimental phenomena erroneously described as palingenesis need only read and reread Johann Otto von Hellwig's *Curiosa physica, oder Lehre von unterschiedlichen Natur-Geheimnissen.* Friedrich Christoph Oetinger's ★lemon balm leaf should also be included among such products of palingenetic fancy (see Oetinger, *Die Philosophie der Alten widerkommend in der güldenen Zeit,* 2:2–3).

> Let no philosopher wonder here that out of a certain earth in which an herb is essentially born, before it is incorporated, all the virtues may be extracted, that the virtues may be kept or preserved, and that the earth may be again put into its place in such a manner that it is thenceforth but a mere earth without any fruitfulness in it, its first *ens* now being sequestered from it: from thence it is wont to come to pass that the virtue of such a first *ens* may be shut up in a glass vessel and brought to that state, so that the form of that same herb may grow in itself without any earth, and when this plant reaches the end of its growth it may have no body, but be notwithstanding a formed thing like a body, the reason whereof is this: because it has no liquor of earth (*liquidum terrae*), from whence it follows that its stalk is nothing other than a certain apparition to the sight, a smoke affecting the form of a substance, but not perceptible by touch.[6]

Paracelsus provides no specific instructions on the method of shadow palingenesis. He does, however, provide relatively detailed instructions on the method of corporeal palingenesis.

> Take a living bird that has just hatched, hermetically seal it in a matrass, and reduce it to ashes over the third degree of fire. Then, still shut up in the matrass, let the ashes of the cremated bird putrefy at the highest degree of putrefaction in horse manure (*venter equinus*) until it becomes a mucilaginous phlegm [produced from ash and empyreumatic oils]. This phlegm can be brought to maturity again and, thus renovated and restored, can become a living bird, provided the phlegm is enclosed once more in its jar or receptacle [which is to say, provided this phlegm is put in an eggshell, then sealed with the utmost care and subjected to the appropriate heat]. The revivification of the dead by regeneration and clarification is indeed a great and profound miracle of nature.[7]

Paracelsus extends this type of palingenesis to all animal species, and Sir Kenelm Digby claims he reconstituted a burnt crayfish in a similar manner.[8] Heinrich Cornelius Agrippa, a contemporary of Paracelsus, also appears to have been familiar with this process, for he avers that "there is an art by which inside an egg being incubated by a hen may be generated a form like unto a man, which I have seen and know how to make and which the magicians say has marvelous virtues—this they call 'true mandrake' (*verus mandragoras*)."[9]

Inspired by their master's characteristically brisk statements, the Paracelsists wrote a great deal on the subject of palingenesis, among them Joseph Duchesne (alias Josephus Quercetanus), Gaston LeDoux de Claves (alias Gasto Clavaeus), Jean Béguin (alias Johannes Beguinus), Pierre Borel (alias Petrus Borellius), Otto Tachenius, Daniel Sennert, Sir Kenelm Digby, David von der Becke, William Maxwell, and Adam Friedrich Pezoldt.[10] The work of Georg Franck von Frankenau, professor of medicine at the University of Heidelberg, is far from exhaustive, and from the experimental point of view it is based primarily on the virtually indistinguishable instructions of Borellius, Tachenius, and Becke.[11] As far as we have been able to determine, the last known testimony of palingenetic practice comes from Karl von Eckartshausen, who reports that two of his friends witnessed successful experiments using different methods, after which they were able to perform the same experiments themselves. The one resuscitated a ⋆buttercup, the other a ⋆rose, and they also successfully performed these experiments on animals.[12] These are the principles and practices we wish to elaborate here.

William Maxwell, the Gustav Jäger of the seventeenth century, speaks of palingenesis in several of his works. Unfortunately, however, he does so in a confused and unnecessarily mysterious fashion. He gives the following instructions to perform operations of shadow palingenesis:

> Take a sufficient quantity of ⋆rose leaves, desiccate them with fire, and calcine them using a bellows until they are reduced to a very white ash [this result may be obtained by the simple combustion of the desiccated rose leaves in a crucible fired to a red heat]. Afterward, extract the salt with ordinary water and introduce the salt into a colatorium [one of those useless devices of old chemistry—any flask stopped with emery will serve the same purpose], having sealed the apertures as tightly as possible. Leave the colatorium over a fire for three months [this fire is obviously the sweet heat of digestion], then bury it in manure [as directed above] and leave it there for another three months [to stimulate putrefaction, practitioners would bury their preparations in horse manure, which they renewed whenever the heat generated by decay diminished]. At the end of this

> period of time, remove the container and place it once more over the fire until the figure appears in the bottle.[13]

Maxwell is of the opinion that the procedures of plant palingenesis and human palingenesis are one and the same. In the same work, he claims that "just as the salts of plants are forced in this manner to cause the figures of whichever plants have rendered these salts to appear in a flask, so too, and this is an indisputable fact, the salt of the blood [which is to say, the salt derived from the ash of the blood] is capable of reproducing, under the influence of an extremely subtle heat, a humanoid figure. And we must see in this the true homunculus of Paracelsus."[14] Maxwell conceives of corporeal palingenesis as the counterpart of umbral or shadow palingenesis, and he provides the following instructions to re-create Agrippa's *verus mandragoras:*

> In a natural, non-artificial, and well-sealed container [that is, in an eggshell voided by suction] mix some blood with the noblest particles of the body as thoroughly as possible and in the proper proportions, then coax a hen to incubate it. Sometime after it hatches you will find the mass coagulated into a humanlike form, and with it you will be able to accomplish marvelous things. You will also see an oil or liquor swimming all about this mass. By mixing this oil or liquor with your own sweat, you will precipitate, through the merest contact with it, various modifications in the perceptions of your senses.[15]

David von der Becke calls the plant's astral body the *idea seminalis vegetabilis,* or "seminal vegetative archetype," and gives the following instructions for the operation of plant palingenesis:

> On a serene day, collect the mature seed of a plant, grind it in a mortar, and place it in a matrass that is roughly the same size as the original plant and has a narrow neck that may be hermetically sealed. Keep the matrass sealed until the arrival of an evening that portends the production of an abundance of dew during the night. Then introduce the seed into a glass vessel, place a tray beneath it so that nothing is lost, and expose it to the air in a meadow or in a garden so that it is well-saturated with dew, but be

> sure to return it back into the matrass before sunrise. Filter the collected dew and distill until the distillate is completely devoid of feces. Calcine the feces and separate the salt by washing it several times, then dissolve the salt in the distilled dew. Pour three fingers of the distilled dew mixed with its salt over the dew-saturated seed and carefully lute the aperture so that the dew does not evaporate. Keep the matrass in a place where a moderate heat prevails. After a few days, the seed will gradually begin to turn into a kind of mucilaginous earth. The supernatant spirit will be marked with striations, a membrane will form on its surface, and the mucilaginous earth will turn green. Expose the matrass to the rays of the Sun and the Moon, but keep it in a warm room during bouts of rainy weather, until all of these signs are fully manifest. If you then subject the matrass to a gentle heat, you will see the image of the plant corresponding to the seed appear, and you will see it disappear when the matrass is cooled. This method of conjuring the *idea seminalis* is employed with few variations by all those who practice *palingenesia*.[16]

Becke also refers to a method of palingenesis by means of ashes. He never gives explicit instructions, but he suggests that by this method it is possible to practice a "lawful necromancy" of one's ancestors, provided, however, their ashes have been preserved.[17] We should note, however, that Becke's method of palingenesis is remarkably similar to that of the anonymous author of the German work titled *The Artificial Resurrection of Plants, Humans, and Animals from Their Ashes.* Here is how this author describes the "palingenesis from ashes":

> Collect the seed of any plant so long as it is mature and was gathered during good weather and under a calm sky. Grind four pounds* of seed in a glass mortar and place it in a matrass that is the size of the entire plant. Hermetically seal the vessel so that the seed does not oxidize, put it in a warm place, and wait for a night when the sky is clear, for it is at such times that dew may be collected in greater abundance. Put the seed

*Practitioners may alter the quantity of plant seed as needed, so long as they use two liters of dew water per pound of seed.

in a bowl and expose it to the air in a garden or meadow. Be sure to place a large tray beneath the bowl so that none of the dew is lost. The dew will saturate the seed and communicate to it its nature. At the same time, drape some clean sheets over some stakes in the ground in order to collect large quantities of dew and transfer eight liters by wringing out the sheets into a glass container. Once the seed is impregnated with dew, transfer it back into the matrass before sunrise so that the daystar does not evaporate the dew. Filter and distill the dew water several times, then calcine the feces to extract the salt. Dissolve the salt in the distilled dew water and pour two fingers over the crushed seed in the matrass, then hermetically seal the vessel with wax. Bury the matrass two feet deep in a damp place or in horse manure and leave it there for an entire month. When you remove it, you will see the seed transformed: a multicolored membrane will form above it and the dew will take on a grass-green coloring according to the nature of the seed. Suspend the vessel for the whole summer in such a place that it receives the rays of the Sun during the day and those of the Moon and stars at night. In case of rain or inclement weather, put it in a dry place until the weather becomes calm and return it to the same place. This work may require as little as two months or as many as two years, depending on the weather. Here are the signs of a successful operation: the mucilaginous matter rises noticeably; the alcohol and the membrane gradually diminish day by day; and all the matter forms into a single mass. Due to the reflection of the Sun, at this point you should also be able to see periodically within the glass a subtle vapor in form of the plant, which will appear colorless and as though it were made of a spiderweb.* The figure will rise and fall frequently according to the energy of the Sun and the phase of the Moon. Finally, the mixture will form into a white ash, which, when subjected to a gentle heat, will give rise to an image of the stem, leaves, and flowers of the plant in their particular shapes and colors. Once the vessel is removed from the heat, the image will disappear and transform back into its material earth. If the vessel is properly sealed, the reappearance of such figures may be achieved indefinitely.[18]

*It is no coincidence that ghosts often manifest before human eyes in this same filmy, spiderweb-like appearance.

The influential chemist Johann Joachim Becher offers two noticeably different methods of plant regeneration. The following method is similar to but much more complete than the one related by Friedrich Christoph Oetinger nearly a century later:

> Pick any plant at the appropriate time, or rather pick each part of the plant at the appropriate times, the root in November, after deseeding the plant, the flower in full bloom, and the leaf before flowering. Take a sizeable portion of each and dry them in a shady place away from the Sun and any other heat source. Calcine the plant matter in an earthenware pot with well-luted joints and extract the salt with hot water. Transfer the juice of the roots, leaves, and flowers into an earthenware vessel and dissolve the salt in this juice. Next, take some virgin earth, which is to say, soil that has been neither plowed nor sown such as you would find in the mountains. This earth should be red, pure, and unmixed. Grind it and pass it through a sieve, then transfer it to a glass or earthenware container and pour the aforementioned juice over it until it absorbs the juice and beings to turn green. Place over this container another of a size that corresponds to the natural size of the plant. Lute the joints carefully so that no draft of air encumbers the image of the plant. The bottom container, however, should have a small aperture at its base so that air can penetrate the earth. Expose the double vessel to the Sun or to a gentle heat and after a half an hour you will see the image of the plant appear in a pearl-gray color.[19]

Becher also provides the following alternative method:

> Grind the plant along with its roots and flowers in a mortar, transfer it into a matrass or similar vessel, and allow it to ferment on its own and produce its own heat. Express the juice, purify it by filtering, and pour it back over the residue in order to resume the process of putrefaction. When the juice turns the natural color of the plant, express and filter it once more, then set it to digest in an alembic until all the impurities precipitate and the juice becomes clear and pure but without losing the natural color of the plant. Pour the juice into another alembic and distill the phlegm and volatile spirits in a water bath using a gentle heat. The sulfur, which is

to say, the solid mass of the extract, will remain at the bottom—set this aside. Distill the phlegm with a slow fire (*ignis lentus*) in order to remove the volatile ammoniacal products of the fermentation—set this aside, too. Calcine the residue (*caput mortuum*) and extract the volatile salt [that is, the ammoniacal salts united to the acid products of the combustion] with the phlegm. Distill the phlegm once more in the water bath or bain-marie to extract the volatile salt and calcine the residue until it becomes white like ash. Pour the phlegm over this residue and extract the fixed salt by depuration. Pour over these two salts—the volatile salt and the fixed salt—the volatile spirits along with the sulfur and the spirits of fire that appeared first during the distillation and allow them to unite intimately. Instead of the phlegm, you can use distilled rainwater and in place of the fixed salt [carbonate of potash], you can use any plant salt, then add the sulfur and coagulate [or desiccate] with a slow fire, thus combining each of the three constituent principles. Put these three principles in a large matrass and add either the distilled water of the plant or spirit of rainwater or of May dew. Only one of these liquids will suffice. Subject the sealed vessel to a slow fire and you will see the immaterial plant grow in this water together with its flowers. The plant will remain visible for as long as the heat lasts. It will disappear by cooling and reappear by reheating. This is truly a great miracle of nature and art.[20]

ROSICRUCIAN OPERATIONS OF PLANT PALINGENESIS

Two similar sets of instructions for operations of corporeal and umbral palingenesis are found in the Rosicrucian manuscripts that belonged to our great-grandfather.[21] The first, which is attributed to Albertus Magnus, appears in a manuscript titled *Güldenes A. B. C. Alberti Magni von den Geheimnissen der Natur* (Albertus Magnus's Golden ABC's of the Secrets of Nature), doubtless a German translation of an older Latin original. The authenticity of this attribution seems likely for two reasons:

1. It is evident from the printed works of Albertus Magnus that this great scholar was familiar with palingenesis.[22]

2. It is quite possible that some authentic manuscripts were not admitted into his corpus because editors were simply unaware of their existence.*

Operation 1. How to Prepare Spiritus Universalis *with* Minera Bismuthi

As some minerals contain the *Spiritus Universi,* or Spirit of the Universe, so from other minerals practitioners may extract a *Spiritus Universalis,* or Universal Spirit. Of all the minerals there are two that produce this *spiritus* by themselves. The first is a *minera bismuthi,* which is found in the mountains;† the second is a brown mineral earth that is found alongside silver ores and contains a most wonderful life-giving spirit. Pebbles from streams can also render such a liquor, but this liquor is suitable solely for ennobling metals.

Here is how the *spiritus* is obtained from bismuth. Collect some *minera bismuthi* from the mountains and grind it, reducing it to a fine powder. Put the powder in a well-luted retort. Immerse the retort in a vessel filled with iron filings in such a manner that it is completely covered by filings and adapt a coil to it. Set the vessel over a gradual fire (*gradus ignis*), and in forty-eight hours you will extract the *spiritus,* which will emerge like tears flowing from a weeping eye. We do not recommend using water at this point, but since dew also furnishes the *Spiritus Universi,* which I call in my writings "spirit of May dew" (*spiritus roris majalis*), you may add a half pound, for it is not at all contrary to the work.‡ Next, introduce the *spiritus bismuthi* and let the fire go out. When the mixture becomes cold, pour the liquor extracted during the distillation into a large alembic and place it in the bain-marie after covering the alembic with its *alambicum,* or capital. Lute the joints well and distill. Using this method you will obtain a *spiritus* that is as pure as crystal and as sweet as honey. This *spiritus* is a living spirit and is the exclusive domain of magic.

This living spirit transformed me into a magician. It is the only active spirit endowed with magical properties that receives the forces it possesses directly from God Almighty, for it can assume all kinds of forms. It is *animal,* for it created Animalia. It is *vegetal,* for it created Vegetabilia. Trees,

*[Kiesewetter's argument fails to convince. Surely this manuscript belongs to the Pseudo-Albertan tradition, which includes some thirty alchemical tractates in Latin. —*Trans.*]

†Bismuth(III) oxide, apparently.

‡See the second set of instructions provided on p. 94.

foliage, herbs, flowers, and, indeed, all vegetal life grow through its agency. It is *mineral,* for it is the principle of all minerals and all metals. It is *astral,* for it descends from on high and from the stars with which it is impregnated. It is *universal,* for it was created at the beginning. It is the Word that issued forth from God, and it is therefore intelligible, the *primum mobile,* or "first mover," of all things. It is the pure nature that surged forth from light and fire and was transplanted and insufflated into inferior things. Hermes Trismegistus refers to it when he says: "Wind carries it in its belly. Earth is its nurse and its receptacle."[23] This spirit gives life and takes it away, and with it alone you may work miracles. Here's how:

Gather some plants, flowers, or fruits before nature has brought them to full maturity, some grape clusters, pears, apples, cherries, plums, or almonds, for example. After selecting the desired plants, hang them whole in the shade and allow them to dry along with their flowers. You will then rebloom and regreen them in winter to the point that they ripen their fruits and acquire the most succulent flavor. Here's how to proceed: Take a matrass with a narrow neck and large belly and pour in one pound of Universal Spirit. Next, add to the container some branches with their flowers and fruits. Be sure to stop the matrass with wax so that the spirit remains inside the container, and then let nature take its course. In twenty-four hours, everything will begin to turn green and grow in height, the fruits will ripen, the flowers will take on their characteristic fragrance, and everything will have a pleasant odor and good flavor. Recognize the presence of the power of God in this process, despite the fact that the bishop of Passau,* because he is ignorant of the divine power, considers it a diabolical working. This spirit is capable of achieving many other ends, as the Holy Father himself could easily observe. We must praise and pray to God for all the benefits and miracles that he gives us, poor wretches that we are. In truth, and it matters not who denies it, revivification by means of the spirit of dead things is a supernatural act. Moreover, it serves as ample proof that this spirit has the power to bring back to life all that is dead. Thus, having taken a bird and cremated it in a vase,

*This must be either Rüdiger von Radeck, bishop of Passau from 1233–1250, or Otto von Lonsdort, bishop of Passau from 1254–1265, both contemporaries of Albertus Magnus (ca. 1193–1280).

I put its ashes in a container like this [in the Rosicrucian manuscript there is an illustration of a blind alembic containing a liquid in which the face of a child is visible]. In another vessel I put the ashes of the decaying corpse of a small child, having first reddened its earth, and in another the ashes of a plant incinerated with its flowers. I filled each of these vessels completely with *spiritus,* and then I let nature take its course. The spirits [or astral bodies] of the child and the plant, which developed in twenty-four hours, showed themselves to me within the *spiritus* and with every appearance of reality. Is this not a true resurrection of these beings? The form awakened by the spirit [or *spiritus*] allows us to see how we ourselves will appear as spirits with pure bodies—which is to say, transparent and in a transfigured form. Just as the body is given new life through the soul and spirit that belong to it, so we shall transfigure in a state of godly contemplation, for this is a luminous force. I would like to add here that I once had a spirit with which I could converse for a few hours a day, but this spirit was only an immaterial simulation of the resurrection from the dead.

Furthermore, during the course of my investigations, Thomas* gave me a vessel in which he had preserved in this liquor a drop of his blood, and I in turn gave him a vessel in which I had preserved in the same liquor a drop of my own blood. If you want a beloved friend to be near you at all times, such a vessel can inform you day or night of his condition. If your friend has fallen ill, the small spark in the middle of the vessel, instead of being brilliant, will cast only a feeble glare. If he becomes ill, the light will become dull. If he becomes angry, the vessel will grow warm. If he is very active, the light will move. But if he dies, the light will go out and the vessel will burst. Moreover, since this unique spirit is capable of anything, you can even use these lights to communicate secretly with your friend, for this spirit is all powerful.

The Paracelsists and Rosicrucians took great care of these "vital lamps" or "lamps of life," and Johann Ernst Burggrav even wrote a monograph on the subject titled *Biolychnium or the Lamp of Life and Death.*[24] To conclude this historical survey of palingenesis, we reproduce one final palingenetic experiment from another of our great-grandfather's manuscripts titled *Testamentum*

*That is, Thomas Aquinas (1225–1274), who was the disciple of Albertus Magnus.

der Fraternitet Roseae et Aureae Crucis (Testament of the Fraternity of the Rosy and Golden Cross). It is comparable to the previous experiment, and for the chemist with a laboratory it is rather easy to perform.

Operation 2. How to Prepare **Spiritus Universalis** *with Dew, Rainwater, Hoarfrost, and Snow*

My dear children, let your zeal for the work animate you at the start of the year. Collect as much white hoarfrost, snow, fog, dew, and rainwater as you can in a large barrel, and allow the contents of the barrel to decay and putrefy until July. You will know that the contents are ready for further work when you see the following signs: the earthy mass will cease to be homogenous; a green membrane will form on the surface; and the verdurous force of the vegetation will be revealed by the appearance of some worms. My children, when you see these signs, stir and mix everything thoroughly, and then pour the mixture into a serpentine alembic and distill ten pounds at a time with a gentle fire until you obtain one hundred pounds or until the putrefied water runs out. Return the product of the first distillation to the serpentine alembic and redistill ten pounds at a time. Next, discard the residue and distill the product of the second distillation ten pounds at a time. Repeat until you have only ten pounds of distillate* and pour it into a strong retort capable of withstanding the fire. Reduce these ten pounds in the ash bath to six pounds. Return the *spiritus* to the retort, plunge it into the water bath, and reduce the six pounds to three pounds. At this point, the seventh distillation, an extremely volatile spirit will rise: pure AIR. This is a live-giving spirit, for if you absorb so much as a spoonful, you will feel the effects of its power in all your limbs. It exhilarates the heart and surges through the entire body as though it had been pierced instantaneously by a wind or a spirit. You must rectify this spirit seven times and push it to its final retrenchment. You can then make it serve different purposes and use it to perform miracles, for this spirit awakens and calls all things to life.

Now, take the ashes of the leaves, flowers, and roots of a plant or the ashes of an animal such as a bird or lizard or even the ashes of the decay-

*As in the rectification of alcohols, the products of all these distillations are reduced by successive distillations to only ten pounds.

ing corpse of a small child and redden them and then place them in a tall matrass or another suitable glass vessel. Pour the wonderful life-giving spirit over these red ashes to the height of one hand and carefully seal the vessel and then place it, without disturbing it, in a warm place. At the end of three twenty-four-hour periods, the plant will appear with all its flowers or the animal or child with all their members, a result that some wrongly use for elaborate juggleries. These beings, however, are purely spiritual creatures, for they are not long in disappearing if one agitates them even a little. They will reappear if the vessel is left at rest, and this is truly a marvel to behold. Whoever witnesses this spectacle witnesses nothing less than the resurrection of the dead. It shows us how all things in nature will reawaken in the universal resurrection.

My friends, I can even revivify in a matter of hours a dried or wilted flower or any other herb, a blade of grass or a grape cluster cut with its vine and leaves, or fruits picked at various stages of development and dried in the shade. Whenever I wanted my pupils to witness a miracle, I would put these twigs, flowers, or fruits in containers—that is, long-necked matrasses, which are wide at the bottom and narrow at the top—and pour over them the necessary quantity of spirit. I would seal the neck of the receptacle with wax and let it rest for four hours. At the end of this period of time, whatever I had placed in the matrasses would start to regreen and rebloom, to such an extent that dried fruits would revivify even in the middle of winter, mature after three to four days and nights, and acquire an exquisite flavor. I even offered such fruits to acquaintances who were ignorant of such matters and told them they had come from such and such a country, and they were none the wiser.

My friends, I even put some of this spirit in a beautiful translucent flask and added a few drops of the blood of a beloved friend and sealed the flask tightly. This way I could constantly observe my friend's condition, whether he was in good or bad health and whether he was happy or unhappy. When he was happy, the light would shine in the flask and everything around it would be alive. When he was unhappy, the light would dim and everything around it would be dull. When he was ill, the bottle would be filled with darkness and agitation. And if he should die a natural or violent death, the vessel will burst. Thus, with this life-giving spirit you can accomplish many great wonders.

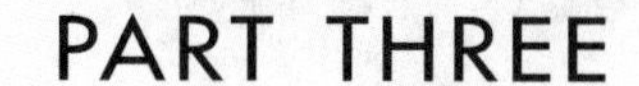

PART THREE

A Concise Dictionary of Magical Plants

Each entry in this dictionary contains the following information:

1. The plant's scientific, common, and, if any, magical, alchemical, and Hermetic names.*
2. The plant's elemental, zodiacal, and planetary signatures.
3. A brief description of the plant's occult properties, medicinal or magical uses, and, if any, special preparations with instructions for their administration.

Unless otherwise noted, the ideal time to harvest each plant is indicated by the ruling planet and the zodiacal sign—which is to say, the plant should be harvested when its ruling planet appears in its corresponding zodiacal sign or, if there are multiple planetary and zodiacal signatures, when at least one of its ruling planets appears in one of its corresponding zodiacal signs. In all cases, however, plant harvests should be avoided if the ruling planet is unfavorably aspected.

We include entries for only a minority of plants, since the principal object of this dictionary is to provide practical examples of the botanical theories expounded in parts 1 and 2 of this book.

*[As stated in the foreword, the first name in each series of French common names, unless otherwise noted, is the name under which the entry appears in the original French publication (the rest proceed in alphabetical order). —*Trans.*]

A

Acacia–Asphodel

Fig. A.1. Mattioli, *Discorsi,* 141.

ACACIA

Latin binomial: *Vachellia nilotica* (L.) P.J.H. Hurter & Mabb.

French names: acacia, acacia à gomme, acacia d'Arabie, gommier blanc, gommier rouge

Elemental qualities: moderately hot and wet (🜁)

Ruling planet: ☿

Zodiacal signature: ♍ (*apud* Mario)

Occult properties: Gum arabic, the exudate or resin of acacia, has demulcent and stomachic properties. Acacia, otherwise known as the gum arabic tree, babul, and thorn mimosa, was sacred to the ancient Egyptians, who used its wood to build the sacred barge of Osiris at the temple of Thebes. Similarly, Moses built the tabernacle, the ark of the covenant, and the table of showbread out of acacia, or *shittim* (שטים), wood. In Freemasonry, acacia is a tripartite symbol for the *immortality of the soul,* because it is an evergreen; *innocence,* because the Greek word ἀκακία also means "guilelessness"; and *initiation,* because a sprig of acacia marks the grave of Hiram Abiff. It is an ideal choice of wood for ritual furniture in rites of ceremonial magic.[1]

AGARIC

Fig. A.2. Chambers and Chambers, *Chambers's Encyclopædia,* 1:185.

Latin binomial (type sp.): *Amanita muscaria* (L.) Lam.

French names: agaric, agaric aux mouches, agaric moucheté, amanite tue-mouches, fausse oronge

Elemental qualities: hot and moderately dry (△)

Ruling planet: ☿ (*apud* Mavéric)

Zodiacal signature: ♈. Harvest at the end of July and the beginning of August.

Occult properties: The type species of the genus *Amanita*—namely, fly agaric or fly amanita—has psychotropic and hallucinogenic properties. Shamans the world over have consumed this mushroom in magical rituals to induce visions, ecstatic experiences, and altered states of consciousness. It is especially good for oneiromancy, foretelling the future through dream interpretation.[2]

AGRIMONY

Fig. A.3. Fuchs, *Plantarum effigies,* 136.

Latin binomial (type sp.): *Agrimonia eupatoria* L.

French names: aigremoine, agrimoine, eupatoire, francormier, grimoino, grimouèno (Languedocian), herbe de Sainte Madeleine, herbe de Saint Guillaume, ingremoine, soubeirette

Elemental qualities: cold and dry (🜃)

Ruling planet: ♃ (*apud* Culpeper, Lenain, Piobb, and Mavéric)

Zodiacal signatures: ♉ and ♒

Occult properties: Agrimonies have astringent, antiedemic, styptic, and tonic properties. They grow among hedges and bushes and in wooded areas. Their astringent leaves combat tonsillitis and nephritis. If placed under someone's pillow, this herb will prevent the sleeper from waking. When burned as a suffumigant, it drives out evil spirits. In lotions, it is good for leucoma, dislocations, and sprains.[3] It also works well as a vermifuge, heals snakebites, and cures sheep's cough.[4]

Fig. A.4. Mattioli, *Discorsi,* 111.

ALDER

Latin binomial (type sp.): *Alnus glutinosa* (L.) Gaertn.

French names: aulne, alnois, aulne glutineux, aulne grisâtre, aulne noir, aulnet, aune, bèrgne (Languedocian), bouleau vergne, ônê, vergne, verne

Elemental qualities: cold and wet (▽)

Ruling planet: ☾

Zodiacal signature: ♓

Occult properties: Alder barks possess astringent, febrifugal, purgative, restorative, styptic, and tonic properties. The wood of the type species of the *Alnus* genus, the black alder or European alder, is frequently used to make magical wands. Its wood charcoal is also useful in rituals of evocation.

Fig. A.5. Fuchs, *Plantarum effigies,* 186.

ALEXANDERS

Latin binomial (type sp.): *Smyrnium olusatrum* L.*

French names: grande ache, ache large, grand persil, maceron, persil d'Alexandre

Properties: Alexanders, or horse parsley, has the same elemental, astral, and occult properties as ★celery.

*[Sédir provides the species name in its uncompounded form, *olus atrum,* meaning "black potherb," so named because of its black, spicy seeds. —*Trans.*]

ALKANET

Fig. A.6. Fuchs, *Plantarum effigies,* 150.

Latin binomial (type sp.): *Anchusa officinalis* L.

French names: buglosse, buglose, langue de bœuf, orcanette

Elemental qualities: cold and dry (🜃)

Ruling planet: ♃

Zodiacal signature: ♉

Occult properties: Alkanet possesses astringent, cordial, demulcent, tonic, and vulnerary properties. It is most commonly used as a blood purifier and in the treatment of coughs and bronchial catarrh. Its root is a natural diuretic. The shape of its leaf has long been thought to resemble a bull's tongue, hence its ancient Greek name βούγλωσσον and the modern derivation "bugloss." According to Pliny, when macerated in wine and served as a drink at parties, it promotes Taurean pleasures, general mirth, and joviality.[5]

ALMOND

Fig. A.7. Mattioli, *Discorsi,* 184.

Latin binomial: *Prunus dulcis* (Mill.) D.A. Webb

French name: amandier

Elemental qualities: cold and dry (🜃)

Ruling planets: ♀ and ♃

Zodiacal signature: ♉ (*apud* Mario)

Occult properties: Almonds possess demulcent and nutritive properties. Five or six bitter almonds taken on an empty stomach prevent the onset of drunkenness.[6] Almond drupes are also good for phthisics, wet nurses, and impotence.

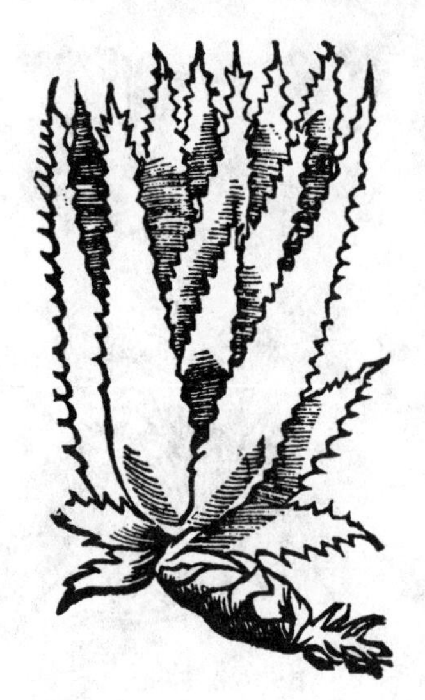

Fig. A.8. Fuchs, *Plantarum effigies,* 75.

ALOE

Latin binomial: *Aloe vera* (L.) Burm.f.
French names: aloès, lys du désert, perroquet*
Elemental qualities: cold and wet (▽)
Ruling planet: ☉
Zodiacal signature: ♋ (*apud* Jollivet-Castelot)
Occult properties: Aloe has antibilious, emollient, hepatic, vermifugal, and vulnerary properties. Its dried powder may be used as a perfume to attract influences from ♃. It serves as a remarkable cure for many hot and dry ailments. Aloe juice mixed with vinegar is a well-known health and beauty elixir. Lotions composed of these two ingredients not only heal sunburns, but they also prevent hair loss.

*[Sédir mistakenly merges *aloès* (aloe) and *bois d'aloès* (aloeswood) in a single entry with the astral signatures ☉ and ♐. See *Les plantes magiques,* 142–43. These have been separated here to form two independent entries, both of which have been revised and augmented following François Jollivet-Castelot's subsequent correction, with ⋆aloe signed by ♋ and ⋆aloeswood by ♐ (*La médecine spagyrique,* 7, 12, 15). Although Jollivet-Castelot retains Sédir's correspondence of the ☉ as the ruling planet for both species, it is interesting to note that both Nicholas Culpeper and Jean Mavéric place aloe under the rulership of ♂. See Culpeper, *Culpeper's English Physician and Complete Herbal,* 71; Mavéric, *La médecine hermétique des plantes,* 51 (English edition: *Hermetic Herbalism,* 44); and Bélus (pseudonym of Mavéric), *Traité des recherches,* 36. The scientific name Sédir provides, *Sempervivum marinum,* is an outdated binomial for ⋆aloe first coined by Leonhart Fuchs (*De historia stirpium commentarii insignes,* 138). —*Trans.*]

ALOESWOOD

Fig. A.9. Pomet, *Histoire générale des drogues,* 103.

Latin binomial (type sp.): *Aquilaria malaccensis* Lam.

French names: bois d'aloès, bois d'agar, bois d'argile, calambac, gaharu*

Elemental qualities: hot and dry (△)

Ruling planet: ☉

Zodiacal signature: ♐

Occult properties: The so-called *lignum aloes* or *xylaloes,* literally "wood of the aloe," otherwise known as agarwood or agalloch, has strong aromatic properties. It produces a resin from which perfumers extract the aphrodisiac fragrance known as *oud* or *oudh* (from عود, the Arabic word for "rod" or "stick"). Decoctions of aloeswood facilitate conception.

ALYSSUM, SWEET

Fig. A.10. Mattioli, *Discorsi,* 477.

Latin binomial: *Lobularia maritima* (L.) Desv.

French names: alysson maritime, corbeille d'argent

Elemental qualities: hot and wet (🜁)

Ruling planet: ☉

Zodiacal signature: ♎ (*apud* Mario)

Occult properties: Sweet alyssum, which is native to the Mediterranean littoral, has diuretic and antiscorbutic properties. When hung in the home as a talisman, it promotes peace, harmony, and well-being.[7]

*[See the preceding note. —*Trans.*]

Fig. A.11. Fuchs, *Plantarum effigies*, 55.

AMARANTH

Latin binomial (type sp.): *Amaranthus caudatus* L.

French names: amaranthe, amarante, crête de coq, mèco-de-piot (Languedocian)

Elemental qualities: moderately hot and wet (🜁)

Ruling planet: ♃

Zodiacal signature: ♊ (*apud* Mario)

Occult properties: Amaranths possess alterative, astringent, diaphoretic, diuretic, and emmenagogic properties. Their leaves are rich in vitamins and minerals. In the Hermetic language of flowers, the amaranth is emblematic of immortality. Its Greek name ἀμάραντον means "never-fading flower," and hence in antiquity it functioned as a funereal herb like ⋆celery and ⋆mint.[8] Crowns made of amaranths bestow fortune, fame, and the favor of powerful figures upon those who wear them.[9]

Fig. A.12. Fuchs, *Plantarum effigies*, 68.

ANGELICA, GARDEN

Latin binomial: *Angelica archangelica* L.

French names: angélique, angéline officinale, archangélique, herbe aux anges, herbe du Saint Esprit, herbe impériale

Elemental qualities: hot and dry (🜂)

Ruling planet: ☉

Zodiacal signatures: ♌ and ♒. Harvest when the ☾ is in ♉ or ♊ or at the end of August.

Occult properties: Garden angelica has alexitary, carminative, diaphoretic, stimulant, stomachic, and tonic properties. It is an effective herb for breaking hexes and curses. When hung as amulets around the necks of children, it keeps them from mischief. It has the same virtues as ⋆vervain with respect to rabies. Its leaves, which are dominated by ♄ and should be harvested when ♄ is in his domicile of ♒, serve as a good remedy for gout. Its roots, which are dominated by the ☉ and ♂ and should be harvested when these planets are in ♌, cure gangrene and venomous bites. When rubbed on decayed teeth, the juice of its leaves soothes toothaches. A decoction of its root taken in the morning on an empty stomach dispels inveterate coughs. An infusion in wine heals internal ulcerations.

ANGELICA, WILD

Latin binomial: *Angelica sylvestris* L.

French names: angélique sauvage, angelico (Languedocian), angélique des bois, angélique des pres, faux panais, herbe à la fièvre, imperatoire sauvage, panais sauvage*

Elemental qualities: hot and dry (△)

Ruling planet: ☉

Zodiacal signature: ♐

Occult properties: Wild angelica has the same occult properties as ★garden angelica but is somewhat less effective.

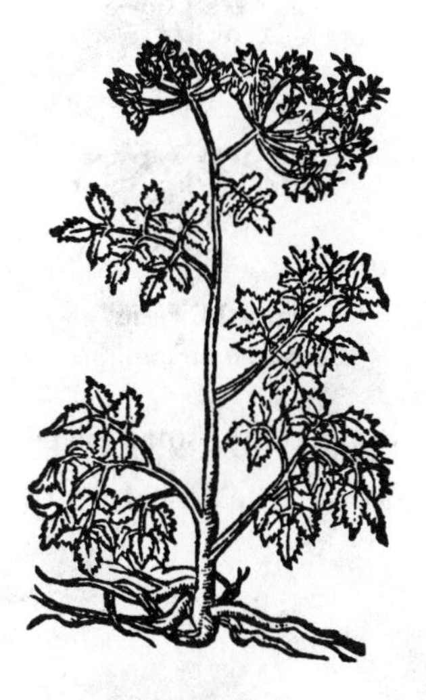

Fig. A.13. Fuchs, *Plantarum effigies,* 69.

*[Sédir briefly describes *angélique sauvage* (wild angelica) in his entry for *angélique* (garden angelica). See *Les plantes magiques,* 143. These have been separated here to form two distinct entries. —*Trans.*]

Fig. A.14. Fuchs, *Plantarum effigies*, 36.

ANISE

Latin binomial: *Pimpinella anisum* L.

French names: anis, anis boucage, anis vert, boucage odorant, fenoul d'anis (Languedocian), herbe odorante, pimpinelle

Elemental qualities: hot and wet (🜁)

Ruling planet: ☿

Zodiacal signatures: ♊ and ♍

Occult properties: Anise has carminative, digestive, emmenagogic, and purgative properties. In baths, it acts as a vermifuge. In lotions, it improves the eyesight. When infused in wine with *saffron, it cures ophthalmia. When chopped into small pieces, macerated in water, and snorted like snuff, it heals ulcers of the nose. Its oil and water are equally useful in treating infantile colic.

Fig. A.15. Mattioli, *Discorsi*, 166.

APPLE

Latin binomial (type sp.): *Malus sylvestris* Mill.

French names: pommier, boquettier, pommier sauvage, poumastre, poumiè (Languedocian)

Elemental qualities: cold and moderately dry (🜃)

Ruling planets: ♃ and ♀

Zodiacal signature: ♍

Occult properties: Apples have antibilious, aperient, carminative, depurative, and tonic properties. The wood of the apple tree is governed by ♃ and signed by ♏. The fruit is governed by ♀ and signed by ♍. It is sacred to the goddess Ceres, the Roman equivalent of the Greek goddess Demeter, who was known by the name Malophoros, or "Apple-Bearer," at the ancient port of Megara in Greece.[10] Whenever a lover dreams about eating an apple, it portends that he or she will be happy in love. The apple bears the sign of the Fall of Adam and Eve.

ARNICA, MOUNTAIN

Fig. A.16. Parkinson, *Theatrum botanicum,* 319.

Latin binomial: *Arnica montana* L.

French names: arnica des montagnes, arnique, bétoine des montagnes, herbe à éternuer, herbe à tous les maux, herbe aux prêcheurs, panacée des chutes, souci des Alpes, tabac des Vosges*

Elemental qualities: hot and dry (△)

Ruling planet: ☉

Zodiacal signatures: ♌, ♈, and ♒ (*apud* Mavéric)

Occult properties: Mountain arnica, otherwise known as mountain tobacco and wolfsbane, but not to be confused or grouped with ★wolfsbane (*Aconitum napellus* L.), is a well-known anti-inflammatory and vulnerary plant. It is a choice herb for healing internal congestions and contusions.[11] Its spagyric essence takes after the nature of ★wormwood, ★lemon balm, and ★saffron.[12] It is one of the twelve magical plants of the Rosicrucians.[13]

*[Sédir's entry appears under the accepted Latin binomial *Arnica montana.* See *Les plantes magiques,* 144. —*Trans.*]

Fig. A.17. Fuchs, *Plantarum effigies,* 33.

ARTICHOKE

Latin binomial: *Cynara scolymus* L.

French names: artichaut, artixaut (Languedocian)

Elemental qualities: hot and dry (△)

Ruling planet: ♂

Zodiacal signature: ♏

Occult properties: The artichoke has aperient, diuretic, digestive, and stimulant properties. It has long been revered as a powerful aphrodisiac. The root or seed harvested when the ☉ is in the fifth degree of ♎ heals diarrhea and dysentery.

Fig. A.18. Mattioli, *Discorsi,* 107.

ASH

Latin binomial (type sp.): *Fraxinus excelsior* L.

French names: frêne, frêne élevé, fraisse (Languedocian)

Elemental qualities: cold and dry (🜃)

Ruling planets: ♃ and ☉

Zodiacal signatures: ♉ and ♏. Its flowers are signed by ♒.

Occult properties: The leaves of the ash tree possess astringent, cathartic, diaphoretic, and laxative properties. A poultice of chewed ash leaves is good for bites from poisonous animals. Its light-gray bark has natural febrifugal properties. According to Paracelsus, a branch collected by a virgin boy when ♄ is in ♍ can cure arthritis and gout and even desiccate wounds. The ancient Greeks called the manna ash (*Fraxinus ornus* L.) μέλια, from which the Meliae or Μελιάδες—the ash-tree nymphs who nursed the infant Zeus (the Greek equivalent of Jupiter)—got their name.[14]

ASPARAGUS

Latin binomial: *Asparagus officinalis* L.

French names: asperge, asparge, esparge, espargue (Languedocian)

Elemental qualities: hot, dry, and moderately wet (△)

Ruling planet: ♂ (*apud* Piobb)

Zodiacal signature: ♈. Harvest when the ☉ and ☾ are in ♋.

Occult properties: Asparagus has aperitive, diuretic, and mild sedative properties (see Table 6.1 on p. 58). It serves not only as a natural diuretic but also as an excellent aphrodisiac. When worn as an amulet, it has contraceptive properties. Ancient botanical lore confirms its Arien provenance. According to tradition, if a gardener pounds a ram's horn into pieces and buries some of the shards, asparagus will sprout up in that very spot.[15]

Fig. A.19. Fuchs, *Plantarum effigies,* 34.

Fig. A.20. Fuchs, *Plantarum effigies,* 86.

ASPHODEL

Latin binomial (type sp.): *Asphodelus ramosus* L.

French names: asphodèle, aledo (Languedocian), asphodèle rameux, asphodèle ramifié, bâton blanc, bâton royal*

Elemental qualities: cold and dry (🜃)

Ruling planet: ♄

Zodiacal signature: ♉ (*apud* Mario)

Occult properties: Asphodel possesses emollient and vulnerary properties. The ancient Greeks associated the asphodel with the dead and the underworld. Homer speaks of a famous parcel in Hades called the Asphodel Meadows, where spirits and "the shades of those whose work is done" dwell.[16] Because of their connection to the underworld, asphodels are frequently used in magical evocations.

*[Sédir also provides the French common name *bâton de Jacob* (Jacob's rod), but this is usually reserved for the yellow asphodel, *Asphodeline lutea* (L.) Rchb., otherwise known as king's spear or yellow Jacob's rod. See *Les plantes magiques,* 144. —*Trans.*]

B

Bachelor's Button–Buttercup

BACHELOR'S BUTTON

Fig. B.1. Fuchs, *Plantarum effigies,* 241.

Latin binomial: *Centaurea cyanus* L.

French names: bleuet, barbeau, barbereau, blavette, bluet (Languedocian), brenecaille, brènefoin, centaurée bleuet

Elemental qualities: cold and wet (∇)

Ruling planet: ♃ (*apud* Piobb)

Zodiacal signature: ♓

Occult properties: The flowers of the bachelor's button, otherwise known as cornflower, have antibilious, emmenagogic, stimulant, and tonic properties. Its leaves, when infused in wine, help stave off infectious diseases and, when boiled in beer, make a good apéritif or digestif.[1]

Fig. B.2. Fuchs, *Plantarum effigies,* 283.

BALM, LEMON

Latin binomial: *Melissa officinalis* L.

French names: mélisse, céline, citronnade, citrounèlo (Languedocian), herbe au citron, mélisse citronelle, piment des ruches, ponchirade, thé de France

Elemental qualities: hot, dry, and moderately wet (△)

Ruling planets: ☉ and ♃

Zodiacal signatures: ♌, ♋, and ♍ (*apud* Mavéric)

Occult properties: Lemon balm has aromatic, carminative, diaphoretic, and febrifugal properties. Oracular priestesses in ancient temples used to consume this herb in the form of a stimulant drink. Its water or liquor mixed with *southernwood and powdered emerald eases the process of parturition and aids in the expulsion of the afterbirth.[2] Its flowers are cordial, hepatic, ophthalmic, and spasmolytic. This "herb of rejuvenation" was sacred to medieval magicians, who gave it the Chaldean name *celeivos.** According to the *Grand Albert,* wearing this herb as an amulet makes the bearer amiable, and attaching it to the neck of an ox will make him follow you around wherever you go.[3] Alchemists and Hermeticists also refer to it by the names *meliphyllum* and *melissophyllum,* which derive from the ancient Greek plant names μελίφυλλον and μελισσόφυλλον.[4†]

*[The variant Sédir provides for Chaldean herb no. 14 is the same variant given by Stanislas de Guaïta (*Le serpent de la Genèse,* 1:372). Here, most Latin editions of the *Grand Albert* read *celayos,* whereas most French editions read *celeyos.* Cf. Table 0.1 on p. xxiii. —*Trans.*]

†[The Greek plant names μελίφυλλον, "honey-leaf" (μέλι, "honey" + φύλλον, "leaf"), and μελισσόφυλλον, "bee-leaf" (μέλισσα, "bee" + φύλλον, "leaf"), the latter of which appears as *melifophilos* in early Latin editions of the *Grand Albert,* were synonymous with μελισσοβότανον, "bee-plant" (μέλισσα, "bee" + βοτάνη, "plant"), another common name for the lemon balm in antiquity. —*Trans.*]

BAMBOO, BLACK

Fig. B.3. Bailey, *Cyclopedia,* 1:130.

Latin binomial: *Phyllostachys nigra* (Lodd. ex Lindl.) Munro
French name: bambou noir
Elemental qualities: cold and wet (🜄)
Ruling planets: ♄ and ♀
Zodiacal signature: ♏
Occult properties: Black bamboo is rich in tabasheer, or tabashir, a translucent white substance otherwise known as bamboo manna or bamboo silica, which has aphrodisiac, astringent, febrifugal, stimulant, and tonic properties. This species is native to the West Indes, where it is commonly used in the place of *vervain to make love potions and magical philters.[5]

BANYAN

Fig. B.4. Linocier, *L'histoire des plantes,* 703.

Latin binomial: *Ficus bengalensis* L.
French names: banian, banian de l'Inde, figuier des banians
Elemental qualities: cold and dry (🜃)
Ruling planet: ♃
Zodiacal signature: ♍
Occult properties: The banyan tree, *vaṭa vṛkṣa* (वटवृक्ष) in Sanskrit, has emollient and vulnerary properties. It has been revered in India since before the Vedic period. It is often equated with the sacred fig, or *aśvattha* (अश्वत्थ), otherwise identified with the species *Ficus religiosa* L., and held to be emblematic of the macrocosm. Krishna describes the cosmic *aśvattha* as a banyan having its roots pointing upward, its branches pointing downward, and the Vedic hymns for its leaves. It is believed that a great many spirits inhabit banyan trees.[6]

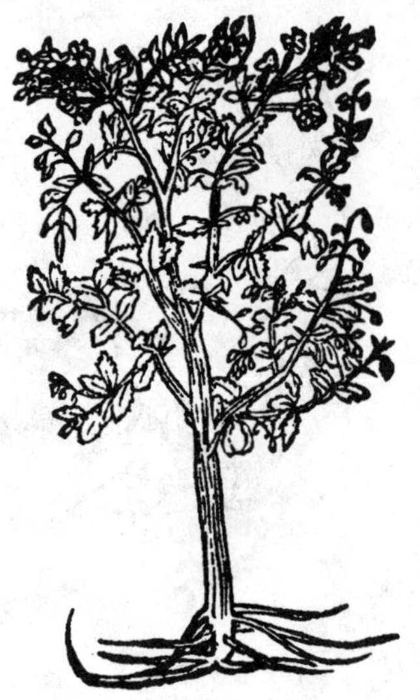

Fig. B.5. Fuchs, *Plantarum effigies*, 309.

BARBERRY

Latin binomial: *Berberis vulgaris* L.

French names: épine-vinette, vinettier

Elemental qualities: cold and dry (🜃)

Ruling planets: ♃ and ♂

Zodiacal signatures: ♉ and ♏ (*apud* Mavéric)

Occult properties: Barberry possesses antiseptic, stomachic, purgative, and tonic properties. Its bitterness derives from its Martian influence, but it is predominantly a Jupiterian shrub. It is most commonly used to heal diarrhea, dysentery, quinsy, jaundice, and menstrual irregularities. Its berries serve as an especially good remedy for hangovers.

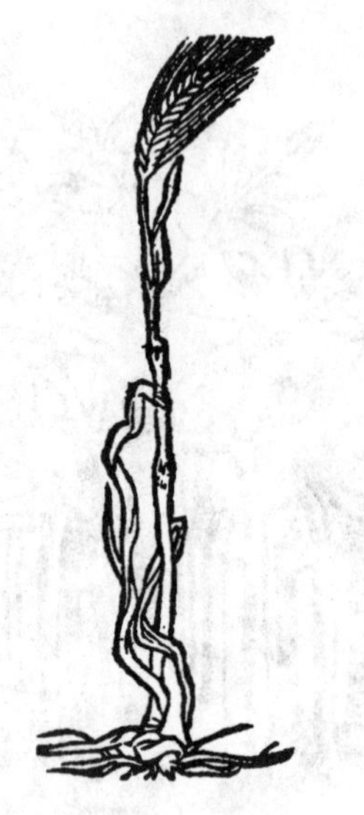

Fig. B.6. Fuchs, *Plantarum effigies*, 247.

BARLEY

Latin binomial (type sp.): *Hordeum vulgare* L.

French names: orge, ordi (Languedocian)

Elemental qualities: cold and dry (🜃)

Ruling planet: ☉

Zodiacal signature: ♍ (*apud* Mario)

Occult properties: Barley grains possess demulcent, emollient, febrifugal, nutritive, and stomachic properties. Brahmins offer the ear or shoot of barley, *yavapraroha* (यवप्ररोह) in Sanskrit, in sacrifices to the gods and the seven spiritual princes, or Manus.

BASIL

Fig. B.7. Fuchs, *Plantarum effigies*, 313.

Latin binomial: *Ocimum basilicum* L.

French names: basilic, basèli (Languedocian), basilie, frambasin, herbe royale, oranger des savetiers

Elemental qualities: hot and dry (△)

Ruling planet: ♂

Zodiacal signature: ♌. Harvest when the ☉ is in ♓ and the ☾ is in ♋.

Occult properties: Basil has carminative, demulcent, diaphoretic, diuretic, stimulant, and stomachic properties. Traditionally known as the herb of love, it has long been considered a powerful aphrodisiac. In the Hermetic language of flowers, it is emblematic of anger. Within this aromatic herb, ♂ is opposed to ♄, and their combat is activated by ☿. Influences from ♀ and ♃ are also present but secondary. Magicians often make use of it because it produces a pestilential Lunar menstruum or solvent, which can be worked in such a way that ☿ forces the toxin under the dominion of ♀. This alters the sphere of influence from ♂ to the ☉ and causes the plant's choleric fire to transmute into an amorous fire.

[BASIL THYME ⚕ *See* THYME, BASIL]

Fig. B.8. Mattioli, *Discorsi,* 105.

BAY

Latin binomial: *Laurus nobilis* L.

French names: laurier, laurier d'Apollon, laurier noble, laurier-sauce, laurier vrai, rampan (Languedocian)

Elemental qualities: hot and dry (△)

Ruling planet: ☉

Zodiacal signature: ♌. Harvest when the ☉ is in ♓ and the ☾ is in ♒.

Occult properties: All parts of this aromatic evergreen are antimicrobial. Its berries, which possess powerful vermifugal properties, are signed by ♍. Its leaves, when chewed and administered in a poultice, are useful against bites from poisonous animals. Ancient diviners wore crowns of bay, otherwise known as laurel and bay laurel, and chewed its leaves in botanomantic rituals. For this reason, they were given the name δαφνηφάγοι, or "bay-eaters," a term that later acquired the figurative meaning of "divinely inspired." The bay became the instrument of the divinatory art known as daphnomancy, by which omens were drawn from the cracks, sparks, and smoke produced during the sacral burning of its branches and leaves. The whole shrub also has the magical virtue of providing protection from lightning, and occult herbalists should study the myth of Daphne with this datum in mind. The juice of its leaves, taken at a dose of three to four drops in some water, provokes menstruation, relieves stomach aches caused by digestive problems, improves hearing loss, cures earaches, and clears up acne. When worn as an amulet, its leaves prevent nightmares. Its berries, if harvested during the hour of ♂ or the hour of ☿ and then powdered and mixed with wine, serve as a good remedy for diarrhea.

BEAR'S BREECHES

Fig. B.9. Fuchs, *Plantarum effigies*, 30.

Latin binomial (type sp.): *Acanthus mollis* L.

French names: acanthe, acanthe à feuilles molles, acanto (Languedocian), branche-ursine, patte-d'ours

Elemental qualities: cold and wet (∇)

Ruling planet: ♂

Zodiacal signature: ♋

Occult properties: The leaves and roots of this spinose genus are powerful emollients and demulcents. Greek and Roman architects decorated the capitals of Corinthian columns with scrolls of acanthus leaves to "soften" the hardness of the stone pillars. When worn as an amulet, it protects the bearer from poisonous creatures.[7]

BEAR'S EAR

Fig. B.10. Mattioli, *Discorsi*, 551.

Latin binomial: *Primula auricula* L.

French names: oreille d'ours, auricule, primevère auricule

Elemental qualities: hot and dry (Δ)

Ruling planet: ♂ (*apud* Piobb)

Zodiacal signature: ♈ (*apud* Mario). Harvest when ♂ is in a beneficent aspect to ♃.

Occult properties: The roots of bear's ear, otherwise known as auricula or mountain cowslip, possess emetic, expectorant, and purgative properties. Its leaves are vulnerary and restorative in action. It is an especially good herb in poultices for cicatrizing wounds.

Fig. B.11. Fuchs, *Plantarum effigies,* 465.

BEET GREENS

Latin binomial (type sp.): *Beta vulgaris* L.

French names: bette, bledo (Languedocian), carde, poirée

Elemental qualities: cold and moderately dry (🜃)

Ruling planet: ☿ (*apud* Piobb)

Zodiacal signature: ♍. Harvest when the ☉ is in ♋ and the ☾ passes ♏.

Occult properties: Beet greens possess aperient, emollient, and vulnerary properties. According to Hippocrates, when soaked in wine and applied topically to wounds, they have remarkable healing and cicatrizant properties. According to Pliny, a poultice made of beet leaves is also good for curing erysipelas.[8]

Fig. B.12. Fuchs, *Plantarum effigies,* 466.

BEETROOT

Latin binomial (type sp.): *Beta vulgaris* L.

French names: betterave, bledorabo (Languedocian)

Elemental qualities: cold and wet (🜄)

Ruling planet: ♃ (*apud* Piobb)

Zodiacal signatures: ♏ and ♐

Occult properties: Beetroot has aperient, nutritive, depurative, and digestive properties. Its juice is hypotensive in action and particularly useful in the treatment of blood disorders.[9]

BELLADONNA

Fig. B.13. Fuchs, *Plantarum effigies,* 395.

Latin binomial: *Atropa belladonna* L.

French names: belladone, belle-dame, bouton noir, guigne des côtes, mandragore baccifère, morelle de jardin, morelle furieuse, parmenton

Elemental qualities: cold and wet (▽)

Ruling planet: ♄ (*apud* Mavéric)

Zodiacal signature: ♏

Occult properties: Belladonna, otherwise known as deadly nightshade, is a powerful narcotic, soporific, and hallucinogenic in the Solanaceae family (see Table 6.1 on p. 58). This Saturnian herb is highly toxic, but in expert hands it makes one of the most remarkable homeopathic remedies. It is the medicament of choice for all nocturnal disorders and is particularly useful in treating eye conditions and spasmodic contractions.[10]

BETONY

Fig. B.14. Fuchs, *Plantarum effigies,* 197.

Latin binomial: *Stachys officinalis* (L.) Trevis. ex Briq.

French names: bétoine (officinale), broutonica (Languedocian), épiaire officinale

Elemental qualities: hot and dry (△)

Ruling planet: ♃

Zodiacal signature: ♈ (*apud* Mario). Harvest with ♃ in ♓ or after the full ☾ terminating the dog days of summer.

Occult properties: Betony, otherwise known as common hedge nettle, wood betony, and bishop's-wort, is a powerful sternutator. Its leaves purify the blood and cure jaundice and edema. When worn as an amulet, it protects the bearer against bewitchment.

Fig. B.15. Fuchs, *Plantarum effigies,* 417.

BINDWEED

Latin binomial (type sp.): *Convolvulus arvensis* L.

French names: liseron, courrejolo (Languedocian), liseron des champs, petite campanette, vrillée

Elemental qualities: hot and wet (🜁)

Ruling planet: ☿

Zodiacal signatures: ♊ and ♒ (*apud* Mavéric)

Occult properties: Bindweed possesses cathartic, diuretic, and purgative properties. The juice of its leaves is typically used in laxative medicines. It is an especially good herb for consecrating altars and other ritual implements.[11]

Fig. B.16. Mattioli, *Discorsi,* 112.

BIRCH

Latin binomial (type sp.): *Betula alba* L.

French names: bouleau, arbre de la sagesse, biole, bois à balais, bois néphrétique, bouillard, boule, sceptre des maîtres d'école

Elemental qualities: hot and dry (🜂)

Ruling planet: ♃

Zodiacal signature: ♐

Occult properties: Birch has anti-inflammatory, cholagogic, and diaphoretic properties. Its bark is aperient and diuretic in action. The Kamchadals of the Russian Far East used birch in a sacred ceremony called the Festival of Brooms.[12] The witches of the Middle Ages are alleged to have used its branches to fly off to their sabbats and to perform weather magic. Its aroma is especially beneficial for melancholics and victims of black magic. The juice of its leaves keeps cheeses sweet and preserves them from mites and maggots.

BIRTHWORT

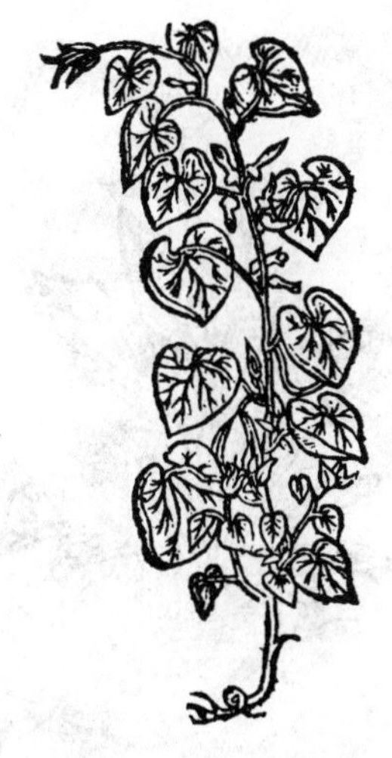

Fig. B.17. Fuchs, *Plantarum effigies,* 50.

Latin binomial (type sp.): *Aristolochia rotunda* L.

French names: aristoloche (ronde), batalu, brigbog, fautèrno (Languedocian), guillebaude, pomerasse, ratalie, ratelaire, sarrasine (ronde)

Elemental qualities: cold and dry (🜃)

Ruling planet: ♃

Zodiacal signatures: ♍ (especially the leaves and roots) and ♓

Occult properties: All species of birthwort have detersive and vulnerary properties. Paracelsus administered birthwort mixed with *comfrey and *aloe in poultices and combined it with essence of turpentine (*oleum terebinthinae*) and earthworms in distilled medicinal waters. It also has pulmonary, diuretic, and emmenagogic properties. When made into a lotion with wine, it dries up scabies and washes out wounds. The smoke of its seeds relieves epileptics, demoniacs, and those made impotent by the *aiguillette,* or magical knot.*

*[The *aiguillette* is a form of sympathetic "castration magic" consisting of a knotted cord or ligature. There were many different ways of tying the *aiguillette.* One early description comes from Albertus Magnus: "If the penis of a wolf has the name of a man or woman tied to it, he or she will not be able to have intercourse until the knot is untied" (Albertus Magnus, *De animalibus libri XXVI,* 22.2.1). A more popular variation of this spell appears in the eighteenth-century grimoire known as the *Petit Albert,* which involves approaching a man's door, calling out his name, and, as soon as he responds, tying the knot around the penis of a freshly killed wolf (Pseudo-Albertus, *Les solide tresor des merveilleux secrets,* 14). Sédir's primary source here, however, is most certainly Jacques Collin de Plancy, who likewise cites the herbal of Pseudo-Apuleius as his source (Collin de Plancy, *Dictionnaire infernal,* 1:230), but the spell involving the suffumigation of birthwort described in Pseudo-Apuleius, *Herbarium* 7.1, concerns deliverance from bewitchment and makes no mention of impotence, natural or magical. —*Trans.*]

Fig. B.18. Fuchs, *Plantarum effigies,* 449.

BISTORT

Latin binomial: *Bistorta officinalis* Delarbre

French names: serpentaire, bistorte, faux épinard, feuillote, renouée bistorte, serpentaire rouge, serpentine*

Elemental qualities: cold and dry (🜃)

Ruling planets: ♄ and ☿

Zodiacal signatures: ♍ and ♒

Occult properties: Bistort is an extremely powerful astringent with febrifugal, laxative, styptic, and vulnerary properties. Placing its chewed leaves or its juice on a wound can save victims of venomous snakebites from certain death. The scent of its root is also useful in the practice of snake charming. Its juice can be gargled to allay sore throats and assuage respiratory arrests. It is one of the most qualified of all plants, in all of its forms, as a condenser of astral fluids. This "herb of fluids" was sacred to medieval magicians, who gave it the Chaldean name *cartulin.* According to the *Grand Albert,* if placed under someone's pillow, it will keep the sleeper awake all night long or until it is removed, and if powdered and burned in a lamp, it will produce visions of astral beings.[13]

*[Sédir provides separate entries for *serpentaire* and *serpentine,* which are variant names for bistort. See *Les plantes magiques,* 168–69. These have been merged together here to form a single entry. Sédir was clearly confused, and justifiably so, by the French common names *serpentaire* and *serpentine* (see further the entry ⋆nettle and the corresponding footnote). More often than not, French *serpentaire* and *serpentine,* both meaning "snake's herb" or "snakeweed," are used to designate the herb bistort (*Bistorta officinalis* Delarbre), and the same is true of their ancient Latin etymons *serpentaria* and *serpentina*. But whereas the French *serpentaire* is also used to refer to the dragon lily (*Dracunculus vulgaris* Schott), French *serpentine* is otherwise used to refer to the herb tarragon (*Artemisia dracunculus* L.). The English equivalent *dragon's wort* is equally ambiguous and may refer to bistort, dragon lily, or tarragon; ultimately, readers must decide its true referrent on a case-by-case basis. Sédir's description of *serpentaire* as a powerful condenser of astral fluids derives from Stanislas de Guaïta's characterization of *serpentaire* as "the herb of fluids" (Guaïta, *Le serpent de la Genèse,* 1:372), but here Guaïta is referring to Chaldean herb no. 16 in the *Grand Albert* (cf. Table 0.1 on p. xxiii). In the Latin edition of 1493, this herb is called *serpentina,* and in the French edition of 1703, it is called *serpentine.* To make matters still more confusing, the Latin and French editions of the *Grand Albert* erroneously supply *quinquefolium* (⋆cinquefoil) as a "Greek" synonym for *serpentina* and

[BITTER CRESS ⚕ *See* CRESS, BITTER]

[BLACK BAMBOO ⚕ *See* BAMBOO, BLACK]

BLACKBERRY

Latin binomial (type sp.): *Rubus allegheniensis* Porter

French names: ronce, ronce à fruits, ronce frutescente, roume (Languedocian)

Elemental qualities: cold and dry (🜃)

Ruling planet: ♄

Zodiacal signature: ♒

Occult properties: Blackberry has antirheumatic, astringent, and tonic properties. It is particularly effective in infusions against pains in the knee and ankle joints. It is sacred to Saturn, as evidenced by its dark-colored fruit. In the Hermetic language of flowers, the blackberry is emblematic of envy.

Fig. B.19. Fuchs, *Plantarum effigies*, 83.

[BLACK HELLEBORE ⚕ *See* HELLEBORE, BLACK]

[BLACK FALSE HELLEBORE ⚕ *See* HELLEBORE, BLACK FALSE]

(*cont.*) *serpentine.* Doubtless Sédir recognized the disparity between *serpentaire* and *serpentine,* but he nonetheless included separate entries for both. These entries, however, are strikingly similar: both herbs are ruled by ♄ and ☿, and both are said to be good remedies for respiratory illnesses. Suffice it to say that bistort is the most likely candidate for both entries. The Pseudo-Albertan tradition of placing *serpentine* under one's pillow logically follows Sédir's description of *serpentaire* as a fluid condenser. It was this very tradition, in fact, that led Guaïta to characterize Chaldean herb no. 16 as "the herb of fluids." In other words, when placed surreptitiously under someone's pillow, the herb bistort will keep the sleeper awake all night long precisely because it is such a powerful fluid condenser and holds such a strong etheric charge. Cf. Sédir's entry ⋆agrimony, which, when placed under someone's pillow, has the opposite effect. —*Trans.*]

Fig. B.20. Fuchs, *Plantarum effigies*, 356.

BLACKTHORN

Latin binomial: *Prunus spinosa* L.

French names: prunellier, agrunelië (Languedocian), épine noire, prunelle, prunier sauvage

Elemental qualities: hot and dry (🜂)

Ruling planet: ♂ (*apud* Piobb and Bélus)

Zodiacal signatures: ♏ (tree) and ♊ (fruit or sloe)

Occult properties: The flowers and fruits of the blackthorn have diaphoretic, depurative, and febrifugal properties. Its bark and roots are highly astringent and possess aperient and diuretic properties. Its juice is invigorating and makes hangovers disappear.

[BLADDER SENNA ❦ *See* SENNA, BLADDER]

[BLESSED THISTLE ❦ *See* THISTLE, BLESSED]

[BOG ROSEMARY ❦ *See* ROSEMARY, BOG]

Fig. B.21. Fuchs, *Plantarum effigies*, 77.

BORAGE

Latin binomial: *Borago officinalis* L.

French names: bourrache, bourracho (Languedocian)

Elemental qualities: hot and wet (🜁)

Ruling planet: ♃

Zodiacal signatures: ♊ and ♒. Harvest under ♊ or when ♃ is in ♒.

Occult properties: Borage, otherwise known as starflower, has cordial, cardiotonic, and diuretic properties. Its chief virtue is its alterative action, which effectively purifies the blood.

BOXWOOD

Fig. B.22. Fuchs, *Plantarum effigies,* 369.

Latin binomial (type sp.): *Buxus sempervirens* L.

French names: buis, arbre vert, bois-béni, bois d'Artois, bouis (Languedocian), buis bénit, guézette, housenic, ozanne

Elemental qualities: hot and dry (△)

Ruling planet: ♀ (*apud* Lenain and Piobb)

Zodiacal signatures: ♌ and ♎. Harvest when the ☉ is in ♓ and the ☾ is in ♒.

Occult properties: The leaves and bark of boxwood possess alterative, diaphoretic, febrifugal, purgative, and vermifugal properties. This evergreen shrub is sacred to Cybele. In antiquity, its wood was used to make the flutes and musical pipes that sounded out the exotic Phrygian melodies of the sacred rites of Magna Mater.[14]

BROOM

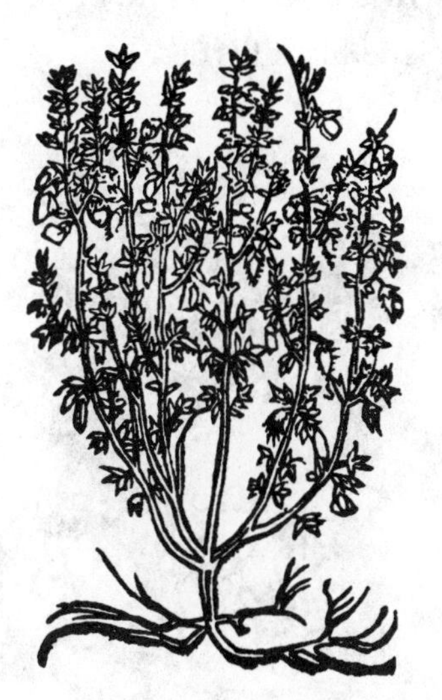

Fig. B.23. Fuchs, *Plantarum effigies,* 124.

Latin binomial (type sp.): *Genista tinctoria* L.

French names: genêt, fleur à tiendre, genestrolle, genêt des teinturiers, herbe à jaunier, herbe aux teinturiers

Elemental qualities: hot and dry (△)

Ruling planet: ♂ (*apud* Piobb)

Zodiacal signature: ♌. Harvest when the ☉ is in ♓ and the ☾ is in ♒.

Occult properties: Broom, otherwise known as dyer's broom or dyer's greenweed, has diuretic, cathartic, and emetic properties. It is known to be an effective remedy for gout, rheumatism, and edema. Its fruits, which are long, shiny pods shaped like green bean pods, are signed by ♍ and therefore have vermifugal properties.

Fig. B.24. Mattioli, *Discorsi,* 702.

BRYONY

Latin binomial: *Bryonia dioica* Jacq.

French names: bryone (dioïque), couleuvrée, feu ardent, herbe de feu, mandragore grimpante, navet du diable, rave de serpent, tuquiè (Languedocian), vigne du diable

Elemental qualities: cold and dry (🜃)

Ruling planet: ♂

Zodiacal signature: ♑

Occult properties: Bryony has powerful diuretic, emetic, and purgative properties. It is an extremely acrid and bitter-tasting herb and is commonly used in rituals of black magic. This species—which is more accurately described by the common name red bryony—produces white flowers and red berries, and for this reason it is often erroneously called white bryony. As a result, occult herbalists sometimes confuse it with ⋆white bryony (*Bryonia alba* L.), which produces white flowers and black berries.[15]

Fig. B.25. Fuchs, *Plantarum effigies,* 52.

BRYONY, WHITE

Latin binomial: *Bryonia alba* L.

French names: bryone blanche, colubrine, coulevrée blanche, racine vierge, vigne blanche

Elemental qualities: hot and dry (△)

Ruling planet: ☿

Zodiacal signature: ♌

Occult properties: White bryony has aperient, cathartic, pectoral, and purgative properties. This climbing plant also has the unique virtue of ensuring protection from lightning.[16] It was known among alchemists and Hermeticists by the name *psilothron,* which is a transliteration of the ancient Greek plant name ψίλωθρον.*

*[Sédir includes the "Hermetic" name *psilothron* as a variant for *bryone* (⋆bryony), following

BUCKTHORN

Fig. B.26. Mattioli, *Discorsi,* 105.

Latin binomial: *Rhamnus catharticus* L.

French names: nerprun, bourg-épine, épine de cerf, nerprun purgatif, noirprun

Elemental qualities: hot and dry (△)

Ruling planet: ♂

Zodiacal signature: ♎

Occult properties: Buckthorn has cathartic, depurative, laxative, and purgative properties. This plant was allegedly used to weave Christ's crown of thorns and hence, like the Christ's thorn jujube (*R. spina-christi* L.) and Jerusalem thorn (*Paliurus spina-christi* Mill.), it is sometimes called Christ's thorn or Crown of Thorns. It is sacred to Saturn and emblematic of virginity, sin, the devil, and humility.[17] When hung as talismans from the doors and windows of a home, its branches become powerful apotropaics that thwart the efforts of black magicians and demons.

(*cont.*) Antoine-Joseph Pernety's *Dictionnaire mytho-hermétique,* 410, but its proper place is under *bryone blanche* (*white bryony). See *Les plantes magiques,* 146. The Greeks adopted the term ψίλωθρον, which literally means "depilatory," as a variant name for white bryony (otherwise known as ἄμπελος λευκή, or "white vine") because its root was chiefly used as a depilatory to remove unwanted hair (see Dioscorides, *On Medical Materials* 4.182). —*Trans.*]

BUGLE

Fig. B.27. Fuchs, *Plantarum effigies,* 221.

Latin binomial (type sp.): *Ajuga reptans* L.

French names: petite consoude, bugle rampante, herbe de Saint Laurent

Elemental qualities: hot and dry (△)

Ruling planets: ♃, ♂, and ☿

Zodiacal signature: ♈

Occult properties: Bugles possess astringent and styptic properties. They cure all kinds of Saturnian maladies such as wounds, bruises, and ulcers, especially those around the region of the mouth. To relieve toothaches, rub a dried bugle root harvested during the month of August on the diseased tooth until it starts to bleed. Afterward, plug the cavity with ⋆willow.

BURDOCK, GREATER

Fig. B.28. Fuchs, *Plantarum effigies,* 41.

Latin binomial: *Arctium lappa* L.

French names: bardane (grande), bouillon noir, chou d'âne, gleteron, glouteron, gouteron, gratteron, herbe aux pouilleux, herbe aux teigneux, houyou, lapparasso (Languedocian), napolier, oreille de géant, peignerolle, pignet, teigne*

Elemental qualities: cold and dry (🜃)

Ruling planet: ♄

Zodiacal signatures: ♒ and ♊

*[Sédir provides separate entries for *grande bardane* and *teigne,* which are variant names for greater burdock. See *Les plantes magiques,* 145, 169. These have been merged here to form a single entry. —*Trans.*]

Occult properties: Greater burdock is an ancient alterative with diuretic and depurative properties. Its root, which is edible, helps unblock obstructions and serves as a common ingredient in remedies for venereal diseases. Suffumigations of its leaves* under the private parts of a girl have the same black magical properties as decoctions of ★lily pollen, which, when mixed in a girl's drink, will reveal that she is no longer a virgin if shortly after consuming it she is overcome with an irresistible urge to urinate.[18]

BURDOCK, LESSER

Latin binomial: *Arctium minus* (Hill) Bernh.

French names: bardane (petite), bardane à petits capitules†

Properties: Lesser burdock has the same elemental, astral, and occult properties as ★greater burdock.

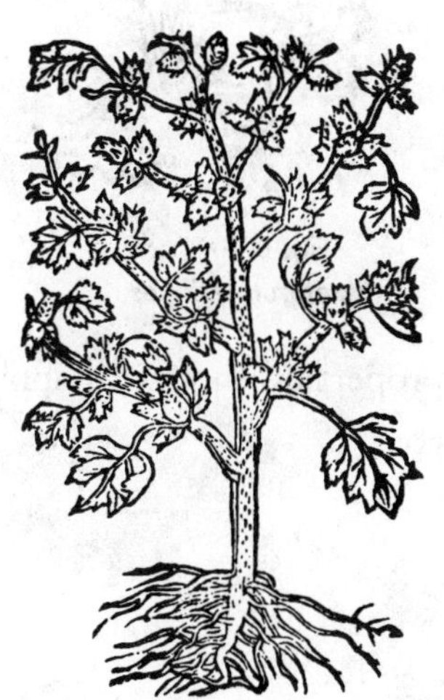

Fig. B.29. Fuchs, *Plantarum effigies,* 333.

*[Sédir's *semences* (seeds) has been corrected here to read *feuilles* (leaves). He identifies his source as Johann Jacob Wecker, but Wecker describes the suffumigant as "the smoke of the seeds of the ★purslane or of the leaves of the personata called burdock" (*fumus seminum portulacae, vel foliorum personatae, quam bardanum vocant*). Wecker in turn identifies his source as Giambattista della Porta, who describes the suffumigant as "the seed of the ★purslane or the leaves of the personata our compatriots call *lapatium*" (*portulacae semen, vel personatae folia, quam vocant nostrates lapatium*). Since the plant name personata is well attested as a variant of greater burdock, Porta's *lapatium* must be a variant for (or misspelling of) *lappatium,* another, but much less common, name for greater burdock (rather than *lapathium,* a known variant of *lapathum,* from Greek λάπαθον—that is, ★sorrel). In any case, Sédir has mistaken his source, which was most likely Pierre Meyssonier's augmented French translation of Wecker's *De secretis libri XVII,* to read "seeds of greater burdock." It is to be noted as well that Porta's magical suffumigants first appear in the 1558 edition of *Magiae naturalis,* but just like his recipes for witches' ointments, these are nowhere to be found in the censored versions of his work, which include the English translation of 1658. —*Trans.*]

†[Sédir includes ★lesser burdock and ★greater burdock in a single entry. See *Les plantes magiques,* 146 s.v. *bardane (petite ou grande).* These have been separated here to form two distinct entries. —*Trans.*]

Fig. B.30. Fuchs, *Plantarum effigies*, 85.

BUTTERCUP

Latin binomial (type sp.): *Ranunculus auricomus* L.

French names: renoncule, bassinet, jaunet, renoncule tête d'or, renounculo (Languedocian)

Elemental qualities: hot and dry (△)

Ruling planet: ♂ (*apud* Culpeper)

Zodiacal signature: ♌. Harvest when the ☉ is in ♌ and the ☾ is in ♍ or when the ☉ is in ♉ and the ☾ is in ♊.

Occult properties: Flowering plants of the family Ranunculaceae, including buttercups, spearworts, and water crowfoots, have antirheumatic, febrifugal, and rubefacient properties. In topical applications, buttercups help remove warts and relieve gout.[19]

C

Cabbage–Cypress

CABBAGE

Fig. C.1. Fuchs, *Plantarum effigies*, 234.

Latin binomial: *Brassica oleracea* L.
French names: chou, caulet (Languedocian)
Elemental qualities: cold and wet (∇)
Ruling planet: ☾ (*apud* Jollivet-Castelot)
Zodiacal signatures: ♋ and, if harvested at the end of October, ♏
Occult properties: Cabbage has antiscorbutic, cardiotonic, stomachic, and vermifugal properties. It is particularly effective against inflammations of the stomach.

Fig. C.2. Fuchs, *Plantarum effigies,* 235.

CABBAGE, RED

Latin binomial: *Brassica oleracea* L. (var. *capitata* f. *rubra*)

French names: chou rouge, caulet rouxe (Languedocian)

Elemental qualities: cold and wet (🜄)

Ruling planets: ☾ and ♃

Zodiacal signature: ♋

Occult properties: Red cabbage is the best species of *cabbage. When consumed before a party, it lessens the effects of drinking too much. It is also beneficial against jaundice and bile. Its expressed juice is often employed as a bechic or pulmonary to treat common colds and bronchitis.[1] Its essence is a universal medicine.

Fig. C.3. Mattioli, *Discorsi,* 692.

CAMELLIA, OILSEED

Latin binomial: *Camellia oleifera* C.Abel.

French name: camélia

Elemental qualities: cold and dry (🜃)

Ruling planet: ☉

Zodiacal signature: ♍

Occult properties: Oilseed camellia, also known as tea oil camellia, has anti-inflammatory and antimicrobial properties. When distilled, its seeds produce an oil that can be used to fuel lamps of adoration.

CAMPHOR

Fig. C.4. Pomet, *Histoire générale des drogues,* 246.

Latin binomial: *Cinnamomum camphora* (L.) J.Presl.

French names: camphrier, camphre.

Elemental qualities: cold and wet (▽)

Ruling planet: ☾

Zodiacal signature: ♋ (*apud* Jollivet-Castelot)

Occult properties: Camphor, the distilled bark of the camphor tree, has anodyne, antipruritic, and sedative properties (see Table 6.1 on p. 58). Its resin, when burned, emits a potent Lunar perfume.

[CAPER SPURGE ❦ *See* SPURGE, CAPER]

CARAWAY

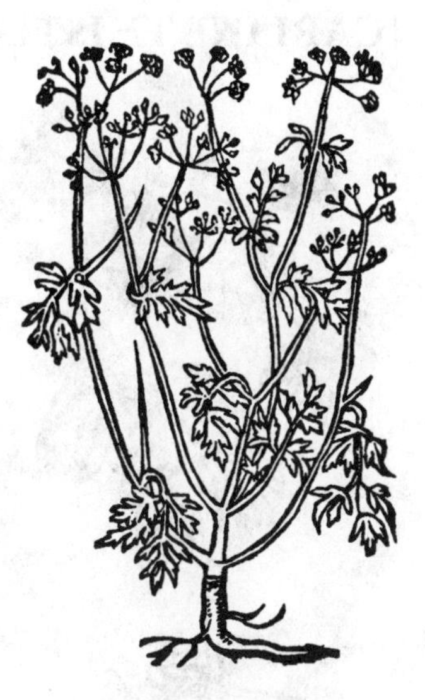

Fig. C.5. Fuchs, *Plantarum effigies,* 224.

Latin binomial: *Carum carvi* L.

French names: carvi, anis bâtard, anis des Vosges, cumin des prés

Elemental qualities: hot and wet (🜁)

Ruling planet: ☉

Zodiacal signature: ♊

Occult properties: Caraway, otherwise known as meridian fennel and Persian cumin, has strong carminative and stomachic properties. The smoke of its seeds serves as an excellent magical perfume not only for magical consecrations but also for psychic well-being.

Fig. C.6. Mattioli, *Discorsi,* 24.

CARDAMOM

Latin binomial (type sp.): *Elettaria cardamomum* (L.) Maton.

French names: cardamome, cardamome aromatique, cardamome verte, cardamomier

Elemental qualities: hot and dry (🜂)

Ruling planet: ☉

Zodiacal signature: ♈

Occult properties: Cardamom has carminative, diuretic, stimulant, and tonic properties. Both the smaller green pods of the genus *Elettaria* and the larger black pods of the genus *Amomum* are aromatic and stomachic in action.

[CARLINE THISTLE ⚘ *See* THISTLE, CARLINE]

Fig. C.7. Fuchs, *Plantarum effigies,* 194.

CASTOR BEAN

Latin binomial: *Ricinus communis* L.

French names: ricin, palme de Christ, ricinier

Elemental qualities: hot and wet (🜁)

Ruling planet: ♃ (*apud* Agrippa)

Zodiacal signatures: ♊ and ♈. Harvest under ♌.

Occult properties: The castor bean or castor-oil plant has cathartic, emollient, laxative, and purgative properties. It also prevents obsession, bewitchment, and sudden fright. It is commonly referred to by the Latin name *palma Christi,* or "palm of Christ," because of its ability to heal wounds and cure a variety of ailments, but some alchemists and Hermeticists refer to it by the name *pentadactylon,* from the Greek πενταδάκτυλον, meaning "five-fingered herb," an epithet ancient herbalists traditionally reserve for ⋆cinquefoil.[2]

CATNIP

Fig. C.8. Fuchs, *Plantarum effigies,* 246.

Latin binomial: *Nepeta cataria* L.

French names: cataire, catàrri (Languedocian), chataire, herbe aux chats, menthe des chats

Elemental qualities: hot and wet (🜁)

Ruling planet: ☿

Zodiacal signature: ♎

Occult properties: Catnip, or catmint, has carminative, emmenagogic, spasmolytic, and stimulant properties. When harvested under a favorable aspect, it can, if the occult herbalist knows how to extract its arcanum, boost a patient's energy levels. This "herb of vitality" was sacred to medieval magicians, who gave it the Chaldean name *bieith.* According to the *Grand Albert,* when held under the nose of some animals, it has a narcotic and stimulant effect, and when drowned insects are placed in its warm ashes, they revivify in a short amount of time.[3]

CEDAR

Fig. C.9. Mattioli, *Discorsi,* 101.

Latin binomial (type sp.): *Cedrus libani* A.Rich.

French name: cèdre

Elemental qualities: hot and dry (△)

Ruling planet: ♃

Zodiacal signatures: ♌, ♒, and ♊ (*apud* Mavéric)

Occult properties: The leaves and wood of cedar possess antiseptic and expectorant properties. The tree's resinous liquor, known as *cedria* and *liquor cedrinus,* is primarily desiccant in action.[4] In the Hebrew scriptures, the cedar is emblematic of pride: "The great cedar towered and set its top among the clouds, and its heart was proud of its height."[5]

Fig. C.10. Fuchs, *Plantarum effigies,* 499.

CELANDINE, GREATER

Latin binomial: *Chelidonium majus* Tourn. ex L.

French names: chélidoine (grande), éclaire, félougène, herbe aux hirondelles, herbe aux verrues, herbe dentaire, herbe de Sainte Claire, sologne, salaranio (Languedocian)

Elemental qualities: cold and dry (🜃)

Ruling planet: ☉

Zodiacal signatures: ♉ and ♎

Occult properties: Greater celandine has alterative, cathartic, diaphoretic, diuretic, and expectorant properties. It remains one of the best-known folk remedies for jaundice and liver diseases. Paracelsus considered the spagyric essences of greater celandine and ★lemon balm to be richest in virtue.[6] This "herb of triumph" was sacred to medieval magicians, who gave it the Chaldean name *aquilaris.* According to the *Grand Albert,* when harvested at the appropriate time and worn as an amulet, it serves as an effective complement to all magical operations for ensuring success in business and especially in trials, and if placed on the heads of patients suffering from a life-threatening illness, it will cause them either to sing, which means they will die, or to cry, which means they will live.[7]*

*[The information concerning the wearing of greater celandine as an amulet, its use in medical prognostication, and its Chaldean name *aquilaris* derive from the *Grand Albert,* which supplies the names *celidonia* and *chélidoine,* but Sédir mistakenly places this information under *chélidoine (petite),* ★lesser celandine, instead of *chélidoine (grande),* greater celandine. See *Les plantes magiques,* 147–48. The French common name *chélidoine* is frequently employed by itself, without the use of the adjective *grande,* to designate the species *Chelidonium majus,* and the same is true of its ancient Greek etymon χελιδόνιον (from χελιδών, "swallow") and the Latin derivative *chelidonia,* both meaning "swallowwort," a name it allegedly received because it blossoms around the time swallows make their nests. For this reason, the information derived from the *Grand Albert* has been moved to its proper place under ★greater celandine. —*Trans.*]

CELANDINE, LESSER

Fig. C.11. Fuchs, *Plantarum effigies,* 500.

Latin binomial: *Ficaria verna* Huds.

French names: chélidoine (petite), billonée, éclairette, ganille, grenouillette, herbe au fic, herbe aux hémorrhoïdes, jauneau, petite éclaire, petite scrofulaire, pissenlit rond

Elemental qualities: hot and moderately dry (△)

Ruling planet: ☉

Zodiacal signature: ♐

Occult properties: Lesser celandine, otherwise known as pilewort, has anti-inflammatory, astringent, and demulcent properties. The root, which is hot and dry and is signed by ♈, is useful in the treatment of gangrene.[8]

CELERY

Fig. C.12. Fuchs, *Plantarum effigies,* 153.

Latin binomial: *Apium graveolens* L.

French names: ache, api (Languedocian), céleri, céleri des marais, persil odorant, persil sauvage

Elemental qualities: hot and dry (△)

Ruling planet: ☿

Zodiacal signature: ♌

Occult properties: Celery has emmenagogic, diuretic, and stomachic properties. It was classed in antiquity among the five great aperitive roots together with *asparagus, butcher's-broom (*Ruscus aculeatus* L.), *fennel, and *parsley.[9] Its seeds make for excellent digestives and carminatives. It was the sacred funerary herb of the ancient Greeks, who gave it the name σέλινον.[10] Alchemists and Hermeticists refer to it by the secret name *oscieum.*[11]*

*[The name *oscieum,* if it is not a typographical error, probably means something like "mouth-mover," from Latin *os,* "mouth," + *ciere,* "to put in motion." —*Trans.*]

Fig. C.13. Mattioli, *Discorsi,* 396.

CENTAURY

Latin binomial (type sp.): *Centaurium erythraea* Rafn

French names: centaurée, erythrée, gentianelle, herb au centaure, herbe au Chiron, hèrbo de santurèo (Languedocian), petite centaurée

Elemental qualities: hot and dry (△)

Ruling planet: ♃

Zodiacal signature: ♌. Harvest when the ☉ is in ♉ and the ☾ is in ♊, at the end of August, or when ♄ and ♂ are in ♐.

Occult properties: Centaury possesses powerful stomachic and tonic properties. According to ancient botanical lore, this herb was first discovered by the centaur Chiron.[12] It heals jaundice, colic, bilious fevers, gout, scurvy, worms, and menstrual irregularities. In antiquity, it was also employed as an antidemoniac. The Duke of Portland's celebrated powder for gout, which consisted of common centaury (*Centaurium erythraea* Rafn, formerly known as *Chironea centaurium*) mixed with the roots of ★gentian and ★birthwort and the tops and leaves of ★germander and ground pine (*Ajuga chamaepitys* L.), was the same formula as Caelius Aurelianus's *diacentaureon* and Aëtius's *antidotos ex duobus centaureae generibus.*[13] From the magical point of view, it is a plant whose virtue is exalted when incantatory words are recited before harvesting it. This "herb of enchantment" was sacred to medieval magicians, who gave it the Chaldean name *siphilon.** According to the *Grand Albert,* when placed in the oil of a lamp with a little "blood of female hoopoe,"† it induces visionary hallucinations. If it is thrown into a fire while you are staring at the night sky, the stars will appear to move. And if someone is made to inhale its fumes, it will make them tremble with fear.[14]

*[In Latin and French editions of the *Grand Albert,* the name of Chaldean herb no. 11 is *isiphilon.* Sédir's variant *siphilon* appears otherwise unattested. Cf. Table 0.1 on p. xxiii. —*Trans.*]

†[Cf. Sédir's discussion of the witches' ointments in chapter 6, especially diabolical herb no. 4 (hoopoe's blood) and the corresponding footnote. —*Trans.*]

CHAMOMILE

Fig. C.14. Fuchs, *Plantarum effigies,* 14.

Latin binomial (type sp.): *Chamaemelum nobile* (L.) All.

French names: camomille (noble), anthémis noble, anthémis odorante, camomille blanche, camomille odorante, camomille romaine, camoumilo (Languedocian)

Elemental qualities: moderately hot and wet (🜁)

Ruling planet: ☿

Zodiacal signatures: ♊ and ♎. Harvest when ♂ conjuncts the ☾ and ☉.

Occult properties: Chamomile flowers are anodyne, nervine, stomachic, and tonic in action. They are particularly effective against congestions of humors in the thoracic organs.

CHASTE TREE

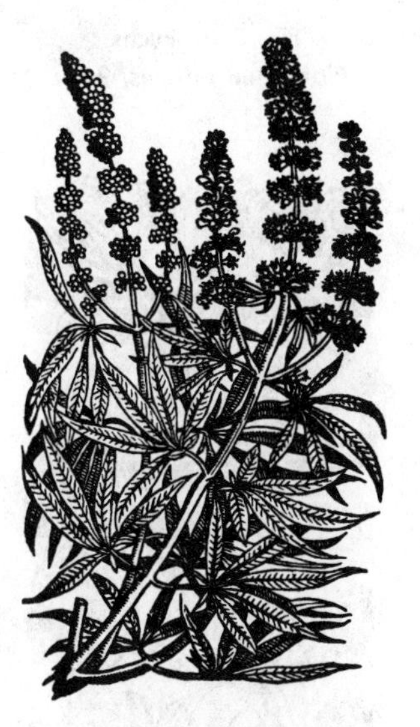

Fig. C.15. Mattioli, *Discorsi,* 143.

Latin binomial: *Vitex agnus-castus* L.

French names: arbre chaste, arbre au poivre, gattilier, petit poivre, poivre sauvage*

Elemental qualities: cold and wet (🜄)

Ruling planet: ♄

Zodiacal signature: ♋

Occult properties: Chaste tree has calmative, emmenagogic, galactagogic, vermifugal, and vulnerary properties. Its calmative properties were already well known to ancient Athenian matrons, who would strew their beds with its leaves to help them preserve their chastity during the Eleusinian Thesmophoria, the yearly festival celebrated in honor of Demeter.[15] Its seeds, when prepared by infusion, serve as a valuable remedy for venereal diseases. A strong

*[Sédir's entry appears under the Latin binomial in the form *Agnus Castus, Vitex.* See *Les plantes magiques,* 142. —*Trans.*]

decoction of chaste tree, ⋆celery, and ⋆sage made in salty water and applied topically to the back of the head cures lethargy.[16] Paracelsus refers to its flowers by the secret names *zatanea, zuccar,* and *zuccaiar.*[17]

Fig. C.16. Fuchs, *Plantarum effigies,* 240.

CHERRY

Latin binomial: *Prunus avium* L.

French names: cerisier, bigarreau, cerièis (Languedocian), cerisier des oiseaux, guignier sauvage, merisier

Elemental qualities: cold and wet (🜄)

Ruling planet: ♃

Zodiacal signature: ♓

Occult properties: The fruit stalks of cherry trees have astringent, bechic, diuretic, and tonic properties. The fruits, or drupes, are purifying, refreshing, and good for combatting hangovers.

Fig. C.17. Mattioli, *Discorsi,* 177.

CHERRY, CORNELIAN

Latin binomial: *Cornus mas* L.

French names: cornouiller (mâle), bois de fer, cornier, courgelier, cournoulhè (Languedocian), fuselier

Elemental qualities: hot and wet (🜁)

Ruling planets: ♃ and ♂

Zodiacal signature: ♏

Occult properties: The cornelian cherry has the same occult properties as the ⋆cherry tree. The wood of the cornelian cherry, however, is much harder and denser. According to Roman mythology, it sprang from the shaft of the javelin that Romulus hurled from the Aventine. It is also sacred to Ares, the Greek god of war.[18]

CHERRY, WINTER

Fig. C.18. Fuchs, *Plantarum effigies*, 396.

Latin binomial: *Physalis alkekengi* L.

French names: alkékenge, alicacabut, cloqueret, herbe à cloques, vésicaire

Elemental qualities: cold and dry (🜃)

Ruling planet: ♀ (*apud* Piobb)

Zodiacal signatures: ♉ and ♎

Occult properties: The winter cherry, or Chinese lantern, has bechic, expectorant, and febrifugal properties. Its fruits, known as bladder cherries, are diuretic in action and are particularly effective in the treatment of edema.

CHICKWEED

Fig. C.19. Mattioli, *Discorsi*, 622.

Latin binomial: *Stellaria media* (L.) Vill.

French names: mouron, bec de moineau, herbe à l'oiseau, morgeline, mouron blanc, mouron des oiseaux, mouron d'hiver, mourrelou (Languedocian)

Elemental qualities: cold and wet (🜄)

Ruling planets: ☾ (*apud* Culpeper) and ♃

Zodiacal signatures: ♋, ♊, and ♎

Occult properties: Chickweed has aperient, astringent, carminative, refrigerant, and vulnerary properties. When harvested at the end of October under ♏, the influence of ♃ predominates.

Fig. C.20. Fuchs, *Plantarum effigies*, 388.

CHICORY

Latin binomial: *Cichorium intybus* L.

French names: chicorée, chicorée amère, chicorée sauvage, herbe à café, intybe, xicourèo (Languedocian)

Elemental qualities: hot and dry (△)

Ruling planet: ☿ (*apud* Piobb)

Zodiacal signatures: ♈ and ♍. Harvest after the full ☾ terminating the dog days of summer.

Occult properties: Chicory has aperient, cholagogic, depurative, digestive, diuretic, and tonic properties. Its root becomes a powerful remedy against evil spirits and spells when the practitioner, on bended knee, exposes it in the ground by means of gold or silver agricultural implements and then uproots it with an iron tool shortly before sunrise on the feast day of Saint John the Baptist (June 24) while swearing an oath and performing ceremonies and exorcisms with the sword of Judas Maccabeus.[19]* If harvested during the hour of ♀ when ♃ is in ♐ and the ☉ is in ♌, this herb will acquire potent vulnerary and healing properties.

*[The German physician Georg Pictorius briefly describes this ritual, which he claims to have witnessed as a young boy, in his treatise *On the Species of Ceremonial Magic,* first published in his *Pantopōlion* in 1563 and subsequently reprinted in most editions of Pseudo-Agrippa's *Fourth Book of Occult Philosophy* but regrettably omitted from Robert Turner's English translation. Pictorius, just like Pseudo-Agrippa, uses the golden sword of Judas Maccabeus (cf. Maccabees 15:15–16) as a metaphor for the consecrated sword of ceremonial magic (cf. Pseudo-Agrippa, *Fourth Book of Occult Philosophy,* 53). Heinrich Cornelius Agrippa himself mentions the golden sword of Judas Maccabeus in his Pavian oration on Hermes Trismegistus (see Agrippa, *Henrici Cornelii Agrippae ab Nettesheym . . . opera,* 2:1076; cf. Morely, *The Life of Henry Cornelius Agrippa von Nettesheim,* 1:282). —*Trans.*]

CHINABERRY

Latin binomial: *Melia azedarach* L.

French names: azédarac, laurier grec, lilas des Indes, margousier*

Elemental qualities: cold and dry (🜃)

Ruling planet: ♄

Zodiacal signature: ♉

Occult properties: The chinaberry tree has cathartic, emetic, and emmenagogic properties. Its grayish-brown bark is employed chiefly as a parasiticide and vermifuge among practitioners of Eastern herbal medicine.[20]

Fig. C.21. Pomet, *Histoire générale des drogues,* 24.

*[Sédir's entry appears under the species name *azedarach* alongside two otherwise unattested variants, *azevarac* and *cetarach.* The latter is an alternate French spelling for *céterach,* a gallicization of the species or genus name—depending on which Latin binomial you prefer—of the fern species known as rustyback (*Asplenium ceterach* L. syn. *Ceterach officinarum* Willd.). It is unclear, however, whether Sédir means to suggest here that rustyback bears the same astral signatures as chinaberry. If so, this is a fascinating example of using Green Language to decipher plant signatures. It is to be noted, moreover, that the other fern species listed in Sédir's dictionary, ★hart's-tongue fern, ★maidenhair fern, and ★male fern, are all placed under the dominion of ♄ and that rustyback is best known as a bechic, or antitussive, which would similarly suggest a Taurean signature. —*Trans.*]

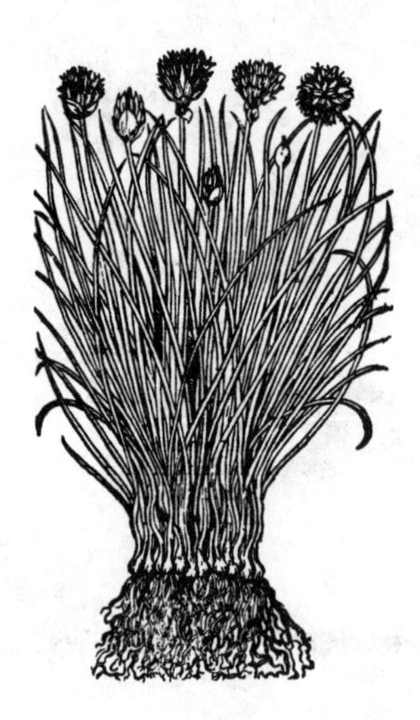

CHIVE

Latin binomial: *Allium schoenoprasum* L.

French names: cive, civette, ciboulette*

Properties: Chive has the same elemental, astral, and occult properties as ★garlic but is less potent and effective.

Fig. C.22. Gerard, *The Herball,* 139.

Fig. C.23. Fuchs, *Plantarum effigies,* 507.

CHRYSANTHEMUM

Latin binomial (type sp.): *Chrysanthemum indicum* L.

French names: chrysanthème, chrysanthème d'automn, chrysanthème vivace, pyrèthre des Indes, renonculier

Elemental qualities: hot and dry (△)

Ruling planet: ☿

Zodiacal signature: ♌ (*apud* Haatan)

Occult properties: Chrysanthemum leaves have depurative properties, whereas the flowers are aperient and stomachic in action. These flowering plants serve as excellent apotropaics against black magic.

*[Sédir's original entry for *cive* and *civette* refers readers to an entry for *ciboulette,* but unfortunately no entry for *ciboulette* is to be found in Sédir's dictionary. For this reason, a new entry has been created for chive. It is possible Sédir had intended to refer readers to his entry for *ciboule* (spring onion), and indeed, it would be equally reasonable to suggest that chive has same elemental, astral, and occult properties as ★spring onion, another close relative of the garlic plant, but both magically and medicinally, chives are more often characterized as a "lesser garlic." Readers will note, moreover, that all species of the *Allium* genus in Sédir's dictionary are Martian herbs. See further the entries ★garlic, ★leek, ★sand leek, ★onion, and ★spring onion. —*Trans.*]

CINNAMON

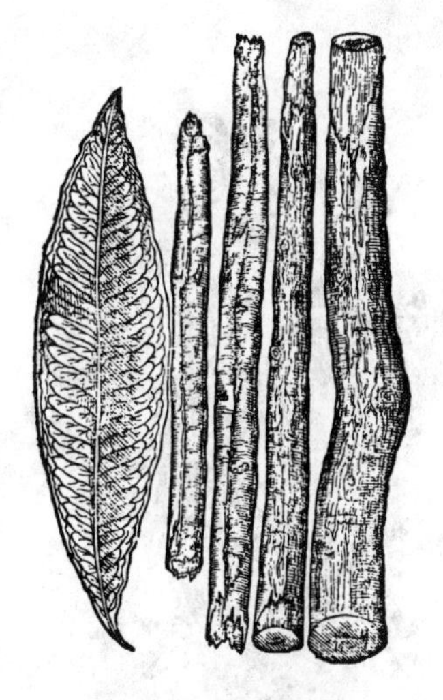

Fig. C.24. Gerard, *The Herball,* 1348.

Latin binomial: *Cinnamomum verum* J.Presl.

French names: cannelle, canelle de Ceylan, canellier

Elemental qualities: hot and dry (△)

Ruling planet: ☉

Zodiacal signature: ♌ (*apud* Mario)

Occult properties: Cinnamon, the spice obtained from the inner bark of the branches of several species of the genus *Cinnamomum,* has astringent, carminative, emmenagogic, stimulant, stomachic, and vermifugal properties. It emits a potent Solar perfume. Its reddish oil, or spagyric essence, when extracted by distillation, has a very pungent taste and serves as a marvelous tonic.

CINQUEFOIL

Fig. C.25. Fuchs, *Plantarum effigies,* 358.

Latin binomial (type sp.): *Potentilla reptans* L.

French names: quintefeuille, cinquefeuille, main de Mars

Elemental qualities: hot and wet (🜁)

Ruling planet: ☿

Zodiacal signatures: ♍ and ♋ (*apud* Mario)

Occult properties: Cinquefoil has astringent and febrifugal properties. The *Grand Albert* calls it *pedactilius* and *pentaphilon,* the former a corruption of the Greek πενταδάκτυλον, "five-fingered herb," the latter a latinization of the Greek name πεντάφυλλον, "five-leaved herb."* When ground and administered in the form of a plaster, its root heals wounds and scabs. Its juice dissolved in water may be drunk or gargled to cure scrofula and soothe toothaches. When worn as an amulet, it bestows good luck upon the bearer, enhances the hearing, and opens the understanding.[21]

*[The corrupt variant *pedactilius* derives from French editions of the *Grand Albert.* In the Latin incunabulum of 1493, the plant name is spelled *pe(n)tadactilius.* —*Trans.*]

Fig. C.26. Fuchs, *Plantarum effigies,* 29.

CLEAVERS

Latin binomial: *Galium aparine* L.

French names: glouteron, aparine, capille à teigneux, gaillet gratteron, gleton, grateron, herbe collante, petit peignot, rièble, saigne-langue, traînasse*

Elemental qualities: cold and dry (🜃)

Ruling planet: ♄

Zodiacal signature: ♍

Occult properties: Cleavers, otherwise known as bedstraw, goosegrass, sticky weed, or catchweed, has aperient, diuretic, hepatic, and lymphatic properties. When gathered under the new ☾ with the ☉ in ♍, its root acquires odontalgic properties and can relieve even the most stubborn toothaches. When gathered under a full ☾, it becomes an effective remedy against inflammations. Its powdered leaves are particularly effective in the treatment of old ulcers. Alchemists and Hermeticists refer to it by the name *philadelphus,* meaning "filial love," an epithet likely inspired by its creeping stems, which allow it to cleave, or latch onto, its vegetal neighbors and any human or animal passersby.[22]

*[Sédir's entry appears under the French common name *glouteron,* a name that is also frequently taken to be synonymous with *bardane* (★greater burdock). That cleavers is the appropriate species here, however, is confirmed by Sédir's inclusion of the species name *aparine.* See *Les plantes magiques,* 154. —*Trans.*]

CLOVE

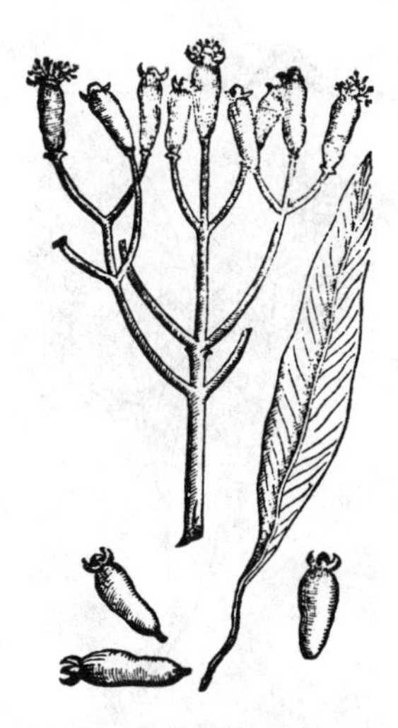

Fig. C.27. Gerard, *The Herball*, 1351.

Latin binomial: *Syzygium aromaticum* (L.) Merrill & Perry

French names: giroflier, clou de giroflé, ginouflado (Languedocian), giroflé

Elemental qualities: hot and dry (△)

Ruling planet: ☉

Zodiacal signature: ♌. Harvest when the ☉ is in ♓ and the ☾ is in ♋.

Occult properties: Cloves possess anesthetic, carminative, diuretic, odontalgic, stimulant, and stomachic properties. Consuming cloves facilitates conception. Its oil is a famous remedy for toothaches. Its essence is particularly useful in certain operations of practical magic: when combined with phosphorus, for example, it nourishes the astral bodies of lemures and larvae (spirits of the restless dead). Holding a clove inside the mouth also serves as a powerful adjuvant for hypnotists.

CLOVER, SWEET

Fig. C.28. Fuchs, *Plantarum effigies*, 299.

Latin binomial (type sp.): *Melilotus officinalis* (L.) Pall.

French names: mélilot, couronne royale, lotier jaune, luzerne bâtarde, petit trèfle jaune, pratelle, trèfle de cheval, trèfle des mouches, trèfle odorant

Elemental qualities: moderately hot and wet (🜁)

Ruling planet: ☉

Zodiacal signature: ♒

Occult properties: Sweet clover has anodyne, aromatic, carminative, emollient, and resolutive properties. Its infusion is often administered for swollen joints, rheumatic pains, and varicose veins.[23] In Roman antiquity, wreaths made of its blossoming branches were de rigueur for dinner parties and banquets. Its festive, decorative, and sartorial applications earned it the Latin sobriquet *sertula Campana,* or "Campanian garland."[24]

COCA

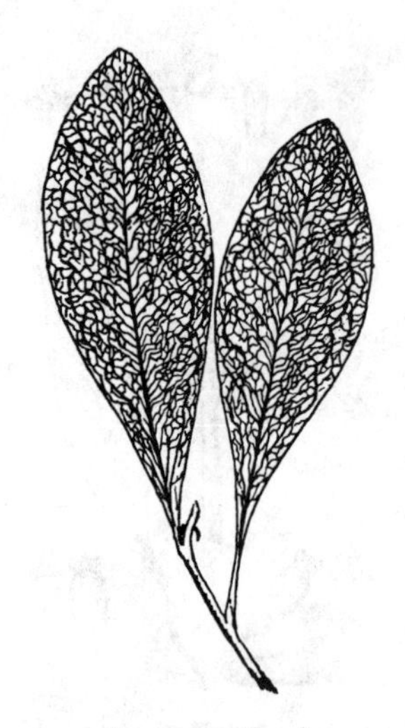

Fig. C.29. Lloyd, "The Mombreros," 5.

Latin binomial: *Erythroxylum coca* Lam.
French name: coca
Elemental qualities: hot and dry (△)
Ruling planets: ♄ and ☉
Zodiacal signatures: ♈, ♌, and ♐
Occult properties: The leaves of this Peruvian plant have potent anesthetic, narcotic, stimulant, and tonic properties. Hypodermic injections of their salt, otherwise known as cocaine, can forge a very real pact with beings in the astral realms. Coca, like hashish but in other respects, exerts a direct and powerful action on the astral body, but its habitual use can be treacherous, as it can lead to the undoing of certain compressive bonds in our hyperphysical nature.[25]

COLOCYNTH

Fig. C.30. Fuchs, *Plantarum effigies,* 212.

Latin binomial: *Citrullus colocynthis* (L.) Schrad.
French names: coloquinte, colokinto (Languedocian), coloquinte vraie
Elemental qualities: hot and dry (△)
Ruling planet: ♂ (*apud* Haatan and Piobb)
Zodiacal signature: ♈. Harvest under ♌.
Occult properties: Colocynth, otherwise known as bitter apple, bitter cucumber, desert gourd, and vine of Sodom, has aperient, cholagogic, demulcent, and diuretic properties. Its fruit is a round gourd, resembling an orange in size and appearance, with numerous seeds embedded in a light, spongy, and extremely bitter pulp. Its fruit extract is most commonly used in purgative medicines.[26] Some alchemists and Hermeticists refer to it by the Arabic name *handal* or *handel.*[27]

COLTSFOOT

Fig. C.31. Fuchs, *Plantarum effigies,* 76.

Latin binomial: *Tussilago farfara* L.

French names: pas d'âne, herbe de Saint Quirin, pas de cheval, pè-pouli (Languedocian), pied de poulan, procheton, taconnet, tussilage*

Elemental qualities: hot, wet, and moderately dry (🜁)

Ruling planet: ♀ (*apud* Piobb)

Zodiacal signatures: ♊ and ♈. Harvest under ♌ or after the full ☾ terminating the dog days of summer.

Occult properties: Coltsfoot has demulcent, expectorant, pectoral, and tonic properties. In antiquity, it was praised primarily for its bechic virtue. According to Dioscorides, the smoking of its leaves cures coughs and other respiratory ailments.[28] The genus name *Tussilago,* in fact, derives from the Latin *tussis,* meaning "cough," and the suffix *-ago,* which forms nouns describing objects, plants, and animals. Alchemists and Hermeticists, however, refer to it by the name *populago* because its leaves are white on one side like those of the ⋆poplar.[29] Coltsfoot is one of the twelve magical plants of the Rosicrucians.[30]

*[Sédir provides separate entries for *pas d'âne* and *farfara,* the latter being the Latin species name for coltsfoot. See *Les plantes magiques,* 152, 163. These have been merged here to form a single entry. In the former entry, Sédir also provides the name *peuplier feuillu,* meaning "leafy poplar," a name otherwise unattested for coltsfoot. It is unclear, however, whether Sédir is here offering his own interpretation of the alchemical name *populago* or, a much more intriguing option, identifying the secret ingredient in witches' ointments that Nynauld calls *feuilles du peuplier* (poplar leaves) but Porta calls *frondes populneas* (poplar-like leaves) as the herb coltsfoot, or both. Cf. Sédir's discussion of witches' ointments in chapter 6, especially diabolical herb no. 13. —*Trans.*]

Fig. C.32. Fuchs, *Plantarum effigies*, 400.

COMFREY

Latin binomial (type sp.): *Symphytum officinale* L.

French names: oreille d'âne, confée, consoude, empes (Languedocian), grande consoude, herbe aux coupures, oreille de vache, pecton

Elemental qualities: moderately hot and wet (🜁)

Ruling planet: ♃

Zodiacal signature: ♊

Occult properties: Comfreys possess anodyne, demulcent, emollient, and vulnerary properties. The roots and leaves of the type species, otherwise known as common comfrey and true comfrey, have powerful styptic and antiemetic properties and quickly stop bleeding wounds, vomiting, and nausea. It is also an especially good herb for pulmonary ulcers, bone fractures, and rheumatism. To get rid of bedbugs without touching them, place some comfrey leaves under the bedside and all the bedbugs will assemble on them.[31]

[COMMON POPPY ⚕ *See* POPPY, COMMON]

Fig. C.33. Mattioli, *Discorsi*, 451.

CORIANDER

Latin binomial: *Coriandrum sativum* L.

French names: coriandre, mari de la punaise, persil arabe

Elemental qualities: hot and dry (△)

Ruling planet: ♀

Zodiacal signature: ♌ (*apud* Mario)

Occult properties: Coriander seeds have carminative, cordial, digestive, and stomachic properties. The crushed seeds of this aromatic herb are often used to give beer a bright citrus flavor.

[CORNELIAN CHERRY ⚕ *See* CHERRY, CORNELIAN]

COSTMARY

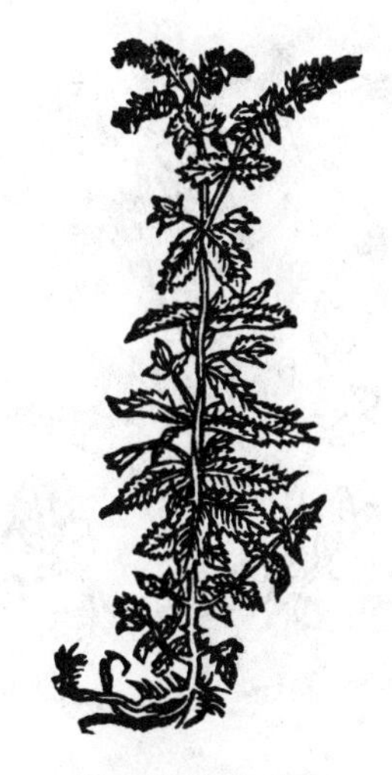

Fig. C.34. Fuchs, *Plantarum effigies*, 165.

Latin binomial: *Tanacetum balsamita* L.

French names: balsamier, baume-coq, baumier, coq (Languedocian), grande balsamite, menthe-coq, menthe de Notre Dame, menthe romaine

Elemental qualities: hot and dry (🜂)

Ruling planet: ☉

Zodiacal signature: ♐

Occult properties: The leaves of the costmary, otherwise known as alecost, have astringent and antiseptic properties. They are commonly used in perfumes and potpourris because of their sweet, balsamic, and minty aroma.

COWSLIP

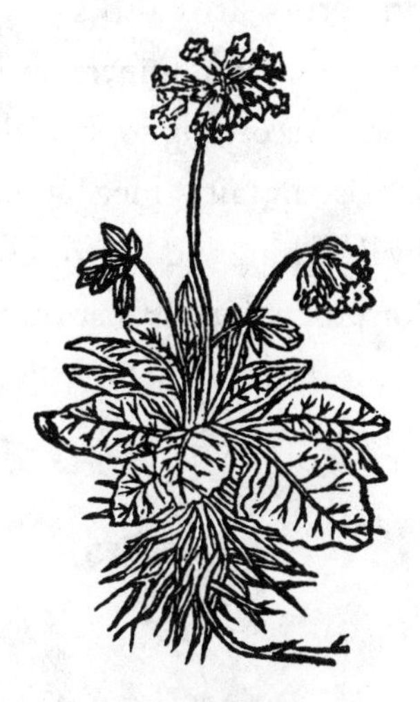

Fig. C.35. Fuchs, *Plantarum effigies*, 492.

Latin binomial: *Primula veris* L.

French names: primevère, coqueluchon, coucou, fleur de coucou, herbe à la paralysie, primevère officinale, printanièiro (Languedocian)

Elemental qualities: hot and wet (🜁)

Ruling planet: ☿

Zodiacal signature: ♎

Occult properties: Cowslip, or primrose, has astringent, sedative, spasmolytic, sternutatory, and vermifugal properties. It serves as an excellent cure for melancholy. According to Hildegard of Bingen, persons suffering from melancholia should place a cowslip, which she calls *Hymelsloszel* or *Himmelschlüssel* (German for "key of heaven"), on the flesh near their heart until their melancholia ceases.[32] Its salt functions as a mild purgative and cures stomatitis and glossitis. It is also a good herb for consecrating altars and other ritual implements.[33] Ancient herbalists used it to remedy palsies, for which reason the Greeks gave it the name παραλύσις, or "paralysis-herb."[34]

Fig. C.36. Fuchs, *Plantarum effigies*, 433.

CRESS, BITTER

Latin binomial (type sp.): *Barbarea vulgaris* W.T. Aiton

French names: vélar (de Sainte Barbe), barbarée, cresson de terre, girarde jaune, gras-capoù (Languedocian), herbe barbara, herbe de Sainte Barbe, roquette des marais

Elemental qualities: hot and dry (△)

Ruling planet: ☉

Zodiacal signature: ♌

Occult properties: Bitter cress, otherwise known as winter cress and yellow rocket, is a cruciferous herb with antiscorbutic and diuretic properties. It is also known by the name herb Barbara, after Saint Barbara, patron saint of miners, military engineers, and persons who work with explosives. According to legend, she was sentenced to death by beheading for her Christian faith. Her father, a pagan, carried out the sentence and was struck by lightning and killed shortly afterward. For this reason, she is often invoked for protection against thunder and lightning. As a Solar herb, bitter cress, or Saint Barbara's herb, may be used as an apotropaic device to achieve the same protective ends.[35]

Fig. C.37. Fuchs, *Plantarum effigies*, 204.

CRESS, GARDEN

Latin binomial: *Lepidium sativum* L.

French names: cresson alénois, cresson de jardin, cressonnette, nanitor (Languedocian), nasitort, passerage cultivée

Elemental qualities: hot and dry (△)

Ruling planet: ♂

Zodiacal signatures: ♈ and ♐. Harvest at the beginning of April or under ♏.

Occult properties: Garden cress has antiscorbutic, diuretic, and stimulant properties. In the East, its seeds are highly prized for their aphrodisiac virtue. Ancient Greek magicians frequently used its seeds in suffumigations, earning it the epithets "cardamom of Hecate" and "cress of Medea."[36]

CRESS, SWINE

Fig. C.38. Fuchs, *Plantarum effigies,* 253.

Latin binomial: *Coronopus squamatus* L.

French names: corne de cerf, coronope, cresson de rivière, pied de corneille

Elemental qualities: hot and dry (△)

Ruling planet: ♂

Zodiacal signature: ♈

Occult properties: Swine cress has alterative, anti-inflammatory, astringent, styptic, and vulnerary properties. When powdered or infused, it can be used to stop hemorrhages and bleeding, for which reason alchemists and Hermeticists gave it the names *sanguinalis* and *sanguinaria,* both meaning "blood-herb."[37] Women who desire to become pregnant can grind this herb into a powder, mix it with cow dung, and form it into an amulet to wear during sexual intercourse.[38]

CUCKOOPINT

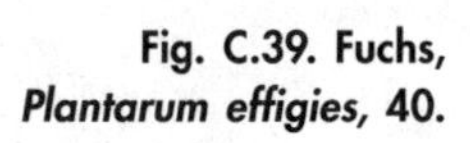

Fig. C.39. Fuchs, *Plantarum effigies,* 40.

Latin binomial (type sp.): *Arum maculatum* L.

French names: arum, arum tachetée, chou pané, glauxòl (Languedocian), gouet, gouet maculé, pied de veau, racine amidonnière*

Elemental qualities: moderately hot and wet (🜁)

Ruling planet: ♂

*[Sédir provides separate entries for *arum* and *gouet,* the latter being the French common name for the type species of the *Arum* genus. See *Les plantes magiques,* 144, 154. These have been merged here to form a single entry. —*Trans.*]

Zodiacal signatures: ♊ (plant) and ♏ (spadix)*

Occult properties: Cuckoopint, a woodland plant species that often grows among shrubs, has diaphoretic, diuretic, expectorant, and purgative properties. Although its red berries are extremely poisonous, its roots and leaves, when cooked well and mixed with vinegar or honey, become antiscorbutic and antiasthmatic in action.[39] Its root is equally toxic in its natural state, and contact with its sap can irritate the skin and cause it to blister, but its extract (arum starch) serves as both a powerful emollient and a useful cosmetic ingredient.

Fig. C.40. Fuchs, *Plantarum effigies*, 401.

CUCUMBER

Latin binomial: *Cucumis sativus* L.

French names: concombre, cornichon

Elemental qualities: cold and wet (∇)

Ruling planet: ☾

Zodiacal signatures: ♋ and ♐

Occult properties: Cucumber has depurative, diuretic, emollient, and resolvent properties. It was classed among the four greater cold seeds of ancient Greek medicine under the name σίκυος or σίκυς together with squash, pumpkin, and melon.[40] Its seeds have natural vermifugal properties. When burned, they also serve to draw down the power and influence of the ☾. A cucumber cut in the shape of a serpent, candied, and then soaked in water will drive away bedbugs.[41]

*[According to Jollivet-Castelot, Sédir's ♏ signature applies exclusively to the cuckoopint's characteristic inflorescence, the spadix, which, in Sédir's system of plant signatures, is hot and dry (△), since it falls under the influence of ♂ in his domicile of ♏. See *La médecine spagyrique,* 15. —*Trans.*]

CUMIN, HORNED

Fig. C.41. Fuchs, *Plantarum effigies,* 188.

Latin binomial: *Hypecoum procumbens* L.

French names: cumin sauvage, cumin couché, cumin cornu*

Elemental qualities: moderately cold and dry (🜃)

Ruling planet: ♄

Zodiacal signatures: ♍ and ♒

Occult properties: Horned cumin seeds have carminative and vermifugal properties. The oil of its seeds is a powerful antirheumatic and febrifuge and should only be taken in very small quantities. Pigeons are very fond of this plant, especially when sprinkled with brine. Steeping the seeds in water and sprinkling the solution around your home will get rid of flies and other noxious insects.[42]

*[Although the French common name *cumin sauvage* (wild cumin) is sometimes used to describe the herb ⋆caraway, Sédir's inclusion of the variant name *hypecoon,* a transliteration of the ancient Greek plant name ὑπήκοον, clearly identifies this species as horned cumin. See *Les plantes magiques,* 150; cf. Pernety, *Dictionnaire mytho-hermétique,* 296 s.v. *hypecoon.* —*Trans.*]

Fig. C.42. Fuchs, *Plantarum effigies,* 254.

CYCLAMEN

Latin binomial: *Cyclamen purpurascens* Mill.

French names: cyclamen, arthanite, coquette, marron de cochon, pain de cochon, pain de pourceau, rave de terre

Elemental qualities: hot and wet (🜁)

Ruling planets: ♀ and ☉ (*apud* Mavéric)

Zodiacal signatures: ♎ (especially the root) and ♌ (especially the leaves)

Occult properties: Cyclamen has emmenagogic, nervine, and digestive properties. It is sacred to the Sun god Apollo. In antiquity, it was known by the name *umbilicus terrae,* or "earth navel," but alchemists and Hermeticists refer to it by the secret name *suffo.*[43] For fistulae, which are caused by ♂, mix its aqueous extract with powdered ⋆ivy root and the ashes of calcined oysters into an unguent and apply it on the skin in the area of the fistula. Do this at the hour of ♄ or ♃, the enemies of ♂.* It is also a common ingredient in love potions and magical philters.

*[In Sédir's original entry, he identifies the two additional ingredients in this remedy by the extremely cryptic phrases *foüs serpentinæ* and *sophiæ Sana.* The remedy for fistula can be traced back to the 1722 edition of the *Petit Albert,* which calls for *la racine de lierre mise en poudre* (powdered ivy root) mixed with *la cendre d'huîtres brûlées* (ash of burnt oysters) and *saindoux de pourceau* (pork fat). See Pseudo-Albertus, *Secrets merveilleux de la magie naturelle et cabalistique du Petit Albert,* 144. In later editions, the ingredient *saindoux de pourceau* reads *sain de pourceau,* which also means "pork fat." Old French *sain,* meaning "lard" or "fat," is now obsolete except in the compound *saindoux,* literally "sweet fat," but *saindoux* by itself signifies "pork fat," and so the change was likely made because *saindoux de pouceau* is a redundancy. Sédir's source for this remedy, however, is most likely the version that appears among the provincial traditions of Hermetic medicine in Papus's *Traité élémentaire de magie pratique,* 427. In Papus's version, the ingredient *sain de pourceau* reads *pain de pourceau*—that is, cyclamen. The French common name *pain de pourceau,* literally "sowbread" or "swinebread," was allegedly given to cyclamen because of the avidity with which pigs would sniff out and consume its tuberous roots. It is unclear whether this change was made because the variant *pain de pourceau* appeared in Papus's edition of the *Petit Albert* or, more likely, because Papus believed *sain de pourceau* to be witches' argot for *pain de pourceau.* In any case, Papus's version also preserves the peculiar variant *lirios* for *lierre,* "ivy." Later editions of the *Petit Albert* read *lieros* or *lireos.* Although Papus's variant *lirios* would appear to be the plural form of Spanish *lirio,* "lily," or a pseudo-Greek form of λείριον, "lily," it is more likely an iotacism, or vowel shift, for λείριος, meaning "lily-like." It is to be noted as well that Spanish *lirio* can also mean "iris," and specifically the yellow iris (*Iris pseudacorus* L.),

CYPRESS

Fig. C.43. Mattioli, *Discorsi*, 96.

Latin binomial (type sp.): *Cupressus sempervirens* L.

French names: cyprès, cyprès pyramidal, cyprès toujours vert

Elemental qualities: hot and dry (△)

Ruling planet: ♄

Zodiacal signature: ♌. Harvest when the ☉ is in ♓ and the ☾ is in ♋.

Occult properties: Cypress trees possess astringent, bechic, pulmonary, and sudorific properties. The cypress has long been asssociated with death and mourning. It is sacred to Pluto, who is often depicted wearing a crown of cypress sprigs and leaves. Its decoction darkens the hair and keeps it healthy.

[CYPRESS SPURGE ⚕ *See* SPURGE, CYPRESS]

(*cont.*) which has lily-like flowers (see Barthés, *Glossaire botanique languedocien,* 131). The roots and rhizomes of the yellow iris are extremely acrid and can be toxic, but they are effective in topical applications for treating wounds, ulcers, and fistulas. It is equally unclear, moreover, whether Sédir means to conceal or "identify" the remedy's true Hermetic ingredients by means of the cryptograms *foüs serpentinæ* and *sophiæ Sana*. In any case, these are unlikely to refer to "ivy root" and "the ashes of calcined oysters." —*Trans.*]

D

Daffodil–Dittany of Crete

Fig. D.1. Mattioli, *Discorsi,* 683.

DAFFODIL

Latin binomial (type sp.): *Narcissus poeticus* L.

French names: narcisse, claudinette, herbe à la vierge, narcisse des poetes, œil de faisan, porillon, vachette

Elemental qualities: cold and dry (🜃)

Ruling planet: ♀

Zodiacal signatures: ♉ and ♏

Occult properties: Daffodil bulbs possess powerful emetic and irritant properties. The ancient Latin name *narcissus* derives from the Greek name νάρκισσος, which was coined from the noun νάρκη, meaning "numbness," in reference to the flower's narcotic fragrance. The Furies wore narcissus flowers in their hair to stupefy their victims, much like Hades used them to enchant Persephone. The distilled water of its root increases sperm motility. When mixed into lotions, it makes the breasts firm. When worn as an amulet, it attracts virgins. Some alchemists and Hermeticists refer to it by the name *keiri* or *keirim,* a name otherwise given to the *wallflower.[1]

DAISY

Fig. D.2. Fuchs, *Plantarum effigies,* 79.

Latin binomial: *Bellis perennis* L.

French names: pâquerette, petite marguerite, margarideto (Languedocian)*

Elemental qualities: moderately hot and wet (🜁)

Ruling planet: ☿

Zodiacal signature: ♒

Occult properties: The daisy has astringent, anti-inflammatory, digestive, emollient, and vulnerary properties. Decoctions of all parts of this beneficent herb reduce stomatitis. Its salt is capable of resolving any engorgement of bile or stone. Eating raw daisies heals various types of fever.

DAISY, OX-EYE

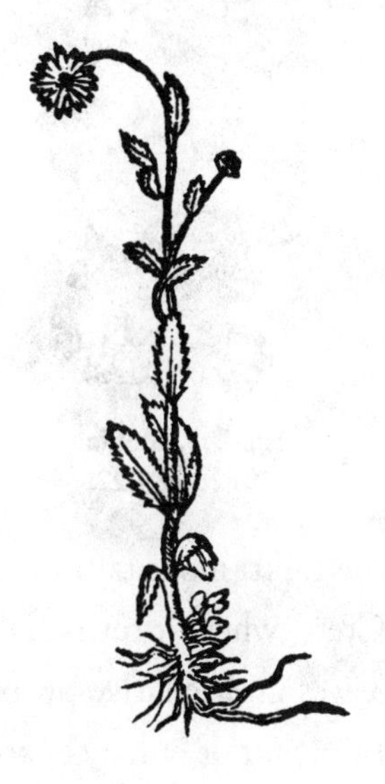

Fig. D.3. Fuchs, *Plantarum effigies,* 81.

Latin binomial: *Leucanthemum vulgare* Lam.

French names: marguerite, grande pâquerette, grand marguerite, grando margarido (Languedocian), œil de bœuf

Properties: The ox-eye daisy has the same elemental, astral, and occult properties as the *daisy.

[DEVIL'S BIT SCABIOUS ❦ *See* SCABIOUS, DEVIL'S BIT]

*[Sédir includes *pâquerette* (daisy) and *marguerite* (ox-eye daisy) in a single entry. See *Les plantes magiques,* 159 s.v. *marguerite.* These have been separated here to form two distinct entries. —*Trans.*]

Fig. D.4. Fuchs, *Plantarum effigies*, 16.

DILL

Latin binomial: *Anethum graveolens* L.

French names: anis sauvage, aneth, aneth odorant, faux anis, fenouil bâtard, fenouil puant

Elemental qualities: hot and wet (🜁)

Ruling planet: ♀ (*apud* Mario)

Zodiacal signature: ♎

Occult properties: Dill has the same occult properties as ★anise but is somewhat less tonic. To cure insomnia, cook some dill stems in oil and apply them on the forehead in the form of a poultice.[2]

Fig. D.5. Mattioli, *Discorsi*, 422.

DITTANY OF CRETE

Latin binomial: *Origanum dictamnus* L.

French names: dictame (de Crète), dictame vrai, origan dictame

Elemental qualities: hot and wet (🜁)

Ruling planet: ☉ (*apud* Agrippa and Lenain)

Zodiacal signatures: ♊ and ♎

Occult properties: The flowering plant known as dittany of Crete, Cretan dittany, and hop marjoram has stomachic and vulnerary properties. The Greek plant name δίκταμον comes from the Dikti mountain range (Δίκτη in Greek) on the eastern side of the island of Crete, where it formerly grew in abundance. This balsamic plant has evergreen leaves and sedative properties. When woven into garlands or burned as a suffumigant, it enhances somnambulistic clairvoyance. When administered in a compress, its leaves are oxytocic in action and hence beneficial for pregnant women. It is sacred to the Roman goddess Lucina, an epithet for Juno in her role as the goddess of childbirth.

[DOG ROSE ⚘ *See* ROSE, DOG]

[DWARF ELDERBERRY ⚘ *See* ELDERBERRY, DWARF]

E

Edelweiss–Eyebright

EDELWEISS

Fig. E.1. Mattioli, *Discorsi,* 656.

Latin binomial: *Leontopodium alpinum* Cass.*

French names: edelweiss (des Allemands), belle étoile, étoile d'argent, étoile des Alpes, pas de lion, pied de lion

Elemental qualities: hot and dry (△)

Ruling planet: ♃

Zodiacal signature: ♌

Occult properties: Edelweiss has astringent, bechic, and discutient properties. Its common name derives from German *Edelweiß,* a compound of *edel,* meaning "noble," and *weiß,* meaning "white." It is one of the twelve magical plants of the Rosicrucians.[1]

*[Sédir provides the Latin binomial *Gnaphalium leontopodium* L., which is synonymous with *Leontopodium nivale* subsp. *alpinum* (Cass.) Greuter or, in its abridged form, *Leontopodium alpinum* Cass. See *Les plantes magiques,* 151. —*Trans.*]

Fig. E.2. Fuchs, *Plantarum effigies,* 37.

ELDERBERRY

Latin binomial (type sp.): *Sambucus nigra* L.

French names: sureau, grand sureau, saüc (Languedocian), seu, sureau noir, suyer, suzeau

Elemental qualities: hot and dry (△)

Ruling planet: ☿

Zodiacal signature: ♈. Harvest under ♌.

Occult properties: Elderberries possess aperient, diaphoretic, diuretic, and emetic properties. The oil drawn from elderberry seeds or an oil in which its seeds have been infused is beneficial in the treatment of gout. The type of *mistletoe that thrives on elderberry trees, especially those that grow next to *willow trees, is particularly effective against epilepsy. Their flowers heal erysipelas and burns, their seeds possess sudorific properties, and their bark is good for edema. A small scion picked just before the new ☾ of October and cut into nine pieces is also useful against edema, so, too, the root when pulled from the earth at noon on the feast day of Saint John the Baptist (June 24). Sprinkling a decoction of its leaves around your home kills flies and other noxious insects.[2] In the Hermetic language of flowers, the elderberry is emblematic of zealousness.

Fig. E.3. Fuchs, *Plantarum effigies,* 38.

ELDERBERRY, DWARF

Latin binomial: *Sambucus ebulus* L.

French names: hièble, eusses (Languedocian), gèble, hyèble, petit sureau, yèble

Elemental qualities: hot and dry (△)

Ruling planet: ♂

Zodiacal signature: ♈. Harvest after the full ☾ terminating the dog days of summer.

Occult properties: The dwarf elderberry has the same occult properties as the *elderberry.

ELECAMPANE

Fig. E.4. Fuchs, *Plantarum effigies,* 135.

Latin binomial (type sp.): *Inula helenium* L.

French names: aromate germanique, aster de chien, aunée officinale, grand aunée, heleniaire, inule, lionne, luno campano (Languedocian), œil de cheval*

Elemental qualities: moderately hot and wet (🜁)

Ruling planet: ♃ (*apud* Agrippa, Lenain, Mario, and Jollivet-Castelot)

Zodiacal signature: ♊ (*apud* Jollivet-Castelot)

Occult properties: Elecampane, otherwise known as horse-heal and elfdock, has antiseptic, astringent, alterative, diaphoretic, and stimulant properties. In decoctions, it is very effective against pruritus and scabies. In infusions, it becomes subject to the influence of the ☉.[3] We celebrate the feast day of Saint Roch (August 16) by hanging tufts of elecampane in stables to keep our horses safe from epidemics. According to mythological tradition, this "flower of Jove" sprung from the tears of Helen of Troy after she was abducted by Paris, hence its Greek name ἑλένιον and the Latin species name *helenium*.[4] For an effective love

*[Sédir's entry for ⋆pennyroyal includes the French names *pouliot sauvage* (pennyroyal), *menthe pouliot* (pennyroyal), *menthe sauvage* (horse mint), and *herbe de Saint Roch* (meadow false fleabane). See *Les plantes magiques,* 165 s.v. *pouliot sauvage.* The confusion appears to stem from Sédir's source, which he does not cite but which is most certainly Huysmans, *La cathédrale,* 292. Clara Bell's English translation of the relevant passage from Huysmans's novel runs as follows: "Then we have the plant or plants dedicated to Saint Roch: the pennyroyal (*menthe pouliot*), and two species of *Inula,* one with bright yellow flowers, a purgative that cures the itch. Formerly on Saint Roch's day branches of this herb were blessed and hung in the cow-houses to preserve the cattle from epidemics" (Huysmans, *The Cathedral,* 202). Since meadow false fleabane, *Pulicaria dysenterica* (L.) Bernh., was formerly classed as *Inula dysenterica* L., and since horse mint, *Mentha longifolia* (L.) Huds., is also known in French by the common name *menthe de cheval* (horse mint), it is reasonable to assume that *menthe sauvage* (wild mint) is a mistake for elecampane, *Inula helenium* L., an herb celebrated for its ability to cure pruritus and equine skin diseases. This herb, moreover, was also, albeit less frequently, called by the French common name *herbe de Saint Roch* (see Azaïs, *Catalogue botanique: Synonymie languedocienne, provençale, gasconne, quercinoise,* 34). For this reason, Sédir's original entry has been made into two separate entries: one for ⋆elecampane, the type species of the *Inula* genus (and an herb frequently mentioned by occult herbalists), and the other for ⋆pennyroyal. —*Trans.*]

charm, gather it before the rising of the ☉ on the eve of the feast day of Saint John the Baptist (June 24), wrap it in fine linen, and wear it over your heart for nine days; then powderize the herb, sprinkle it over a bouquet of flowers or some food, and give the flowers or food to your beloved.[5]

Fig. E.5. Fuchs, ***Plantarum effigies,*** **137.**

EYEBRIGHT

Latin binomial: *Euphrasia officinalis* L.

French names: euphraise, brise lunettes, casse-lunette, herbe aux myopes, luminet, urfrèso (Languedocian)

Elemental qualities: hot and dry (△)

Ruling planet: ☉ (*apud* Mavéric)

Zodiacal signature: ♈ (especially the flowers)

Occult properties: Eyebright has powerful astringent, anti-inflammatory, and ophthalmic properties. It bears a mark on its blossom resembling the human eye, and for this reason it has long been viewed as a powerful natural remedy for all eye diseases.[6]

F

Fava Bean–Fumitory

FAVA BEAN

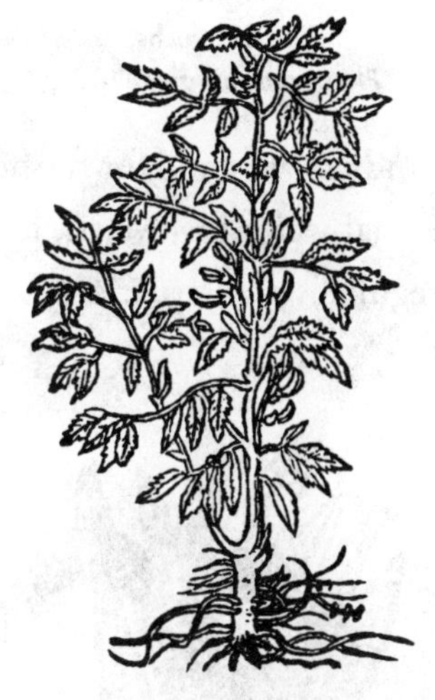

Fig. F.1. Fuchs, *Plantarum effigies,* 220.

Latin binomial: *Vicia faba* L.

French names: fève, fabo (Languedocian), fève des marais, féverole, fèvette, gourgane

Elemental qualities: cold and wet (🜄)

Ruling planets: ♄ and ☿. The fruit is governed by ♄ and the ☾.

Zodiacal signature: ♋. Harvest at the end of October when the plant has submitted to the influence of ♏ (with ☿ in ♏).

Occult properties: Fava or broad beans possess diuretic, depurative, and antirheumatic properties. Decoctions of roasted fava beans are effective against bladder and kidney stones. Plasters made with fava bean flour soften and resolve tumors of the sex organs and relieve sunburn. According to the Pythagorean school, fava bean flowers bear the mark of Hades.*

*[This curious statement appears to be a conflation of two Pythagorean traditions. The first, from Aristotle (as quoted by Diogenes Laertius), maintains that the Pythagorean proscription against fava beans was due to Pythagoras's own view that the shape of the "jointless" bean resembles the human genitals or "the gates of Hades" (Diogenes Laertius, *Lives of Eminent Philosophers* 8.34). The second, from Porphyry, maintains that "if at the time when the beans bloom, you take a little of the flower, which is then black, and place it in an earthenware vessel, seal it tightly, and bury in the ground for ninety days, and afterward dig it up and unseal it, you will find either the head of an infant or the pudenda of a woman" (Porphyry, *Life of Pythagoras,* 44). —*Trans.*]

Fig. F.2. Fuchs, *Plantarum effigies,* 285.

FENNEL

Latin binomial: *Foeniculum vulgare* Mill.

French names: fenouil, aneth doux, aneth fenouil, fenouil des vignes, fenoul (Languedocian)

Elemental qualities: hot and wet (🜁)

Ruling planet: ☿ (*apud* Mavéric)

Zodiacal signatures: ♊ and ♒

Occult properties: Fennel has aperitive, carminative, diuretic, emmenagogic, stimulant, and stomachic properties. The Greeks gave it the name μάραθον, which is the origin of the place-name Marathon (Μαραθών in Greek), meaning "overgrown with fennel." Its umbels, when candied, are good for freshening the breath. When distilled, it produces a beneficial water for uveitis, iritis, and inflammations of the eyes. In infusions, it regulates menstruation.

Fig. F.3. Fuchs, *Plantarum effigies,* 168.

FERN, HART'S-TONGUE

Latin binomial: *Asplenium scolopendrium* L.

French names: langue de cerf, escalapandro (Languedocian), herbe à la rate, herbe hépatique, scolopendre*

Elemental qualities: hot and dry (🜂)

Ruling planet: ♄

Zodiacal signatures: ♈, ♉, ♋, and ♎

Occult properties: Hart's-tongue fern has astringent, cholagogic, diaphoretic, splenic, and vulnerary properties. In ancient Greek medicine, it was classed among the five great capillary herbs—that is, herbs that promote hair growth, together

*[Sédir provides separate entries for *langue de cerf* and *scolopendre,* which are variant names for hart's-tongue fern. In the latter entry, he also provides the genus name *Phyllitis,* in which the hart's-tongue fern was formerly classed: *Phyllitis scolopendrium* (L.) Newman is synonymous with *Asplenium scolopendrium* L. See *Les plantes magiques,* 157, 168. These have been merged here to form a single entry. —*Trans.*]

with *maidenhair fern, black spleenwort (*Asplenium adiantum-nigrum* L.), great golden maidenhair (*Polytrichum commune* Hedw.), and wall rue (*Asplenium ruta-muraria* L.).[1]

FERN, MAIDENHAIR

Fig. F.4. Fuchs, *Plantarum effigies,* 46.

Latin binomial: *Adiantum capillus-veneris* L.

French names: capillaire, capillèro (Languedocian), cheveux de Vénus*

Elemental qualities: cold and dry (🜃)

Ruling planet: ♄

Zodiacal signature: ♉

Occult properties: The maidenhair fern, otherwise known as the southern maidenhair, black maidenhair, and Venus-hair, has astringent, bechic, capillary, demulcent, depurative, and emetic properties. It is sacred to Pluto, who is frequently depicted wearing a wreath of maidenhair stems and leaves, and to his abductee, Proserpine. The ancient Latin name, which is the same as the modern genus name *Adiantum,* derives from the Greek adjective ἀδίαντος, meaning "unwetted," in reference to the plant's unique water-repellent properties. Ancient herbalists noted that maidenhair ferns would remain dry even when fully plunged in water. This same characteristic was attributed to the hair of Venus, which remained dry even as she rose up from the sea.[2]

*[Sédir provides separate entries for *capillaire* and *cheveux de Vénus,* which are variant names for maidenhair fern. See *Les plantes magiques,* 146, 148. These have been merged here to form a single entry. —*Trans.*]

Fig. F.5. Fuchs, *Plantarum effigies,* 341.

FERN, MALE

Latin binomial: *Dryopteris filix-mas* (L.) Schott

French names: fougère mâle, falièiro de crabo (Languedocian), fougère de chèvre

Elemental qualities: hot and dry (△)

Ruling planets: ♄ and, to a lesser extent, ♂

Zodiacal signature: ♐

Occult properties: Male fern has abortifacient, anodyne, anti-inflammatory, astringent, febrifugal, and vulnerary properties. Several ancient physicians also extol its virtue as an anthelmintic and vermifuge.[3] The root, when powdered, is particularly effective against tapeworm. Louis XV purchased a "secret nostrum" for tapeworm from Madame Nouffleur at a considerable sum, which later proved to be nothing more than Galen's male fern remedy.[4] When cooked in wine, it can unblock obstructions of the spleen, cure melancholy, provoke menstruation, and prevent reproduction. When worn as an amulet, it enhances magical spells, dispels nightmares, and protects against lightning, hail, evil spirits, and black magic. A blade of male fern plucked at noon on the eve of the feast day of Saint John the Baptist (June 24) brings good luck in games of chance.[5] Alchemists and Hermeticists refer to ferns by the name *pteris,* from the ancient Greek plant name πτερίς, likely because their leaves so resemble a bird's feathers (πτερόν in Greek).[6] In the Hermetic language of flowers, the male fern is emblematic of humility.

[FIELD WORMWOOD ⚕ *See* WORMWOOD, FIELD]

FIG

Fig. F.6. Fuchs, *Plantarum effigies,* 285.

Latin binomial (type sp.): *Ficus carica* L.

French names: figuier, figuiè (Languedocian), figuier comestible, figuier commun

Elemental qualities: moderately hot and wet (🜁)

Ruling planet: ♃

Zodiacal signature: ♒

Occult properties: Figs have aperient, bechic, emollient, stimulant, and resolvent properties. The fruit of the black fig (*Ficus carica* L.) is governed by ♄, and the fruit of the white fig (*Ficus virens* Aiton) by ♃ and ♀. The fig tree is sacred to a number of gods and goddesses but especially to Hermes and Dionysus. Many of the latter's theonyms make reference to the fig tree (συκέα in Greek), such as Dionysos Sykites, who was revered in Sparta. Saturn is also sometimes depicted wearing a crown of fig leaves. In India, it is sacred to Vishnu. A fig branch cut under a favorable aspect can be used to calm angry bulls. Its mashed fruit is also a natural remedy for corns: simply coat the callused area of the foot for several days in a row. There was also an ancient method of divination by means of fig leaves known as sycomancy. In general, practitioners would write out a question on a fresh fig leaf, and if the leaf did not dry quickly, it augured a bad outcome.

Fig. F.7. Mattioli, *Discorsi,* 629.

FIGWORT

Latin binomial (type sp.): *Scrophularia nodosa* L.

French names: scrofulaire, grande scrofulaire, scrofulaire bâtarde, scrofulaire noueuse

Elemental qualities: cold and dry (🜃)

Ruling planet: ♀ (*apud* Mavéric)

Zodiacal signatures: ♉, ♋, ♎, and, if harvested at the end of October, ♏

Occult properties: Figwort possesses alterative, anti-inflammatory, bechic, diuretic, purgative, and vulnerary properties. Because of the throat-like shape of its flowers, it has long been used to heal scrofula and inflammations of the parotid gland. According to Dioscorides, when administered in a poultice, the nettle-leaved figwort (*S. peregrina* L.) is useful in the treatment of various skin disorders like eczema, psoriasis, and gangrene.[7]

Fig. F.8. Mattioli, *Discorsi,* 48.

FLAG, SWEET

Latin binomial: *Acorus calamus* L.

French names: jonc odorant, acore odorant, acore vraie, roseau aromatique*

Elemental qualities: moderately cold and dry (🜃)

Ruling planet: ☉

Zodiacal signature: ♍

Occult properties: Sweet flag has antirheumatic, carminative, digestive, stomachic, tonic, and vermifugal properties. In most traditions of herbal folk magic, it is considered an herb of prosperity and good luck. In East Prussia, it was ceremonially fed to cows on the evening of the festival of Saint John the Baptist (June 24). In China, its leaves are

*[Sédir provides separate entries for *jonc odorant* and *roseau aromatique,* which are variant names for sweet flag. See *Les plantes magiques,* 156, 166. These have been merged here to form a single entry. —*Trans.*]

bound in bundles and placed as talismanic devices at the bedside and at the four corners of doors and windows.[8]

FLAX

Fig. F.9. Fuchs, *Plantarum effigies,* 267.

Latin binomial (type sp.): *Linum usitatissimum* L.

French names: lin, li (Languedocian), lin cultivé.

Elemental qualities: cold and dry (🜃)

Ruling planet: ♃

Zodiacal signature: ♉ (*apud* Haatan and Jollivet-Castelot)

Occult properties: Flax has astringent, demulcent, pectoral, and vulnerary properties. It also serves as a powerful emollient. In lotions, it relieves pleurisy, resolves ulcers, and mollifies malignant breast tumors.

FORGET-ME-NOT

Fig. F.10. Mattioli, *Discorsi,* 382.

Latin binomial (type sp.): *Myosotis scorpioides* L.

French names: gromillet, ne m'oublez pas, oreille de rat, oreille de souris, regardez-moi, scorpionne*

Elemental qualities: cold and dry (🜃)

Ruling planet: ♀ (*apud* Piobb)

Zodiacal signature: ♉

Occult properties: Forget-me-not has bechic, pulmonary, and pectoral properties. In the Hermetic language of flowers, its blue flowers are emblematic of memory and philosophical anamnesis.[9]

*[Sédir's entry appears under the genus name *Myosotis,* from the Greek μυοσωτίς, literally "mouse-ear" (μυός + ὦτα), a name Dioscorides uses to describe madwort (*Asperugo procumbens* L.) and ⋆pellitory-of-the-wall (see Dioscorides, *On Medical Materials* 2.183 and RV 4.86). —*Trans.*]

Fig. F.11. Fuchs, *Plantarum effigies,* 515.

FOXGLOVE

Latin binomial: *Digitalis purpurea* L.

French names: digitale, bragos-de-coucut (Languedocian), digitale pourprée, doigt de Notre Dame, doigtier, gant de Notre Dame, gantelée

Elemental qualities: cold and wet (∇)

Ruling planet: ♂

Zodiacal signature: ♓

Occult properties: Foxglove has diaphoretic, emetic, stimulant, and purgative properties. All parts of this plant are highly toxic and can be lethal if consumed, even in small quantities (see Table 6.1 on p. 58). When subjected to prolonged distillations, however, it renders a benign liquor that may be used externally in astringent lotions for healing wounds. It may also be used internally in homeopathic doses to regulate the heartbeat, relieve respiratory disorders, and allay uncontrollable vomiting.

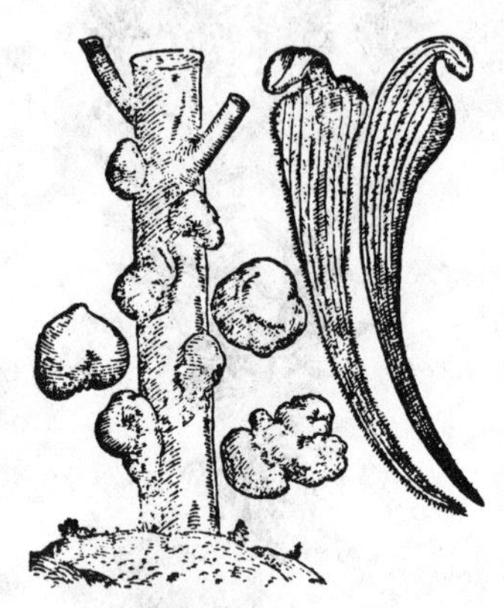

Fig. F.12. Gerard, *The Herball,* 1247.

FRANKINCENSE

Latin binomial: *Boswellia sacra* Flueck.

French names: encens, arbre à encens

Elemental qualities: hot and dry (∆)

Ruling planet: ☉

Zodiacal signature: ♌

Occult properties: Frankincense, or olibanum, the resin extract obtained from trees of the genus *Boswellia,* has anti-inflammatory, nervine, and psychotropic properties. According to Greek mythology, the Sun god Helios changed the lifeless body of the "white goddess" Leucothoë into the frankincense, or olibanum, tree. Its resin emits a powerful Solar perfume that acts directly on the psychic center.

FUCHSIA

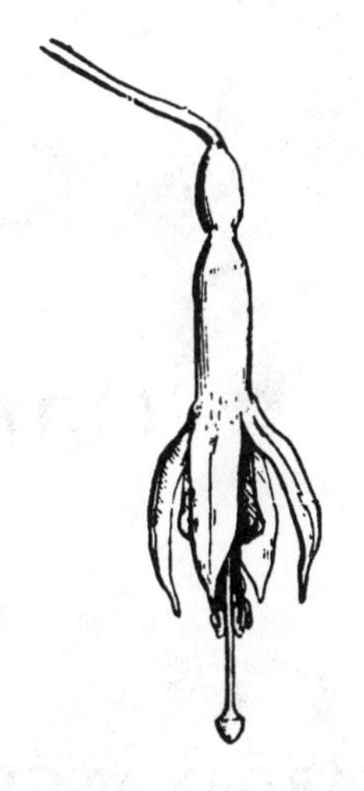

Fig. F.13. Wilson, *Chapters on Evolution,* 332.

Latin binomial (type sp.): *Fuchsia triphylla* L.

French name: fuchsia

Elemental qualities: hot and dry (△)

Ruling planet: ♀

Zodiacal signature: ♌

Occult properties: Fuchsia berries have diuretic and febrifugal properties. The fuchsia is one of the twelve sacred plants of the Rosicrucians.[10]

FUMITORY

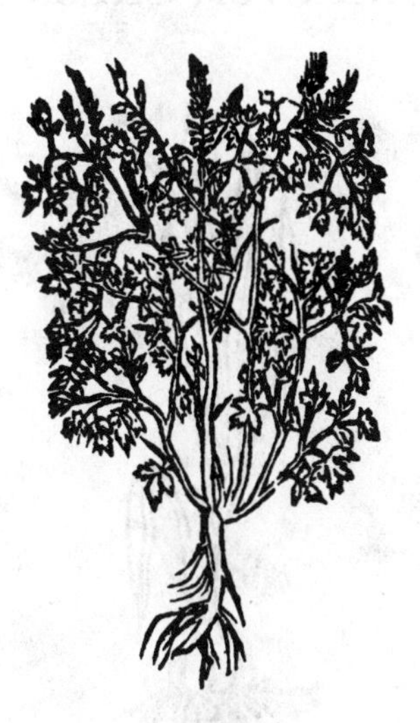

Fig. F.14. Fuchs, *Plantarum effigies,* 193.

Latin binomial (type sp.): *Fumaria officinalis* L.

French names: fumeterre, chausse rouge, fiel de terre, fine terre, fleur de terre, fumoterro (Languedocian), herbe à la jaunisse, lait battu, pied de géline, pisse-sang

Elemental qualities: hot and dry (△)

Ruling planets: ♃, ♄, and ♂

Zodiacal signature: ♈ (*apud* Mario)

Occult properties: Fumitory possesses desiccant, diuretic, purgative, and tonic properties. It is particularly useful in the treatment of scabies and syphilis.

G

Garlic–Grapevine

[GARDEN ANGELICA ❦ *See* ANGELICA, GARDEN]

[GARDEN CRESS ❦ *See* CRESS, GARDEN]

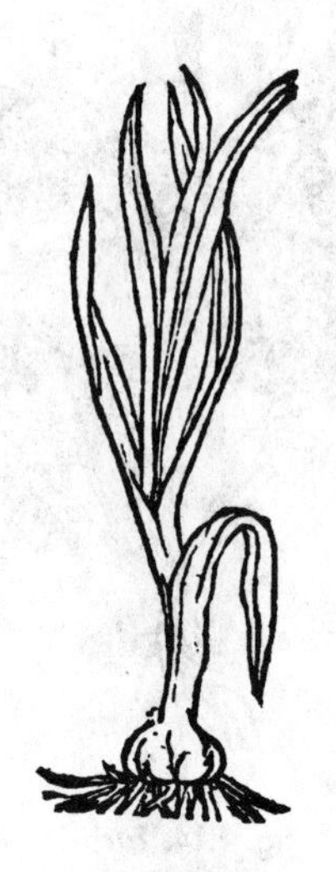

Fig. G.1. Fuchs, *Plantarum effigies*, 428.

GARLIC

Latin binomial: *Allium sativum* L.

French names: ail, ail cultivé, al (Languedocian), chapon, perdrix, thériaque des pauves

Elemental qualities: hot and dry (△)

Ruling planet: ♂

Zodiacal signature: ♐

Occult properties: Garlic has powerful cholagogic, diaphoretic, and stimulant properties. It is also a natural diuretic and promotes menstruation. The ancient Egyptians revered the garlic plant, but the Greeks forbade anyone who had eaten the herb σκόροδον, the Greek name for the garlic plant, from entering the sacred precincts of the temple of Cybele.[1] When taken on an empty stomach, it has an apotropaic function and guards the consumer against evil spells. For maximum effect, consume it mixed with ♄ (vinegar). Hanging a string of garlic bulbs from a tree or cleansing the tree using an instrument that has been rubbed with garlic is known to keep away birds. If you prefer your garlic inodorous, plant and harvest it when the ☾ is below the horizon.

GENTIAN

Fig. G.2. Fuchs, *Plantarum effigies,* 136.

Latin binomial (type sp.): *Gentiana lutea* L.

French names: gentiane, gentiane jaune, gìussano (Languedocian), grande gentiane

Elemental qualities: hot and dry (△)

Ruling planet: ☉

Zodiacal signatures: ♌ and ♈. Harvest under ♌ or under ♉ with the ☾ in ♊.

Occult properties: Gentian possesses aperient, aperitive, febrifugal, stimulant, and stomachic properties. It is also a good herb for consecrating altars and other ritual implements.[2] The species that grows in the mountains, commonly known as great yellow gentian, the type species of the genus, is one of the twelve sacred plants of the Rosicrucians.[3]

GERMANDER

Fig. G.3. Fuchs, *Plantarum effigies,* 501.

Latin binomial: *Teucrium chamaedrys* L.

French names: germandrée, calamandrier, chêneau, chênette, germandrée petit chêne, hèrbo de garroulho (Languedocian), petit chêne, sauge amère, thériaque d'Angleterre

Elemental qualities: cold and dry (🜃)

Ruling planets: ♂ and ♃

Zodiacal signature: ♉

Occult properties: Germander has diuretic, purgative, resolvent, and sudorific properties. When administered in the form of a topical application, it relieves pains caused by hemorrhoids and hyperemia.

Fig. G.4. Mattioli, *Discorsi,* 489.

GERMANDER, WOODLAND

Latin binomial: *Teucrium scorodonia* L.

French names: sauge des bois, faux scordion, germandrée sauvage, hèrbo d'abelho (Languedocian), sauge des montagnes

Elemental qualities: cold and dry (🜃)

Ruling planet: ♀ (*apud* Culpeper)

Zodiacal signatures: ♍ and ♒

Occult properties: Woodland germander, otherwise known as wood sage, has antiseptic, febrifugal, tonic, vermifugal, and vulnerary properties. Decoctions of this herb in wine or vinegar are particularly effective in the treatment of skin conditions such as bruises, wounds, ulcers, and psoriasis.[4]

Fig. G.5. Mattioli, *Discorsi,* 562.

GLADIOLUS

Latin binomial: *Gladiolus communis* L.

French names: glaïeul, lauiol (Languedocian)

Elemental qualities: cold and dry (🜃)

Ruling planet: ♂ (*apud* Piobb)

Zodiacal signature: ♉ (*apud* Mario)

Occult properties: The gladiolus, or sword lily, has aperient, astringent, diuretic, emmenagogic, and purgative properties. Its root has the same virtues as the root of the *iris but is somewhat less effective. The Latin genus name *Gladiolus* is a diminutive of *gladius,* meaning "sword," but alchemists and Hermeticists refer to it by the name *xiphidium* or *xiphium,* derivatives of the Greek diminutive ξίφιον, from ξίφος, also meaning "sword."[5]

GLADIOLUS, MARSH

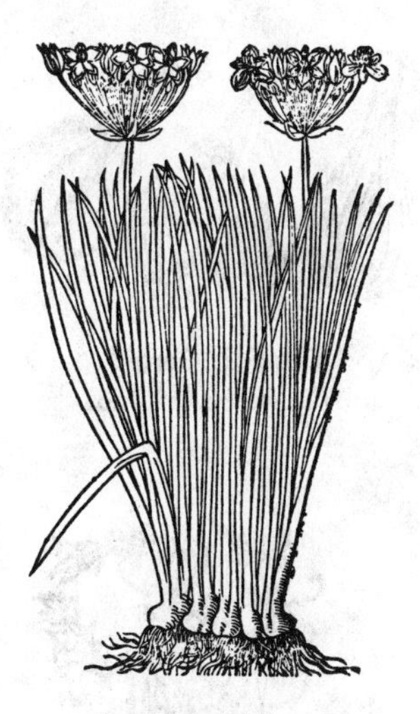

Fig. G.6. Gerard, *The Herball,* 27.

Latin binomial: *Gladiolus palustris* Gaudin

French names: glaïeul de rivière, glaïeul des marais

Elemental qualities: cold and wet (▽)

Ruling planet: ♂

Zodiacal signature: ♋

Occult properties: The marsh gladiolus has properties similar to the ★gladiolus, but with one notable difference: decoctions of the upper parts of the root are said to have aphrodisiac properties, whereas decoctions of the lower parts of the root are held to be anaphrodisiac in action.[6]

GOAT'S-BEARD

Fig. G.7. Fuchs, *Plantarum effigies,* 474.

Latin binomial (type sp.): *Tragopogon porrifolius* L.

French names: barbe de bouc, barberon, cercifix, salsifis, salsifix blanc, sarsifi (Languedocian)

Elemental qualities: hot and wet (🜁)

Ruling planet: ♀ (*apud* Piobb)

Zodiacal signature: ♎

Occult properties: Goat's-beard roots have anti-inflammatory, emollient, and resolvent properties. Alchemists and Hermeticists gave this herb the secret name *ipcacidos* or *ipoacidos.*[7]

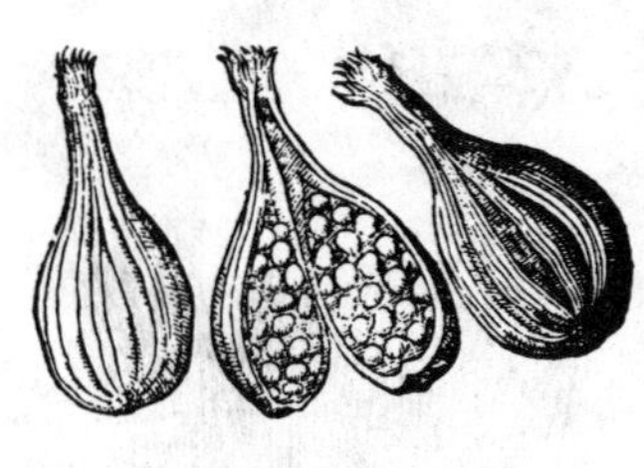

Fig. G.8. Gerard, *The Herball,* 1,358.

GRAINS OF PARADISE

Latin binomial: *Aframomum melegueta* K. Schum.

French names: graine de paradis, maniguette, plante du paradis*

Properties: Grains of paradise has the same elemental, astral, and occult properties as ★cardamom.

Fig. G.9. Fuchs, *Plantarum effigies,* 47.

GRAPEVINE

Latin binomial (type sp.): *Vitis vinifera* L.

French names: vigne, bigno (Languedocian), vigne cultivée, vigne sauvage, vigne vinifère

Elemental qualities: hot and wet (🜁)

Ruling planets: ☉ and ♃ (*apud* Agrippa)

Zodiacal signature: ♎ (*apud* Haatan and Jollivet-Castelot)

Occult properties: Grapevines possess anodyne, anti-inflammatory, aperient, demulcent, and febrifugal properties. The juice of the leaves cures dysentery, stops hemorrhages, and allays vomiting. The seeds, when roasted, crushed, and applied topically on the stomach in the form of a poultice, are another known cure for dysentery. When ground and applied to the stomach in a cataplasm, or poultice, the leaves and filaments are good for newly pregnant women tormented by hunger.[8]

[GREATER BURDOCK ❦ *See* BURDOCK, GREATER]

[GREATER CELANDINE ❦ *See* CELANDINE, GREATER]

[GREEN HELLEBORE ❦ *See* HELLEBORE, GREEN]

---

*[Sédir includes *graine de paradis* and *maniguette* with *cardamome* (★cardamom) in a single entry. See *Les plantes magiques,* 146 s.v. *cardamome.* These have been separated here to form two distinct entries. —*Trans.*]

H
Hazel–Hyacinth

[HART'S-TONGUE FERN ⚕ *See* FERN, HART'S-TONGUE]

HAZEL

Fig. H.1. Mattioli, *Discorsi,* 189.

Latin binomial (type sp.): *Corylus avellana* L.
French names: coudrier, abelaniè (Languedocian), avellinier, coudre, noisetier*
Elemental qualities: hot and wet (🜁)
Ruling planets: ♃ and ☿
Zodiacal signatures: ♎ and ♋
Occult properties: Hazel has astringent, diaphoretic, febrifugal, stimulant, and tonic properties. Spirit of hazel wood extracted under the conjunction of the ☾ and ☿ is excellent for the eyesight. Magic wands cut under favorable aspects are employed in ceremonial magic and rhabdomancy. Hazelnuts can heal dislocations of limbs by sympathy if the practitioner's will is strong enough: simply join two hazelnuts with two ★almond nuts and wear them as an amulet.

*[Sédir provides separate entries for *coudrier* (♃) and *noisetier* (☿), the former, it seems, for the hazel tree and the latter for hazelnuts. See *Les plantes magiques,* 150, 161. These have been merged here to form a single entry. Sédir's signatures apparently derive from Agrippa, who similarly assigns the hazel tree (*corilus*) to the same two planets, ♃ and ☿ (see Agrippa, *De occulta philosophia libri tres,* liber I, capita XXVI et XXIX, xxxiii, xxxv; English edition: *Three Books of Occult Philosophy,* bk. 1, chaps. 26 and 29, 96, 104). In the French translation of 1727, however, Agrippa is made to assign the hazel tree (*coudrier*) to ☿ and hazelnuts (*noisettes*) to ♃ (see Agrippa, *La philosophie occulte de Henr. Corn. Agrippa,* 1:73, 77). —*Trans.*]

Fig. H.2. Fuchs, *Plantarum effigies,* 142.

HEATHER

Latin binomial (type sp.): *Calluna vulgaris* (L.) Hull

French names: bruyère (callune), béruée, brande, brugo (Languedocian), bucane, péterolle

Elemental qualities: cold and wet (▽)

Ruling planet: ☿

Zodiacal signature: ♏ (*apud* Haatan)

Occult properties: Heather has anti-inflammatory, cholagogic, and diuretic properties. It is an especially good herb for divination and botanomancy.

Fig. H.3. Mattioli, *Discorsi,* 710.

HELIOTROPE

Latin binomial (type sp.): *Heliotropium europaeum* L.

French names: héliotrope, hélianthe, herbe aux verrues, herbe du cancer, liotrop (Languedocian), soleil, tournesol

Elemental qualities: hot and dry (△)

Ruling planet: ☉

Zodiacal signature: ♌

Occult properties: Heliotrope has cardiotonic, cholagogic, emmenagogic, and febrifugal properties. Its name derives from the Greek words ἥλιος, "Sun," and τρέπειν, "to turn." As its name would suggest, the heliotrope is sacred to the Sun gods Helios and Apollo. If magicians magnetize themselves with a heliotrope stem harvested at the appropriate time under ♌, they will be granted true and faithful revelations. This "herb of sincerity" was sacred to medieval magicians, who gave it the Chaldean name *ileos.** According

*[Latin and French editions of the *Grand Albert* give the Chaldean name as *ireos.* Once again, Sédir's variant *ileos* is otherwise unattested, but this spelling seems intended to draw a closer connection between the name of Chaldean herb no. 1 and the Greek word ἥλιος, the name of its ruling planet. Cf. Table 0.1 on p. xxiii. —*Trans.*]

to the *Grand Albert,* when wrapped in a ⋆bay leaf with a wolf's tooth and worn as an amulet, the heliotrope makes others speak well of the bearer. When placed under one's pillow at night, it reveals information in dreams about lost or stolen objects. And when placed at the door of a church in which women are present, those who have been unfaithful to their husbands will not be able to exit.[1] The heliotrope is one of the twelve magical plants of the Rosicrucians.[2]

HELLEBORE, BLACK

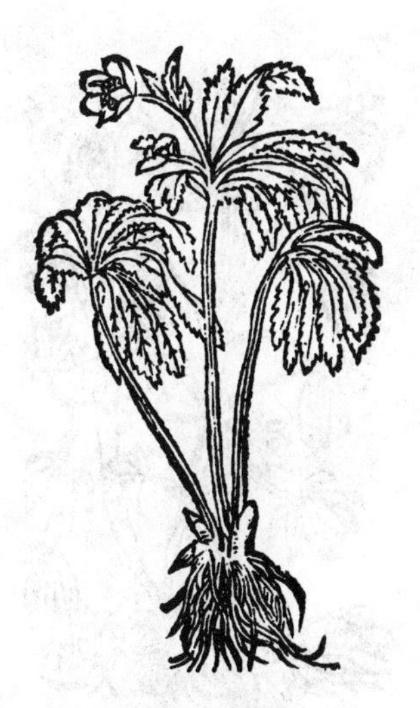

Fig. H.4. Fuchs, *Plantarum effigies,* 155.

Latin binomial: *Helleborus niger* L.
French names: hellébore noir, herbe aux fous, rose de Noël, rose d'hiver*
Elemental qualities: cold and dry (🜃)
Ruling planet: ♄
Zodiacal signature: ♑
Occult properties: Black hellebore has diuretic, emetic, emmenagogic, irritant, narcotic, and purgative properties. When macerated in spirit of wine and distilled over a slow fire (*ignis lentus*), it renders a valuable medicinal liquor, to which sugar candy is often added. When this liquor is administered in pure water or infused with horse tongue lily (*Ruscus hypoglossum* L.), it becomes an especially good remedy for *le mal caduc,* or epilepsy. The oil of its root is equally well suited for this purpose. Its powdered root serves as a useful perfume in operations of Saturnian and Capricornian magic. Occult herbalists sometimes call it *offoditius.*† According to the *Grand Albert,* its juice is beneficial for

*[Sédir's discussions of ⋆black hellebore, ⋆black false hellebore, and ⋆white hellebore all appear in a single entry. See *Les plantes magiques,* 151 s.v. *ellébore.* These have been separated into individual entries to avoid confusion. Cf. Sédir's discussion of hellebores in chapter 5. —*Trans.*]

†[Sédir lists the name *offoditius* alongside the French common name *rose de Noël* (Christmas rose) as a synonym for black hellebore. His source for this information is probably Papus's *Traité élémentaire de magie pratique,* 235. Papus's description of the occult properties of this herb comes straight from the description of planetary herb no. 1 (♄) in the *Grand Albert,* but what reads *affodilius* in the Latin incunabulum of 1493 reads *offodilius* in the French edition of 1703 and *offoditius* in the Lyon edition of 1800. The original reading, *affodilius,* is usually taken to be a variant for medieval Latin *affodilus,* a derivative of the ancient Latin plant name *asphodelus,* itself a loanword from Greek ἀσφόδελος. Latin *affodilius* and *affodilus,* however,

soothing and healing kidney pains and sore legs. It is also good for bladder infections. Its root, when cooked only slightly and then wrapped in a white linen cloth and worn as an amulet, cures patients suffering from demonic possession and melancholy. And if prepared in the same manner but hung in the home as a talisman, the root will drive out evil spirits.[3] Alchemists and Hermeticists refer to its seed by the secret name *mandella*.[4]

Fig. H.5. Fuchs, *Plantarum effigies,* 156.

HELLEBORE, BLACK FALSE

Latin binomial: *Veratrum nigrum* L.

French names: fausse hellébore, vératre noir

Elemental qualities: hot and dry (△)

Ruling planet: ♄

Zodiacal signature: ♈. Harvest at the beginning of April or under ♏.

Occult properties: Black false hellebore possesses emetic, errhine, expectorant, sternutatory, and vermifugal properties. This species of hellebore, which alchemists and Hermeticists call *helebria,* has dark red-black flowers.[5] It is often administered to scabious horses and sheep.

(*cont. from p. 181*) usually refer to the *daffodil, although the Hermetic tractate *On Plants of the Seven Planets,* Pseudo-Albertus's primary source for the archetypal planetary herbs, clearly identifies the herb of Saturn as *asphodel (see Table 0.2 on p. xxvi). It would appear that Papus took *offoditius* to be a secret or Chaldean name for black hellebore on the grounds that the occult properties ascribed to planetary herb no. 1 in the *Grand Albert* appear to have little to no connection to those of the daffodil. —*Trans.*]

HELLEBORE, GREEN

Fig. H.6. Mattioli, *Discorsi*, 670.

Latin binomial: *Helleborus viridis* L.

French names: hellébore vert, herbe à la bosse, herbe à sétons, herbe de Saint Antoine, pommelière*

Elemental qualities: hot and dry (△)

Ruling planet: ♄

Zodiacal signature: ♐

Occult properties: Green hellebore has many of the same occult properties as stinking hellebore (*Helleborus foetidus* L.), which in its natural state is violently narcotic and purgative in action. This species of wild hellebore is most likely the plant named *consiligo* discovered by the Marsi in antiquity.[6] Its root is poisonous but useful as a bait for capturing wolves and foxes.

HELLEBORE, WHITE

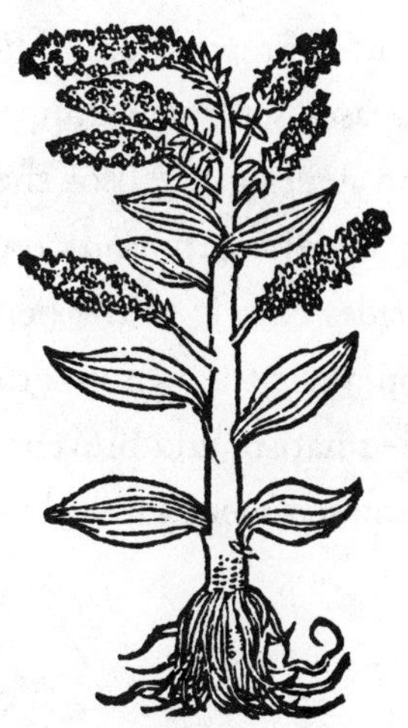

Fig. H.7. Fuchs, *Plantarum effigies*, 154.

Latin binomial: *Veratrum album* L.

French names: hellébore blanc, baraire (Languedocian), varaire blanc, vérâtre blanc

Elemental qualities: hot and dry (△)

Ruling planet: ♄

Zodiacal signature: ♈. Harvest when ♃ favorably aspects the ☾.

Occult properties: White hellebore, otherwise known as false helleborine, has many of the same occult properties as *black hellebore (see Table 6.1 on p. 58). *Veratrum album* is variable in size and flower color, but the best variety has whitish-red flowers. It can be made into a good topical cream for edema, mental illness, and maladies caused by the onset of old age, but it works best in the form of a dry powder. Melancholics will experience relief if they wear its root as an amulet.

---

*[Sédir's entry appears under the ancient Latin plant name *consiligo.* See *Les plantes magiques,* 149. The name *consiligo* was also used in antiquity to identify the herb *lungwort. —*Trans.*]

Fig. H.8. Fuchs, *Plantarum effigies,* 130.

HEMLOCK

Latin binomial: *Conium maculatum* L.

French names: ciguë, ciguë de Socrate, ciguë maculée, grande ciguë, jalbertasso (Languedocian)

Elemental qualities: cold, dry, and moderately wet (🜃)

Ruling planet: ♄

Zodiacal signatures: ♍ and ♒

Occult properties: Hemlock has anaphrodisiac, emetic, narcotic, sedative, soporific, and spasmolytic properties (see Table 6.1 on p. 58). This tall biennial plant is highly poisonous, but in expert hands it can be transformed into beneficent and benign medicines. "Historically," writes Ernest Bosc, "it has had multiple uses in herbal therapeutics. Preparations of hemlock have been deemed effective against cancerous diseases, scrofula, lymphatic congestions of the abdominal organs, ophthalmia, and neuralgia. All of these medicaments owe their medicinal virtues to the chemical compound coniine, an extremely volatile and active principle that resides chiefly in the hemlock fruits."[7] Hemlock must be harvested when ♃ conjuncts the ☉ for it to become anaphrodisiac in action. Its water can cure rheumatism and prevent excessive growth of the breasts. Its juice mixed with wine sediment causes birds to fall into a state of lethargy.[8]

HEMP

Fig. H.9. Mattioli, *Discorsi,* 533.

Latin binomials (type sp.): *Cannabis sativa* L.

French names: chanvre, canbe (Languedocian), chanvre cultivé, chanvre d'Inde, chanvre indien, chanvre sauvage, pantagruélion (Rabelais)

Elemental qualities: cold and dry (🜃)

Ruling planet: ♄

Zodiacal signatures: ♉, ♊, and ♒ (*apud* Mavéric)

Occult properties: Hemp possesses anodyne, hallucinogenic, narcotic, nervine, sedative, and spasmolytic properties (see Table 6.1 on p. 58). It produces a fat extract known as hashish. This unguent, when smoked or swallowed, provokes ecstatic experiences that are still poorly understood in the West, but some Muslim, Buddhist, and Taoist sects use it in regular, methodical doses to study the psychurgic arts.[9]

HENBANE

Fig. H.10. Fuchs, *Plantarum effigies,* 480.

Latin binomial: *Hyoscyamus niger* L.

French names: jusquiame, carelhado (Languedocian), hannebane, herbe aux engelures, jusquiame noire, mort aux poules, porcelet, potelée

Elemental qualities: hot and dry (△)

Ruling planets: ♄ and ♃

Zodiacal signatures: ♈, ♐, and ♑. Harvest when ♄ is in ♈ or ♏.

Occult properties: Henbane, or stinking nightshade, has antirheumatic, bechic, odontalgic, narcotic, nervine, and sedative properties (see Table 6.1 on p. 58). Decoctions of its bark soothe toothaches. Its roots and seeds, when administered in the form of topical creams, dry out buboes. Its root is effective against gout, and its juice is good for liver pain. Poultices of this herb are also beneficial for all types of breast disease. Wearing the entire plant as an amulet prevents the onset of inflammatory swelling and colic. It will also make whoever wears it good-natured and amiable, but burning its seeds as a suffumigant provokes anger. When hidden indoors in its natural state, it can provoke nervous crises, and it can be worked in such a manner that it causes death, even at a distance. This "herb of death," otherwise known by the secret name *octharon,* was sacred to medieval magicians, who gave it the Chaldean name *mansera.** According to the *Grand Albert,* when its juice is mixed with the blood of a young hare and rubbed on the skin, it attracts hares.[10]

HERB

Latin terms: *gramen, herba, ocimum, ocimus, ruta, satureia*

Greek terms: βοτάνη, πόα, λάχανον, φυτόν, χλόη, χόρτος

Sanskrit terms: *auṣadha* (अउषध), *hayuṣā* (हयुषा), *mahāmuni* (महामुनि), *oṣadhi* (ओषधि)

Folk remedy: The following is a generic magical herbal remedy from the Vosges to stop nosebleeds: Gather a fistful of any herb at random with your left hand without looking and say, "I am from Noah, an herb neither planted nor sown, and I shall do as God commands." Place the herb under the nostrils, and the bleeding will stop immediately. For greater efficacy, gather the herb at night under the light of the ☾.[11]

*[Sédir lists *octharan* and *mansera* as variants for the French common name *jusquiame.* Both of these names derive from the *Grand Albert,* and both appear to be otherwise unattested spellings. The former spelling, *octharan,* a variant for planetary herb no. 6 (♃), appears in Papus's *Traité élémentaire de magie pratique,* 235, but here the Latin incunabulum of 1493 reads *acharonia,* while the Latin edition of 1555 and the French editions of 1703 and 1800 read *acharon* (cf. Table 0.2 on p. xxvi). The latter spelling, *mansera,* a variant for Chaldean herb no. 8, reads *ma(n)ſela* (with a horizontal stroke over the letter *a* to indicate the supplemental nasal *n* and with one long *s*) in the Latin incunabulum of 1493, *manſeſa* (with two long *s*'s) in the Latin edition of 1555 and the French edition of 1703, and *mensera* in the French edition of 1800 (cf. Table 0.1 on p. xxiii). —*Trans.*]

HOLLY

Fig. H.11. Mattioli, *Discorsi,* 148.

Latin binomial (type sp.): *Ilex aquifolium* L.

French names: houx, grìfoul (Languedocian), houx commun

Elemental qualities: hot and dry (△)

Ruling planets: ♄ and ♂

Zodiacal signature: ♈ (*apud* Haatan and Jollivet-Castelot)

Occult properties: Holly possesses diaphoretic, febrifugal, and tonic properties. Its red berries, which are emetic and purgative in action, are thought by some to have been used as an ingredient in witches' ointments.[12] Febrile patients who rub themselves against the first holly they come across will be relieved of their fever almost immediately.

HONEYSUCKLE

Fig. H.12. Fuchs, *Plantarum effigies,* 371.

Latin binomial (type sp.): *Lonicera caprifolium* L.*

French names: chèvrefeuille, chèvrefeuille des jardins, chèvrefeuille d'Italie, couteto (Languedocian)

Elemental qualities: cold and dry (🜃)

Ruling planet: ☿

Zodiacal signature: ♍ (*apud* Haatan)

Occult properties: Honeysuckle possesses emollient, spasmolytic, and vulnerary properties. The *lilium inter spinas,* or "lily among thorns," mentioned in Song of Songs 2:2 is commonly thought to be a reference to the honeysuckle, but alchemists, Hermeticists, and early pharmacopeists refer to it by the name *matersylva,* meaning "mother-of-the-forest."[13] It is an especially good herb for consecrating altars and other ritual implements.[14]

*[Sédir also supplies the species name *periclymenum*—that is, the European honeysuckle, *Lonicera periclymenum* L. See *Les plantes magiques,* 148. —*Trans.*]

Fig. H.13. Fuchs, *Plantarum effigies,* 339.

HOREHOUND

Latin binomial (type sp.): *Marrubium vulgare* L.

French names: marrube (blanc), bon-blanc, bonhomme, herbe aux crocs, herbe vierge, marrible (Languedocian), mont-blanc*

Elemental qualities: hot and dry (△)

Ruling planet: ♂ (*apud* Piobb)

Zodiacal signature: ♈. Harvest at the beginning of April or under ♏.

Occult properties: The herb horehound has diaphoretic, diuretic, expectorant, and stimulant properties. In antiquity, its juice was primarily used as a bechic or pectoral to treat respiratory illnesses.[15]

Fig. H.14. Mattioli, *Discorsi,* 114.

HORNBEAM

Latin binomial: *Carpinus betulus* L.

French names: charme, calpre (Languedocian), charme faux bouleau, charmille

Elemental qualities: cold and dry (🜃)

Ruling planets: ☉ and ♃

Zodiacal signatures: ♊ and ♍

Occult properties: The European, or common, hornbeam, the type species of the genus *Carpinus,* has astringent, styptic, and vulnerary properties. Its wood is especially good for making magical wands and for use in divinatory and therapeutic practices.

[HORNED CUMIN ⚘ *See* CUMIN, HORNED]

[HORNED POPPY ⚘ *See* POPPY, HORNED]

*[Sédir's entry appears under the genus name *Marrubium,* which is thought to derive from the name of the ancient Roman village *Maria urbs.* See *Les plantes magiques,* 159. —*Trans.*]

HORSERADISH

Fig. H.15. Fuchs, *Plantarum effigies,* 379.

Latin binomial: *Armoracia rusticana* G.Gaertn., B.Mey. & Scherb.

French names: raifort, cran, cranson, mérédic, moutarde de capucin, rafe (Languedocian), raifort sauvage

Elemental qualities: hot and dry (△)

Ruling planet: ♂

Zodiacal signature: ♐

Occult properties: Horseradish possesses aperient, diuretic, expectorant, irritant, rubefacient, and stimulant properties. According to the *Grand Albert,* to walk through fire or carry hot iron without getting burned, mix together *marshmallow juice, egg white, *parsley seed, and quicklime, reduce the mixture to a powder, and then mix it with egg white and horseradish juice. Rub the mixture on your body or hands, let it dry, then rub the mixture on your body or hands a second time.[16]

HOUND'S TONGUE

Fig. H.16. Fuchs, *Plantarum effigies,* 264.

Latin binomial: *Cynoglossum officinale* L.

French names: cynoglosse, langue de chien, lengo-de-co (Languedocian)

Elemental qualities: hot and dry (△)

Ruling planet: ♂

Zodiacal signature: ♈

Occult properties: Hound's tongue has anodyne, astringent, digestive, diuretic, emollient, and spasmolytic properties. When worn as an amulet around the neck, it destroys prejudice and enmity and promotes empathy. This "herb of sympathy" was sacred to medieval magicians, who gave it the Chaldean name *algeil.* According to the *Grand Albert,* when worn as an amulet under the big toe, it prevents dogs from barking at the bearer.[17]

Fig. H.17. Fuchs, *Plantarum effigies*, 17.

HOUSELEEK

Latin binomial (type sp.): *Sempervivum tectorum* L.

French names: joubarbe, artichaut bâtard, artichaut sauvage, barbe de Jupiter, chou de chèvre, cussòudo (Languedocian), herbe aux cors, ioubarbe, joubarbe sauvage, œil de Jupiter (Rabelais)

Elemental qualities: cold and dry (🜃)

Ruling planet: ♀

Zodiacal signature: ♍

Occult properties: Houseleek possesses astringent, diuretic, refrigerant, and vulnerary properties. When crushed and mixed with *barley flour and oil and administered in the form of a topical application, it quickly heals scabs and skin rashes. Consuming houseleek can break the spell of the *aiguillette* and cure magical castration and impotence.[18]

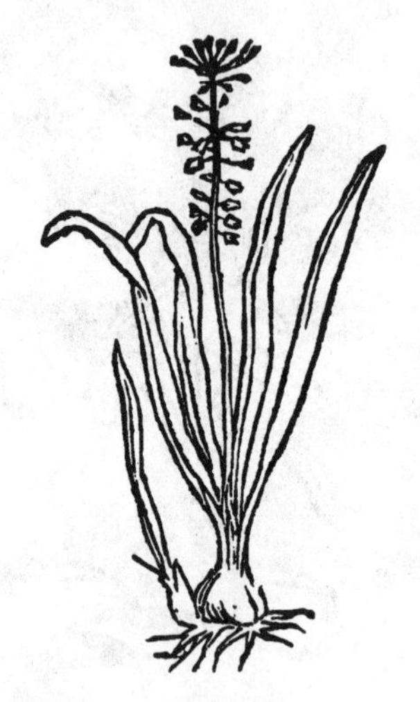

Fig. H.18. Fuchs, *Plantarum effigies*, 481.

HYACINTH

Latin binomial (type sp.): *Hyacinthus orientalis* L.

French names: jacinthe, jacinthe des jardiniers, jacinthe d'Orient, jacinto (Languedocian)

Elemental qualities: hot and wet (🜁)

Ruling planets: ♀ and ☉

Zodiacal signatures: ♎ and ♉

Occult properties: Hyacinth possesses astringent and diuretic properties. When worn as an amulet, hyacinth helps procure friendships with influential and prominent women. The juice of its root prevents the growth of body hair and postpones puberty. The root, when boiled, can also heal testicular tumors. Hyacinth is one of the twelve sacred plants of the Rosicrucians.[19]

I

Iris–Ivy

IRIS

Latin binomial (type sp.): *Iris germanica* L.

French names: iris, courtrai, couteau, coutèlo (Languedocian), flambe, flamme, glaïeul bleu, glais

Elemental qualities: hot and wet (🜁)

Ruling planet: ♀

Zodiacal signature: ♎

Occult properties: Iris roots possess astringent, cathartic, diuretic, emetic, and pectoral properties. They are most often employed in the treatment of bronchitis, phthisis, and edema.[1] In the Hermetic language of flowers, the iris is emblematic of peace.

Fig. I.1. Fuchs, *Plantarum effigies*, 182.

Fig. I.2. Fuchs, *Plantarum effigies*, 238.

IVY

Latin binomial (type sp.): *Hedera helix* L.

French names: lierre, leùno (Languedocian), lierre grimpant

Elemental qualities: cold and dry (🜃)

Ruling planet: ☿

Zodiacal signatures: ♉ and ♐

Occult properties: Ivy possesses anodyne, anti-inflammatory, bechic, spasmolytic, and vermifugal properties. Its leaves not only prevent drunkenness but also cure hangovers. It should come as no surprise, then, that the crown of Bacchus consisted of a wreath of intertwined ivy leaves and stems. To combat sore throat and bad breath, take twenty ivy leaves and put them in a small pot with some old wine and a little salt, bring it slowly to a boil, and gargle a small sip at as hot a temperature as you can stand.[2] Fumigations of ivy can get rid of unwanted bats in your home. In Montenegro, people hang ivy on their doors on Christmas Eve to garner protection for the coming year. In Germany, the first time a cow is milked in the springtime, people perform this same talismanic operation with ivy wreaths.

J

Jimsonweed–Juniper

JIMSONWEED

Fig. J.1. Fuchs, *Plantarum effigies,* 399.

Latin binomial: *Datura stramonium* L.

French names: endormeuse, chasse-taupe, datura officinal, endormie, endourmidouiro (Languedocian), herbe aux sorcières, herbe aux taupes, herbe des magiciens, herbe du diable, pomme endormante, pomme épineuse, pomme poison, stramoine, trompette des anges*

Elemental qualities: cold and wet (▽)

Ruling planets: ♄ and ☾

Zodiacal signature: ♏

Occult properties: Jimsonweed, otherwise known as devil's snare, devil's trumpet, devil's weed, thorn apple, and moon flower, is a member of the nightshade family (Solanaceae). It has extremely powerful soporific, narcotic, and hallucinogenic properties (see Table 6.1 on p. 58), as it is saturated with malefic fluids—namely, daturine, a mixture of the deliriant alkaloids atropine and

*[Sédir provides separate entries for jimsonweed, the first under the genus name *Datura* and the second under *stramoine,* a gallicized form of the species name. See *Les plantes magiques,* 150, 169. In the latter entry, Sédir supplies the French common name *morelle furieuse* and the alchemical names *datel* and *tatel* as variants, but *morelle furieuse* (deadly nightshade) is in fact a well-known common name for ★belladonna. The error is attributable to Sédir's source—namely, Pernety's *Dictionnaire mytho-hermétique.* The alchemical names, at any rate, are appropriately placed. The two entries have been merged here to form a single entry. Cf. Sédir's discussion of witches' ointments in chapter 6. —*Trans.*]

hyoscyamine—for which reason it became a favorite herb of medieval sorcerers and witches. Its seeds have several medicinal applications, as they are anodyne, anti-inflammatory, stomachic, and vermifugal in action.[1] Alchemists and Hermeticists refer to it by the secret names *datel* and *tatel,* which are abridged forms of the Hindi name *dhatura* and the Arabic name *tatorah.*[2]

Fig. J.2. Fuchs, *Plantarum effigies,* 44.

JUNIPER

Latin binomial (type sp.): *Juniperus communis* L.

French names: genièvre, genévrier, genibre (Languedocian)

Elemental qualities: hot and dry (△)

Ruling planet: ☿

Zodiacal signatures: ♈, ♌, and ♐

Occult properties: The juniper tree, especially its berries, possesses aromatic, carminative, diaphoretic, diuretic, stomachic, and tonic properties. Its branches serve as powerful amulets and talismans for warding off or expelling evil spirits and serpents. A suffumigation of the berries, which alchemists and Hermeticists call *harmat* but Martin Ruland calls *ebel,** will purify any room.[3] Juniper seeds can also be used to cure demoniacs. These exorcistic, apotropaic, and purificatory actions are due to the fact that juniper produces fruit that bears the sign of the Trinity: each berry, more often than not, bears three triangular seeds. The triangular seeds as well as the berries are effective against edema, plague, venoms, colic, coughs, asthma, scabies, and gout. Juniper decoctions, especially when mixed with a decoction of ⋆elderberry flowers, are particularly useful in the treatment of hemorrhoids, and the same is true of sandarac, the resin produced by the savin juniper (*Juniperus sabina* L.). The juniper extract, or "honey," known as *mel juniperinum,* is a remarkable curative for asthma. Alchemists and Hermeticists refer to the various juniper species by the secret name *hara.*[4]

*[The alchemical name *ebel* is otherwise used to describe the seeds of the herb ⋆sage. Pernety, Sédir's source, casts doubt over Ruland's attribution of this name to juniper seeds. See Ruland, *Lexicon alchemiae,* 192 s.v. *ebel;* cf. Pernety, *Dictionnaire mytho-hermétique,* 126 s.v. *ebel.* —*Trans.*]

K

Knotweed–Kusha Grass

KNOTWEED

Fig. K.1. Fuchs, *Plantarum effigies,* 352.

Latin binomial (type sp.): *Polygonum aviculare* L.

French names: renouée, herbe de couchons, herbò nousado (Languedocian), persicaire, renouée des oiseaux, traînasse

Elemental qualities: hot and wet (🜁)

Ruling planets: ♃ and ☉

Zodiacal signatures: ♊ and ♐ (*apud* Mavéric)

Occult properties: Knotweed possesses astringent, cardiotonic, cholagogic, diuretic, febrifugal, and vulnerary properties. If the leaves are applied to a contused wound and then stored in a damp place, the wound will heal magnetically. Knotweed has multiple medicinal applications and may be used to treat ailments of the chest, the heart, and the stomach. Its infusion is good not only for sexual arousal and for inciting love but also for healing pulmonary engorgements and melancholy. When worn as an amulet around the neck, its root heals diseases of the eyes. According to the *Grand Albert,* when hung in the home as a talisman, it brings in influences from the owner's ☉ sign, and when worn as an amulet over the chest, it becomes antihysteric and pulmonary in action.[1] Alchemists and Hermeticists refer to knotweed by the names *proserpinaca, seminalis,* and *corrigiole,** but they call a related member of the Polygonaceae family known as spotted lady's thumb (*Persicaria maculosa* S.F.Gray syn. *Polygonum persicaria* L.) by the Saturnian name *molybdaena.*[2]

Fig. K.2. Mattioli, *Discorsi,* 567.

KUSHA GRASS

Latin binomial: *Desmostachya bipinnata* (L.) Stapf.

French names: kousa, herbe dharba, herbe kusa

Elemental qualities: hot and dry (△)

Ruling planet: ☉

Zodiacal signature: ♌

Occult properties: Kusha grass has astringent, diuretic, and stimulant properties. In India, this herb is known by many names, including *kuśa* (कुश) and *darbha* (दर्भ) in Sanskrit.[3] It is the sacred herb of the Hindus, for whom it serves as the material for the seat, asana, or mat of all acts of religious and ascetic devotion. The sacred Hindu scriptures describe it as the ideal seat for all meditation.[4] This sacrificial grass has powerful magnetic properties and serves as a universal vehicle.

*[The latter plant name, *corrigiole,* is a variant for planetary herb no. 2 (☉) in the *Grand Albert.* Here the Latin incunabulum of 1493 reads *corrigiola,* and the French edition of 1703 reads *corrigiale* (cf. Table 0.2 on p. xxvi). This herb should not be confused with any plant of the modern genus *Corrigiola,* whose modern type species is the strapwort (*Corrigiola litoralis* L.). —*Trans.*]

L

Larkspur–Lungwort

LARKSPUR

Fig. L.1. Fuchs, *Plantarum effigies,* 239.

Latin binomial (type sp.): *Consolida regalis* Gray

French names: consoude royale, dauphinelle consoude, dauphinelle des champs, dauphinelle royale, pied d'alouette, pied d'alouette royal

Elemental qualities: hot and dry (△)

Ruling planet: ♀ (*apud* Piobb)

Zodiacal signatures: ♈, ♐, and ♒. Harvest after the full ☾ terminating the dog days of summer.

Occult properties: Larkspur has deobstruent, diuretic, hypnotic, and purgative properties. It is sacred to the Roman goddess Juno in her guise as Lucina. Its powder has vulnerary and antihemorroidal properties. Its decoction in a compress is effective against leucoma.[1] Paracelsus, who calls it *aquilena,*[2] used it frequently in conjunction with ★birthwort, ★aloe, ★Saint John's wort, and ★bay oil.*

*[Sédir's original entry also includes the following statement: "To get rid of bedbugs without touching them, place some leaves from this plant under the bedside, and all the bedbugs will assemble on them." This folk remedy derives from Papus's *Traité élémentaire de magie pratique,* 549 s.v. *punaises,* where the key ingredient is *grande consoude* (comfrey), not *consoude royale* (larkspur). For this reason, the folk remedy has been moved under ★comfrey. See further Martin-Lauzer, "De la grande consoude," 355. Sédir's inclusion of the folk remedy here, however, may well be intended as a correction, since larkspurs, which are extremely poisonous in their natural state, arguably possess stronger insecticidal properties. —*Trans.*]

Fig. L.2. Fuchs, *Plantarum effigies,* 514.

LAVENDER

Latin binomial (type sp.): *Lavandula spica* L.

French names: lavande, aspic, estamous (Languedocian), faux nard

Elemental qualities: hot and dry (△)

Ruling planet: ☉ (*apud* Agrippa, Lenain, and Piobb)

Zodiacal signature: ♌. Harvest when the ☉ is in ♓ and the ☾ is in ♋.

Occult properties: Lavender possesses powerful aromatic, antiseptic, carminative, febrifugal, styptic, and tonic properties. The aroma of the lavender flower has long been celebrated for its powerful nervine action.[3] When employed as a suffumigant, its smoke chases away evil spirits.

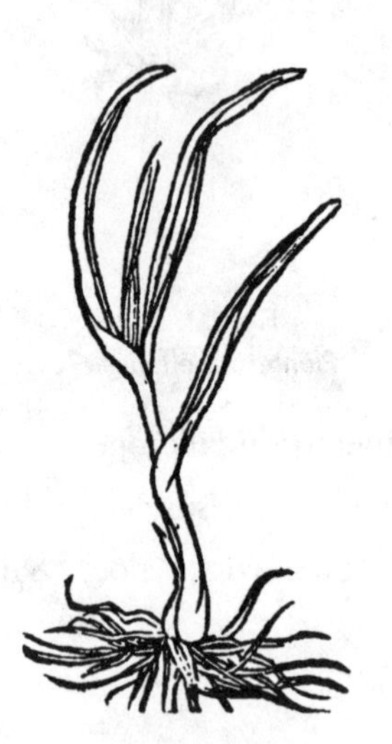

Fig. L.3. Fuchs, *Plantarum effigies,* 364.

LEEK

Latin binomial: *Allium ampeloprasum* L.

French names: poireau, ail porreau, poireau d'été, poireau perlé, poireau perpétuel, porre (Languedocian), porreau

Elemental qualities: hot and dry (△)

Ruling planets: ♂ and ☾

Zodiacal signatures: ♏ and ♊

Occult properties: Leek possesses the same occult properties as *garlic, but its action is much milder due to the influence of the ☾. When mixed with a sympathetic aliment, it has diuretic properties and provokes menstruation. According to a Hermetic tradition preserved in the *Grand Albert,* putting leek seeds in spoiled vinegar will cause the vinegar to regain its former strength.[4]

LEEK, SAND

Fig. L.4. Mattioli, *Discorsi,* 342.

Latin binomial: *Allium scorodoprasum* L.

French names: ail rocambole, échalote d'Espagne, oignon bulbifère, oignon d'Egypte, poireau du Levant, rocambole*

Properties: Sand leek has the same elemental, astral, and occult properties as ★leek.

LEMON

Fig. L.5. Mattioli, *Discorsi,* 171.

Latin binomial: *Citrus limon* (L.) Osbeck

French names: citronnier, citron, citrouniè (Languedocian), limonier

Elemental qualities: hot and wet (🜁)

Ruling planet: ♃

Zodiacal signature: ♎ (*apud* Haatan and Jollivet-Castelot)

Occult properties: The lemon tree possesses astringent, febrifugal, rubefacient, stomachic, and tonic properties. The fruit is ruled by the ☉ and signed by ♓. Its juice is antiscorbutic and diuretic in action and serves as an excellent remedy for hangovers. The juice of the inner bark of the lemon tree makes an ideal plaster for healing uveitis and other eye inflammations.

*[In his entry for ★leek, Sédir includes the species name *scorodoprasum,* which derives from the ancient Greek name for the sand leek, σκοροδόπρασον, literally "garlic leek" (σκόροδον, "garlic," + πράσον, "leek"). See *Les plantes magiques,* 164 s.v. *poireau.* These have been separated here to form two distinct entries. —*Trans.*]

[**LEMON BALM** ❦ *See* BALM, LEMON]

[**LESSER BURDOCK** ❦ *See* BURDOCK, LESSER]

[**LESSER CELANDINE** ❦ *See* CELANDINE, LESSER]

Fig. L.6. Fuchs, *Plantarum effigies*, 171.

LETTUCE

Latin binomial (type sp.): *Lactuca sativa* L.

French names: laitue, laitue cultivée, lachugo (Languedocian), salade, salade des jardins

Elemental qualities: cold and wet (🜄)

Ruling planet: ☾

Zodiacal signatures: ♋ and ♊ (*apud* Jollivet-Castelot)

Occult properties: All species of lettuce, especially wild lettuce (*Lactuca virosa* L.), contain a milky exudate called lactucarium, which has anodyne, hypnotic, narcotic, sedative, and soporific properties (see Table 6.1 on p. 58). As a narcotic, it resembles opium, but it is both weaker and not habit forming. The leaves, being rich in vitamins and minerals, have natural nutritive properties.[5] All parts of the plant are also galactagogic in action and are often prescribed to nursing mothers and wet nurses to increase their breast milk supply.

Fig. L.7. Fuchs, *Plantarum effigies*, 454.

LICEBANE

Latin binomial: *Delphinium staphisagria* L.

French names: herbe aux poux, dauphinelle staphisaigre, herbe aux goutteux, juco-lait (Languedocian), raisin sauvage, staphisaigre

Elemental qualities: cold and wet (🜄)

Ruling planet: ♄

Zodiacal signatures: ♏, ♉, and ♎ (*apud* Mavéric)

Occult properties: Licebane, otherwise known as stavesacre, has emetic, sialagogic, spasmolytic, and vermifugal properties. All parts of this plant are highly toxic, and for this reason it is rarely used internally.[6] When crushed and mixed with oil, however, it may be used topically in the form of a lotion to kill lice, treat warts, and relieve itchy skin.

LICHEN, LUNGWORT

Fig. L.8. Fuchs, *Plantarum effigies,* 268.

Latin binomial: *Lobaria pulmonaria* (L.) Hoffm.

French names: lichen pulmonaire, hépatique des bois, lobaire pulmonaire, mousse pulmonaire, pulmonaire des chênes, pulmonaire en arbre, thé des forêts*

Elemental qualities: cold and wet (∇)

Ruling planet: ♄ (*apud* Piobb and Bélus)

Zodiacal signature: ♋ (*apud* Jollivet-Castelot)

Occult properties: Lungwort lichen, otherwise known as tree lungwort, possesses anodyne, antirheumatic, bechic, pectoral, and pulmonary properties. In the Hermetic language of flowers, lichens are emblematic of solitude.

*[Sédir's entry concerns all lichens. See *Les plantes magiques,* 157 s.v. *lichen.* The original entry has been modified here to reflect the astral signatures supplemented by Sédir's colleagues. See Piobb, *Formulaire de haute magie,* 201; Bélus, *Traité des recherches,* 35; and Jollivet-Castelot, *La médecine spagyrique,* 12. Of the many and various families and genera of lichens, these signatures are best exemplified by the lungwort lichen, the type species of the genus *Lobaria.* The ridges and lobes of the lungwort lichen have long been thought to resemble human lung tissue, and for this reason it has a long history of use in the treatment of pulmonary ailments. —*Trans.*]

Fig. L.9. Fuchs, *Plantarum effigies,* 206.

LILY

Latin binomial (type sp.): *Lilium candidum* L.

French names: lis, lis blanc, lis candide, lìri (Languedocian), lis de la Madone, lys, rose de Junon

Elemental qualities: cold and dry (🜃)

Ruling planets: ☾, ♀, and ♃

Zodiacal signature: ♉. The bulb is hot and dry (△) and signed by ♈.

Occult properties: The lily has anti-inflammatory, astringent, demulcent, emmenagogic, and tonic properties. Its flower is the image of universal generation, preformation, and the action of the primal fire on the mother water. For this reason, the angel Gabriel is depicted carrying a lily in most Renaissance paintings of the Annunciation. In the Hermetic language of flowers, the lily is emblematic of purity. In the Middle Ages, its pollen was used as a diuretic to expose women who did not preserve their chastity. In lotions, it is good for treating burns and whitening the complexion. Hildegard of Bingen used the extremity of its root ground with rancid fat to heal leprosy.[7] The root, harvested when ♀ conjuncts the ☾ in ♉ or ♎ and hung around the neck as an amulet, works well as a love charm. Its distilled water eases pains in childbirth, soreness of the eyes, and stomach aches. Consuming two fragments of its root helps women who have suffered a miscarriage expel the placenta and dead fetus. When crushed and boiled with bread crumbs, its bulbs make abscesses ripen and burst in a short amount of time. Perfumes and suffumigants made from lilies prepare the rooms in which they are burned for astral manifestations. This "herb of manifestation" was sacred to medieval magicians, who gave it the Chaldean name *augoeides.** According to the *Grand Albert,* if reduced to a powder and rubbed on someone's clothes, it will prevent them from falling asleep. Its flower is said to be a good treatment for kidney pain because it, too, increases and decreases with

*[In Latin editions of the *Grand Albert,* the name of Chaldean herb no. 9 is *augo,* but French editions read *ango* (cf. Table 0.1 on p. xxiii). On *augoeides* as a plant name, see the comments in the foreword. —*Trans.*]

the ☾.[8] The author of the *Grand Albert* also refers to the Madonna lily, the type species of the genus *Lilium,* by the secret name *chrynostates.**

LILY, WATER

Fig. L.10. Fuchs, *Plantarum effigies,* 303.

Latin binomial (type sp.): *Nymphaea alba* L.

French names: nénuphar, lis d'eau, lis des étangs, lune d'eau, nénuphar blanc, volant d'eau

Elemental qualities: cold and wet (∇)

Ruling planets: ☾ and ♀

Zodiacal signatures: ♋ and ♊ (*apud* Haatan and Jollivet-Castelot)

Occult properties: Water lily has anodyne, astringent, cardiotonic, demulcent, and sedative properties. If harvested in June or July, it can be used with profit in the treatment of migraines and vertigo. When mixed with a plant governed

*[Although French editions of the *Grand Albert* read *chrynostates,* Latin editions read *chynostates,* and Agrippa similarly refers to this herb by the name *chinostates* (erroneously rendered as *chinosta* in English translations). Angelo de Gubernatis suggests that these names are corruptions of the otherwise unattested form *chrysostates,* or χρυσοστάτης, a compound of Greek χρυσός, meaning "gold," and an agent-nominal suffix derived from the aorist active participial stem of the verb ἵστημι, meaning "to stand." Gubernatis's comparison of the *chrysostates* with legends about herbs that turn objects into gold, however, is unconvincing. If *chrysostates* is indeed an uncorrupted form, then it is essentially a medieval equivalent of modern English *heliostat* (literally, "one who stands in the ☉"), an instrument that holds the image of the ☉ stationary by rotating to reflect its light. However, as the *Grand Albert* points out, the *chrysostat,* if you will, is closely linked to the ☾, and so the initial element χρυσός would refer to the light of the ☾ or, more precisely, to the reflection of golden rays of the ☉ on the ☾. It is to be noted, however, that the lily is not a selenotrope—which is to say, it does not follow the ☾ in the same way a ★heliotrope follows the ☉. In fact, both Pseudo-Albertus and Agrippa assert that it follows *the phases of the* ☾, increasing and decreasing "not only in fluid, but also in virtue" as the ☾ waxes and wanes. But whereas the *Grand Albert* attributes this ebb and flow to the flower, Agrippa attributes it to the leaves. See further Agrippa, *De occulta philosophia libri tres,* liber I, caput XXIV, xxxii (English edition: *Three Books of Occult Philosophy,* bk. 1, chap. 24, 94); and Gubernatis, *La mythologie des plantes,* 1:214. At any rate, if the Latin formation *chynostates* or *chinostates* is not simply a corruption of κυνόσβατος (whiterose), which appears in the much older Hermetic tractate *On Plants of the Seven Planets* (cf. Table 0.2 on p. xxvi), or of κρίνον, the ancient Greek plant name for the Madonna lily, then it would appear to be the equivalent of κρινοστάτης, literally "one who stands in the lily," which would be a fitting secret name for the astral body of the Madonna lily. See further Sédir's comments on the astral bodies of plants in chapters 1 and 9. —*Trans.*]

by ♄, it can cure gonorrhea and stop muscle spasms. Its root heals leucorrhea and menorrhagia. When harvested under favorable influences of the ☾ and ♄, it can be made into a potent anaphrodisiac beverage. In the Hermetic language of flowers, the water lily is emblematic of charity (*caritas*).

Fig. L.11. Fuchs, *Plantarum effigies*, 498.

LINDEN

Latin binomial (type sp.): *Tilia europaea* L.

French names: tilleul, tel (Languedocian), tilleul d'Europe

Elemental qualities: moderately hot and wet (🜁)

Ruling planet: ☾

Zodiacal signatures: ♎. The flower is signed by ♐.

Occult properties: Linden possesses bechic, diaphoretic, febrifugal, and nervine properties. Infusions of linden, which should be made when the ☾ is in ♓, have calmative properties and are particularly effective against menstrual ailments, epilepsy, and colic.

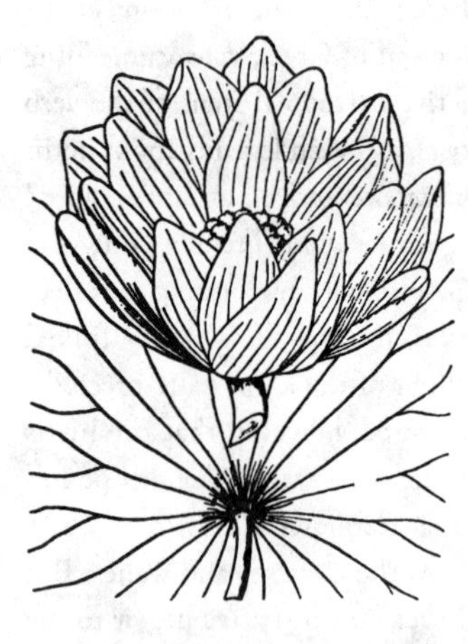

Fig. L.12. Britton, *An Illustrated Flora*, 2:46.

LOTUS

Latin binomial: *Nelumbo nucifera* Gaertn.

French names: lotus, lotier, lotus d'Égypte, lotus des Indes, lotus sacré*

Elemental qualities: cold and wet (🜄)

Ruling planet: ☉

Zodiacal signature: ♓

Occult properties: The sacred lotus has astringent, cardiotonic, resolvent, stomachic, styptic, and tonic properties. According to legend, the Bodhisattva presented a white lotus flower

*[Sédir's entry appears under the spelling *lotos,* for *lōtos,* a transliteration of λωτός, the ancient Greek name for the Egyptian lotus. See *Les plantes magiques,* 158. This extremely versatile herb should not be confused with the other Greek plant named λωτός—namely, the strawberry clover (*Trifolium fragiferum* L.)—or with the modern genus *Lotus,* which includes a variety of trefoils and vetches. —*Trans.*]

to Queen Māyā of Sakya. From the religious point of view, the Indian, or Egyptian, lotus conveys the same esoteric meaning as the *lily. In Christian tradition, the archangel Gabriel appears to the Virgin Mary at the Annunciation holding a spray of lilies. "This spray typifying fire and water, or the idea of creation and generation," writes Helena Petrovna Blavatsky, "symbolizes *precisely the same idea as the lotus* in the hand of the Bodhisattva who announces to Mahā-Māyā, Gautama's mother, the birth of the world's Savior, Buddha."[9] This FIRE-WATER dichotomy is clearly exemplified in the plant's astral signatures.

LUNGWORT

Fig. L.13. Fuchs, *Plantarum effigies,* 366.

Latin binomial: *Pulmonaria officinalis* L.

French names: pulmonaire, herbe au lait de Notre Dame, herbe aux poumons, herbe cardiaque, herbe de cœur, herbò dal paumou (Languedocian)

Elemental qualities: cold and wet (🜄)

Ruling planets: ☿ and ♄

Zodiacal signature: ♋ (*apud* Jollivet-Castelot)

Occult properties: Lungwort, otherwise known as Mary's tears and Our Lady's milk drops, has bechic, demulcent, emollient, pectoral, and pulmonary properties. Its flower has astringent and desiccant properties. When applied topically in the form of a poultice, its leaves are useful for healing wounds.

[LUNGWORT LICHEN *See* LICHEN, LUNGWORT]

M

Madder–Myrtle

Fig. M.1. Fuchs, *Plantarum effigies*, 158.

MADDER

Latin binomial (type sp.): *Rubia tinctorum* L.

French names: garance, garance des teinturiers, rouge des teinturiers

Elemental qualities: hot and dry (△)

Ruling planets: ♃ and ♂

Zodiacal signature: ♐ (*apud* Jollivet-Castelot)

Occult properties: Madder possesses aperient, astringent, cholagogic, diuretic, and emmenagogic properties. It is particularly valuable for its ability to heal hernias, but it is also effective in the treatment of edema, jaundice, and amenorrhea, or suppressed menstruation.

[MAIDENHAIR FERN ⚕ *See* FERN, MAIDENHAIR**]**

[MALE FERN ⚕ *See* FERN, MALE**]**

MALLOW

Fig. M.2. Fuchs, *Plantarum effigies,* 290.

Latin binomial (type sp.): *Malva sylvestris* L.

French names: mauve, fausse guimauve, fromageon, grande mauve, malbo (Languedocian), maulue, maulve, mauve des bois, petit fromage

Elemental qualities: cold and dry (🜃)

Ruling planet: ♀ (*apud* Piobb)

Zodiacal signatures: ♉ and ♎

Occult properties: Mallow possesses aperient, astringent, bechic, demulcent, emollient, and pectoral properties. It is especially common in emollient creams and in remedies for asthma, bronchitis, and sore throats.[1] Our French ancestors, however, who were inveterate debauchees, would burn mallow and use its smoke to ensure the virginity of young lovers. This practice is reminiscent of the black magical application of *greater burdock.

MALLOW, MARSH

Fig. M.3. Fuchs, *Plantarum effigies,* 7.

Latin binomial: *Althaea officinalis* L.

French names: guimauve, bourdon de Saint Jacques, guimauve sauvage, malbo blanco (Languedocian), mauve blanche, sucette

Elemental qualities: hot and wet (🜁)

Ruling planet: ♀ (*apud* Mavéric)

Zodiacal signatures: ♊ and ♉

Occult properties: Marsh mallow possesses marvelous anti-inflammatory, bechic, demulcent, emollient, and vulnerary properties. The *guimauve* is known colloquially as *fromjon* or *fromjom,* from *from,* meaning "brave," "strong," or "virtuous," and *jom,* meaning "misery," "pain," or, in general, "anything caused by evil." The *guimauve,* or *fromjom,* then, is a plant whose virtue consists in combating pain and illness.[2] This etymology demonstrates that marsh mallow removes illnesses and heals wounds by means of a powerful cleansing action. Indeed, all parts of the plant have emollient properties and may be used in herbal teas, poultices, sitz baths, and anti-inflammatory creams. When powdered, kneaded into an ointment, and rubbed on the face and hands, its seeds act as an insect repellent. Its flowers, when kneaded with pork fat and terebinth resin or turpentine and applied topically on the stomach, cures pelvic inflammatory disease and inflammation of the womb. Its roots infused in wine heal urinary retention.[3]

Fig. M.4. Fuchs, *Plantarum effigies,* 301.

MANDRAKE

Latin binomial: *Mandragora officinarum* L.

French names: mandragore, herbe de Circé, main de gloire, pomme d'amour

Elemental qualities: cold and moderately dry (🜃)

Ruling planets: ♄ and ☾

Zodiacal signature: ♑

Occult properties: Mandrake has strong anesthetic, cathartic, emetic, hallucinogenic, and narcotic properties (see Table 6.1 on p. 58). Its malefic properties can cause madness if left uncorrected by the influence of the ☉, which can turn it into a powerful and beneficent narcotic. The Germans used to carve small statues of the gods, called *Abrunes,* from the roots and sell them as charms against all varieties of sickness and misfortune. Witches allegedly used mandrake root as a means of flying to their sabbats. The root is an extremely powerful astral condenser, and the anthropomorphic shape it always assumes reveals its unique properties and singular energy. Our friend Sisera possesses one such root portraying a father, mother, and child in exquisite detail.* Some still entertain the ludicrous theories of

*[Sisera, or Noël Sisera, was the mystical name adopted by Léon Champrenaud (1870–1925) upon entry into the Supreme Council of the Martinist Order, of which Sédir was also a member. Champrenaud, like Papus and many other members of his Martinist Order, took his magical name from Éliphas Lévi's commentary on an excerpt from the *Nuctemeron of Apollonius of Tyana,* in which Sisera appears as the name of the *génie du désire* (genius of desire), the first daemon of the second hour of the night (see Lévi, *Dogme et rituel de la haute magie,* 2:392). Like Sédir, he was also consecrated into the bishopric of the Église gnostique de France, in which he served as bishop of Versailles under the episcopal name Tau Théophane. Together with Tau Simon, bishop of Tyre and the East—namely, Albert de Pouvourville (1861–1939),

wayward magicians who thought they could use mandrake to produce the elixir of life or to create *teraphim,* or tutelar gods.[4] Medieval alchemists call it by many names, including *aroph,* a contraction of the Latin collocation *ar(oma) ph(ilosophorum),* or "aroma of the philosophers"; *dudaim,* a latinization of the Hebrew plant name דודאים (used only in the plural in the Hebrew Bible); and *jabora,* a latinization of the Arabic plant name يبروح or *yabrūḥ.*[5] Mandrake is also one of the twelve sacred plants of the Rosicrucians.[6]

MARJORAM

Fig. M.5. Fuchs, *Plantarum effigies,* 382.

Latin binomial: *Origanum majorana* L.

French names: marjolaine, majourano (Languedocian), origan des jardins

Elemental qualities: hot and dry (△)

Ruling planets: ☉ and ☿

Zodiacal signature: ♈. Harvest at the beginning of April or under ♏.

Occult properties: Marjoram has carminative, cholagogic, diaphoretic, diuretic, and stimulant properties. The oil extracted from this herb is particularly useful in the treatment of lethargy and apoplexy.

[MARSH GLADIOLUS ⚕ *See* GLADIOLUS, MARSH]

[MARSH MALLOW ⚕ *See* MALLOW, MARSH]

[MEADOW SAFFRON ⚕ *See* SAFFRON, MEADOW]

(*cont.*) alias Matgioï—he coauthored a work titled *Les enseignements secrets de la Gnose* (The Secret Teachings of Gnosis), which became the foundational initiatic text of the Église gnostique de France. —*Trans.*]

Fig. M.6. Fuchs, *Plantarum effigies,* 269.

MERCURY

Latin binomial (type sp.): *Mercurialis perennis* L.*

French names: mercuriale, caquenlit, chou de chien, cynocrambe, foirolle, herbe foireuse, mercuriale pérenne, mercuriale vivace, mourtariol (Languedocian), ortie bâtarde, ramberge

Elemental qualities: cold and wet (∇)

Ruling planet: ☿ (*apud* Agrippa, Piobb, Mavéric, and Jollivet-Castelot)

Zodiacal signature: ♏

Occult properties: Mercury, which is dioecious (having male and female reproductive organs in separate plants) and wind pollinated, possesses emollient, emmenagogic, laxative, and purgative properties. When administered to women in the form of a decoction for four consecutive days, the juice of the male plant facilitates the conception of boys, whereas the juice of the female plant facilitates the conception of girls.[7]

*[Sédir provides the outdated genus name *Phyllum* (from Greek φύλλον, meaning "leaf"), following Pernety, *Dictionnaire mytho-hermétique,* 185. Old herbalists and early botanists formerly used the Latin binomials *Phyllum marificum* and *Phyllum feminificum* to distinguish between male and female mercuries, which are comparable to the ἀρρενόγονον and θηλύγονον of the ancient Greek herbalists. See, for example, Franck von Franckenau, *Flora Francica aucta,* 492. —*Trans.*]

MIGNONETTE

Fig. M.7. Parkinson, *Theatrum botanicum*, 822.

Latin binomial (type sp.): *Reseda odorata* L.

French names: réséda, herbe de Saint Luc, mignonnette, resera (Languedocian)

Elemental qualities: cold and dry (🜃)

Ruling planets: ♀ and ☉

Zodiacal signature: ♑ (*apud* Mario)

Occult properties: Mignonette possesses alexitary, aromatic, diaphoretic, and laxative properties. The clairvoyant Catherine Emmerich, who had a profound mystical relationship with the Virgin Mary, claimed that mignonette was the favorite herbal remedy of the evangelist Luke, who was also a practicing physician.[8] Sometimes, Emmerich asserts, he would macerate it in oil and, after blessing it, apply it as an unguent on the brow or mouth of the patient in the form of the cross; other times he would desiccate the herb and administer it in infusions.[9] In the Hermetic language of flowers, the mignonette is emblematic of gentleness.

Fig. M.8. Mattioli, *Discorsi,* 425.

MINT

Latin binomial (type sp.): *Mentha spicata* L.*

French names: menthe, mentastre (Languedocian), menthe crépue, menthe crispée, menthe douce, menthe verte

Elemental qualities: hot and dry (△)

Ruling planets: ♃ and ♂

Zodiacal signatures: ♈ and ♒. Harvest either after the full ☾ terminating the dog days of summer or when the ☉ is in ♌ and the ☾ is in ♍ or, lastly, when the ☉ is in ♉ and the ☾ is in ♊.

Occult properties: Mint possesses antiseptic, carminative, diuretic, spasmolytic, stimulant, stomachic, and restorative properties. According to Greek mythology, Persephone tore her rival, Minthe—a naiad associated with the river Cocytus and Hades's former lover—limb from limb out of jealousy, after which Hades granted that Minthe should become ἡδύοσμος, or "sweet-smelling," and transformed her into garden mint (ἡδύοσμον in Greek). Mint was also offered to the dead in ancient Greek funeral rites, much like ⋆amaranth and ⋆celery.

Fig. M.9. Fuchs, *Plantarum effigies,* 418.

MINT, WATER

Latin binomial: *Mentha aquatica* L.

French names: pouliot aquatique, menthe aquatique

Elemental qualities: hot and dry (△)

Ruling planet: ♂

Zodiacal signature: ♈

Occult properties: Water mint has antiseptic, bechic, and stimulant properties. Its leaves, which have a minty flavor similar to but stronger than ⋆peppermint, may be eaten raw to

*[For other species of *Mentha,* see Sédir's entries ⋆peppermint, ⋆pennyroyal, and ⋆water mint. —*Trans.*]

freshen the breath. It is most frequently used in medicine as a mycteric to decongest and clear the nasal cavity.[10]

MISTLETOE

Fig. M.10. Fuchs, *Plantarum effigies,* 187.

Latin binomial: *Viscum album* L.

French names: gui de chêne, besc de poumiè (Languedocian), gillon, gui commun, guy

Elemental qualities: cold and dry (🜃)

Ruling planet: ☉ (*apud* Bélus)

Zodiacal signature: ♉

Occult properties: Mistletoe has cardiotonic, diuretic, narcotic, nervine, and stimulant properties. Its infusion, when administered after menstruation, facilitates conception.[11] Its dried berries, when powdered and dissolved in a generous wine, are effective against epilepsy. The Druids harvested it with great pomp at Christmastime at a precise astronomical hour. Its leaves, when saturated with the triple magnetism of the host tree, the stars, and a pious crowd, become powerful magnetic capacitors that serve to effect miraculous cures in some rather desperate cases. This "herb of salvation" was sacred to medieval magicians, who gave it the Chaldean name *luperax.* According to the *Grand Albert,* a sprig of mistletoe hung from a tree with the wings of a swallow attracts birds.[12] The leaves of the mistletoe growing on a hawthorn (*Crataegus rhipidophylla* Gand.) furnish a good tincture against all diseases of the chest. In the East, the mistletoe potion called *guytama* has replaced the expressed juice of the *barsom,* the ritual implement of the Zoroastrian priests.[13] Alchemists and Hermeticists refer to the mistletoe by the names *dabat* and *helle* or *hele.*[14]

Fig. M.11. Mattioli, *Discorsi,* 52.

MOSS, SPHAGNUM

Latin binomial (type sp.): *Sphagnum palustre* L.

French names: mousse, sphaigne, tourbe

Elemental qualities: cold and wet (∇)

Ruling planet: ♄

Zodiacal signatures: ♓ and ♉ (*apud* Haatan and Jollivet-Castelot)

Occult properties: Sphagnum moss has capillary, emollient, and vulnerary properties. Decoctions of it are good for regrowing hair, strengthening the teeth, and stopping bleeding. When harvested from trees bearing Lunar signatures and cooked in wine, the moss becomes diuretic and soporific in action. Alchemists and Hermeticists refer to mosses by the name *serpigo,* literally "creeper," a derivative of the Latin verb *serpere,* "to creep."[15]

[MOUNTAIN ARNICA ⚘ *See* ARNICA, MOUNTAIN]

MUGWORT

Fig. M.12. Fuchs, *Plantarum effigies,* 25.

Latin binomial: *Artemisia vulgaris* L.

French names: armoise, artemise, ceinture de Saint Jean, couronne de Jean-Baptiste, fleur de Saint Jean, herbe de feu, herbe de l'armise, remise, tabac de Saint Pierre*

Elemental qualities: cold and dry (🜃)

Ruling planet: ♂ (*apud* Piobb and Mavéric)

Zodiacal signature: ♍, ♎, and ♏ (*apud* Mavéric)

Occult properties: Mugwort has emmenagogic, diuretic, diaphoretic, nervine, stimulant, and tonic properties. In the Bocage of Normandy, when gathered on the eve of Saint John's day in midsummer (June 24), it is believed to have the power to counteract spells preventing cows from giving their milk for butter. In southern Germany and Bohemia, celebrants of the same midsummer festival throw belts made of mugwort into the honorary fires to garner protection for the coming year.[16] If you wish to go on a strenuous hike or take a long walk without ever becoming tired, carry some blades of mugwort in your hand and weave them into a belt as you hike or walk. Afterward, boil the belt and wash your feet with its water and you will never experience any fatigue or soreness. When cooked in wine and drunk regularly in small doses, it prevents miscarriages. Suffumigations of boiled mugwort in a sitz or hip bath help women expel miscarried fetuses.[17] When worn as an amulet, it protects the bearer from the *aiguillette* and other malignant spells and charms. When hung as a talisman above the door of one's home or buried under its threshold, it protects the home and keeps evil spirits and sorcerers from entering.[18]

*[Sédir mistakenly merges *armoise* (★mugwort) and *herbe de Saint Jean* (★Saint John's wort) in a single entry with the astral signatures ☉, ♈, and ♋. See *Les plantes magiques,* 155. The confusion is due to the fact that the French common name *herbe de Saint Jean* was used for both herbs. Since these are two different species, however, they have been separated here to form two independent entries. That Sédir's astral signatures apply solely to Saint John's wort is confirmed by his colleagues; see especially Mario, "La flore mystérieuse," *La vie mystérieuse,* 31, 106 (☉); 32, 125 (♈); and 34, 150 (♋); and Jollivet-Castelot, *La médecine spagyrique,* 7 (☉), 12 (♋). —*Trans.*]

MULBERRY

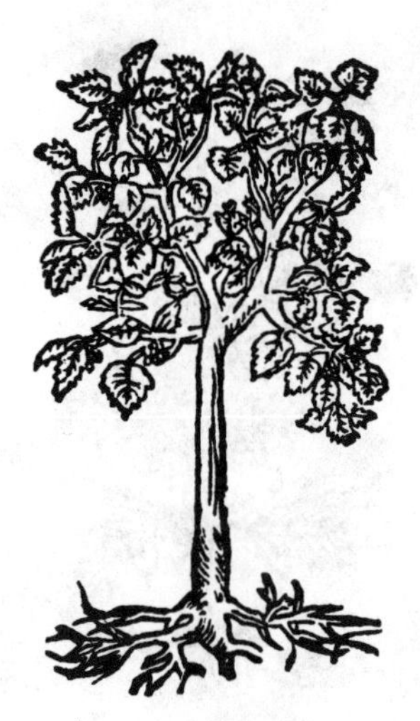

Fig. M.13. Fuchs, *Plantarum effigies,* 297.

Latin binomial (type sp.): *Morus nigra* L.

French names: mûrier, amouriè (Languedocian), mûrier noir, mûrier noir du Moyen Orient

Elemental qualities: cold and dry (🜃)

Ruling planet: ☿

Zodiacal signature: ♑

Occult properties: Mulberry has anti-inflammatory, diuretic, emollient, expectorant, and vermifugal properties. The berries, which start out green but become pink or red while ripening and turn a deep purple or black when fully ripe, are governed by ♃. The red berries are aperitive and purgative in action, whereas the green are good for hyperemia, dysentery, hemoptysis, and stomatitis.

MULLEIN

Fig. M.14. Fuchs, *Plantarum effigies,* 488.

Latin binomial: *Verbascum thapsus* L.

French names: bouillon blanc, brisan (Languedocian), cierge de Notre Dame, herbe de Saint Fiacre, molène, oreille de loup, oreille de Saint Cloud*

Elemental qualities: cold and dry (🜃)

Ruling planets: ♃

Zodiacal signatures: ♍ (especially the leaves) and ♌

Occult properties: Mullein, otherwise known as great mullein, has powerful astringent, emollient, demulcent, pectoral, and spasmolytic properties. When boiled and administered in the form of a poultice, its leaves sooth renal colic. Offerings of this herb are made to Saint Fiacre of Breuil (600–670 CE), the patron saint of gardeners

*[Sédir provides separate entries for *bouillon blanc* and *molène,* which are variant names for mullein. See *Les plantes magiques,* 145, 160. These have been merged here to form a single entry. —*Trans.*]

and medicinal plants.[19] Alchemists and Hermeticists refer to it by the secret name *ploma,* the Latin word for "feather" or "plume."[20]

MUSTARD

Latin binomial (type sp.): *Sinapis alba* L.

French names: moutarde, herbe au beurre, moutarde blanche, moutardin, seneue, sénevé blanc

Elemental qualities: hot and dry (△)

Ruling planet: ♂

Zodiacal signature: ♈ (*apud* Haatan)

Occult properties: Mustard possesses antibacterial, diaphoretic, diuretic, emetic, expectorant, rubefacient, and stimulant properties. The mustard seed, by far the most versatile part of the plant, symbolizes the Christ and omniscience.[21]

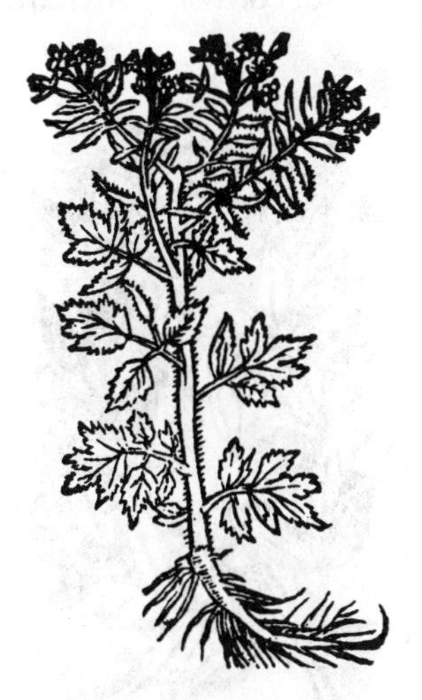

Fig. M.15. Fuchs, *Plantarum effigies,* 307.

MYRRH

Latin binomial: *Commiphora myrrha* (Nees) Engl.

French names: myrrhe, arbre du paradis, mìrro (Languedocian), myrrhier, persil d'asne

Elemental qualities: hot and dry (△)

Ruling planet: ☉ (*apud* Gessmann)

Zodiacal signatures: ♈ and ♒ (*apud* Haatan and Jollivet-Castelot)

Occult properties: Myrrh, the oleo-gum resin of the myrrh tree, has astringent, antiseptic, bechic, emmenagogic, spasmolytic, and vermifugal properties. According to Greek mythology, Myrrha, the mother of Adonis, was transformed into a myrrh tree for tricking her father, Cinyras, into having an incestuous relationship with her. The tree's aromatic, pungent, and woody exudate was thought to originate from Myrrha's tears. The

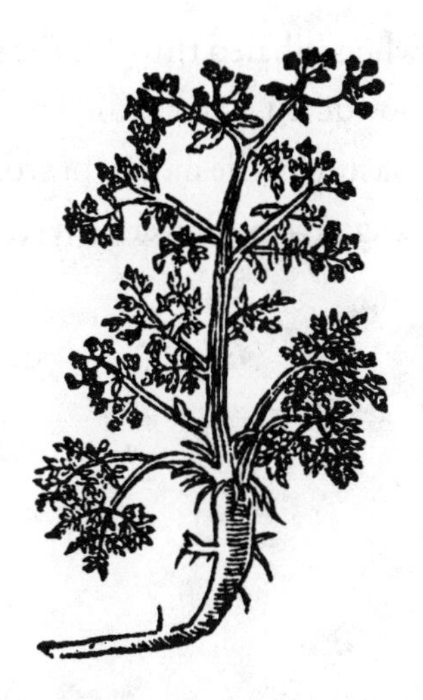

Fig. M.16. Fuchs, *Plantarum effigies,* 298.

Helmontian aphorism "Whoever can make myrrh soluble by the human body holds the secret to prolonging life" was popular among early chemists, pharmacopoeists, and apothecaries. This can be achieved by means of alcohol, which "extracts only the inflammable parts."[22]

Fig. M.17. Mattioli, *Discorsi,* 161.

MYRTLE

Latin binomial (type sp.): *Myrtus communis* L.

French names: myrte, mirto (Languedocian)

Elemental qualities: cold and dry (🜃)

Ruling planet: ♀

Zodiacal signature: ♉

Occult properties: Myrtle possesses aromatic, balsamic, styptic, and tonic properties. It is sacred to Venus and to the ancestral Lares, ancient spirits who, if propitiated, watch over the house or community to which they belong. Its leaves, when woven into crowns, heal cephalic tumors. The vapors of its infusion, when inhaled through the mouth, relieve migraines. Its fruit, when desiccated, powdered, and candied with egg white and then administered in the form of a plaster on the mouth or stomach, cures nausea and vomiting.[23] In the Hermetic language of flowers, myrtle flowers are emblematic of compassion.

N
Needleleaf–Nutmeg

NEEDLELEAF

Fig. N.1. Parkinson, *Theatrum botanicum*, 568.

Latin binomial (type sp.): *Polycnemum arvense* L.

French names: genouillère, herbe à mille nœuds

Elemental qualities: hot and wet (🜁)

Ruling planet: ☿

Zodiacal signature: ♎

Occult properties: Plants of the Polycnemoideae family have diuretic, nervine, and vulnerary properties. Their nonsucculent leaves, when bruised and infused in white wine, make an effective tonic against hallucinations.[1] The French common name *herb à mille nœuds* is the equivalent of the ancient Greek name πολυκνημόν, from which the modern genus name *Polycnemum* derives, meaning "herb with many nodes."

Fig. N.2. Fuchs, *Plantarum effigies,* 58.

NETTLE

Latin binomial (type sp.): *Urtica dioica* L.

French names: ortie, bournigas (Languedocian), grande ortie, ortie dioïque, ortie piquante

Elemental qualities: hot and dry (△)

Ruling planet: ♂

Zodiacal signature: ♌. Harvest the species that is not malodorous—namely, the stinging nettle (*Urtica dioica* L.)—when ♂ is in the east in ♏ or ♑. Harvest the malodorous species—namely, the red dead nettle (*Lamium purpureum* L.)—when the ☉ and ☾ are in ♍ or when the ☉ is in ♉ and the ☾ is in ♊.*

Occult properties: Nettle has astringent, depurative, diuretic, stimulant, and styptic properties. The species that is not malodorous softens tumors and relieves gout and asthma. A nettle placed in the fresh urine of a patient suffering from a life-threatening illness and left for twenty-four hours will indicate, if desiccated, that the patient will die or, if it is still green, that the patient will live.[2] When worn as an amulet, it gives the bearer courage. This "herb of bravery" was sacred to medieval magicians, who gave it the Chaldean name *roybra.*† According to the *Grand Albert,* when held in the hand as a talisman with ⋆yarrow, it cures spectrophobia. To catch fish barehanded, mix its juice with the juice of ⋆bistort,‡ rub your hands

*[The malodorous variety to which Sédir refers, the dead nettle (French: *ortie morte*), so called because it lacks the characteristic trichomes of the stinging nettle that produce a sharp, stinging sensation upon contact with the skin, was otherwise known among the old herbalists as stinking nettle (French: *ortie puante*) because of is fetid odor: *Lamium purpureum* L. is synonymous with *Lamium foetidum* Gilib. Although the stinking nettle is not a true nettle, its herbal actions are strikingly similar to those of the stinging nettle, except that the latter also has rubefacient properties. Most likely Sédir introduces the malodorous species here because the amuletic, talismanic, and ichthyocidal formulae in the *Grand Albert* all involve some form of contact with the skin. —*Trans.*]

†[In Latin and French editions of the *Grand Albert,* the name of Chaldean herb no. 2 is spelled *royb.* Cf. Table 0.1 on p. xxiii. —*Trans.*]

‡[At the end of Sédir's entry for *serpentine,* he refers readers to this entry for additional information. Here, however, Sédir's text reads *serpentaire.* On Sédir's confusion of the French common names *serpentaire* and *serpentine,* both of which may be used to identify the herb bistort,

with the mixture, and then throw the rest into the river.[3] When cooked in wine, its seed is effective against pleurisy and pneumonia. Its bruised leaves administered in a poultice stop gangrene, and a decoction of its seed counters mushroom poisoning. In the Hermetic language of flowers, nettles are emblematic of lust (*cupiditas*).

[NIÇOISE OLIVE ❦ *See* OLIVE, NIÇOISE]

NUTMEG

Fig. N.3. Gerard, *The Herball*, 1353.

Latin binomial: *Myristica fragrans* Houtt.

French names: noix muscade, muscadier, noyer muscade

Elemental qualities: hot and dry (△)

Ruling planets: ☉, ♂, and ♃ (*apud* Mavéric)

Zodiacal signature: ♈

Occult properties: Nutmeg, the spice made from the seeds of the nutmeg tree, and mace, the spice made from the arils, possess aromatic, carminative, digestive, and stimulant properties. The flower, which is even more strongly signed by ♈, facilitates conception. The fruit, when taken on an empty stomach, reduces the effects of drinking too much wine.

(*cont.*) see the entry *bistort and the corresponding footnote. Although the French editions of the *Grand Albert* read *serpentine,* their Latin predecessors read *sempervivum,* the ancient Latin name for *houseleek. —*Trans.*]

O

Oak–Oregano

Fig. O.1. Mattioli, *Discorsi,* 147.

OAK

Latin binomial (type sp.): *Quercus robur* L.

French names: chêne, châgne, chêne à grappe, chêne blanc, chêne pédoncule, gravelin, xaine (Languedocian)

Elemental qualities: cold and dry (🜃)

Ruling planet: ♃

Zodiacal signatures: ♍ and ♉

Occult properties: The oak tree possesses anti-inflammatory, antiseptic, astringent, demulcent, styptic, and tonic properties. It is sacred to Zeus, Jupiter, and Thor. The Druids revered the oak as the Tree of Knowledge. Oaks are both magnetic and attractive and tenacious and hard and hence black or dark in color. In Genesis 35:4–5, Jacob buries the former idols and blasphemies of his household under the oak of Shechem and thereby protects it from the infernal hunger of the wrath of God in the *turba magna.** In the Hermetic language of flowers, oaks are emblematic of strength.

*[Here Sédir reverts back to using Boehmic argot (see chapter 2). Elsewhere he defines the *turba magna* as "any environment delivered unto the ebullition and effervescence of the fire of anger" (Sédir, *De la signature des choses,* 84 n. 1) and, in another work, as "an eighth form that resides in the multiplicity of wills: it is the kingdom of anger or workshop of the devil, and it will be consumed only in the deluge of fire" (Sédir, *Le bienheureux Jacob Boehme,* 36). —*Trans.*]

OAT

Fig. O.2. Fuchs, *Plantarum effigies*, 103.

Latin binomial: *Avena sativa* L.

French names: avoine, avoine byzantine, avoine commune, avoine cultivée, cibado (Languedocian)

Elemental qualities: cold and dry (🜃)

Ruling planets: ☾ and ☉

Zodiacal signature: ♍ (*apud* Mario)

Occult properties: Oat possesses antirheumatic, cardiotonic, emollient, nervine, nutritive, spasmolytic, and stimulant properties. To cure yourself of scabies, roll naked in an oat field, rub your body with fountain water, and then dry yourself in the fresh air beside a tree or hedge and the scabies will dry up.

[OILSEED CAMELLIA ❦ *See* CAMELLIA, OILSEED]

OLD-MAN'S BEARD

Fig. O.3. Parkinson, *Theatrum botanicum*, 382.

Latin binomial: *Clematis vitalba* L.

French names: viorne (des pauvres), aubavis, aubervigne, bildalbo (Languedocian), bois à fumer, bois de pipe, clématite vigne-blanche, cranquillier, herbe aux gueux, vigne de Salomon

Elemental qualities: hot and dry (△)

Ruling planet: ☿

Zodiacal signatures: ♈ and ♋ (*apud* Mavéric)

Occult properties: Old-man's beard, otherwise known as traveler's joy, has astringent, diuretic, rubefacient, and vulnerary properties. Its leaves, when decocted in wine, have the power to cure epilepsy.[1]

Fig. O.4. Mattioli, *Discorsi,* 145.

OLIVE

Latin binomial: *Olea europaea* L.
French names: olivier, oulìou (Languedocian)
Elemental qualities: hot and dry (△)
Ruling planets: ♃ and ☉
Zodiacal signatures: ♈ and ♑ (*apud* Jollivet-Castelot)
Occult properties: The olive tree possesses aperient, astringent, cholagogic, diuretic, emollient, and febrifugal properties. The olive is sacred to Athena and Minerva. Its oil is a powerful condenser of astral light and is frequently used in medicinal magic. Ingesting two fingers of olive oil on an empty stomach prevents inebriation. To relieve migraine headaches, write the name of the goddess Athena (ΑΘΗΝΑ or Ἀθηνᾶ) on an olive leaf and tie it to the head.* In the Hermetic language of flowers, olive tree flowers are emblematic of peace.

*[This remedy derives from the Byzantine agricultural handbook attributed to Cassianus Bassus (*Agricultural Pursuits* 9.1.5), but the practice of inscribing magical names, words, and characters on plant leaves and wearing them as amulets is at least as old as the Greek and Demotic magical papyri, if not much older. Socrates, in fact, employs a similar leaf-and-incantation formula to cure Charmides's headache (Plato, *Charmides* 155b). See further *PGM* I. 262–347 (bay-leaf amulet); *PGM* II. 40 (cinquefoil-leaf amulet); *PGM* IV. 780–88 (persea-leaf amulet); *PGM* VII. 213–14 (olive-leaf amulet); and *PGM* CXIXb. 1–5 (olive-leaf amulet) and *Testament of Solomon* 18.37 (ivy-leaf amulet). —*Trans.*]

OLIVE, NIÇOISE

Fig. O.5. Mattioli, *Discorsi,* 145.

Latin binomial: *Olea europaea* L. (subsp. *europaea* var. Cailletier)

French names: olivette, cailletier, oulibedo (Languedocian)

Elemental qualities: hot and dry (△)

Ruling planets: ♃ and ☉

Zodiacal signature: ♈. Harvest under ♌.

Occult properties: The Niçoise olive comes from France and specifically from the region of Nice. It has many of the same occult properties as the ⋆olive.

ONION

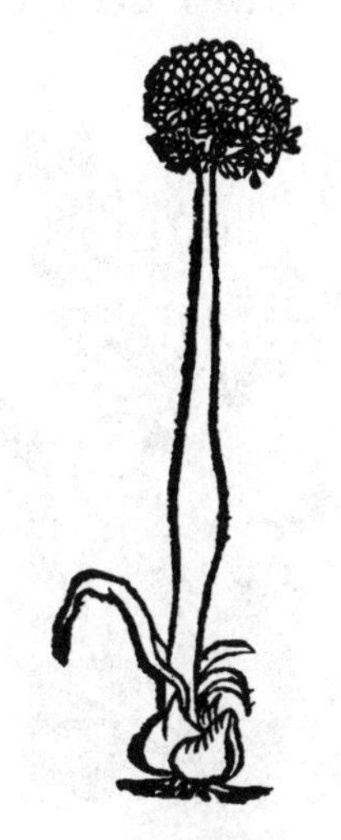

Fig. O.6. Fuchs, *Plantarum effigies,* 242.

Latin binomial: *Allium cepa* L.

French names: oignon, cebo (Languedocian), ognon

Elemental qualities: hot and dry (△)

Ruling planet: ♂

Zodiacal signature: ♐

Occult properties: Onions possess aperitive, anti-inflammatory, stomachic, stimulant, and tonic properties. They become aphrodisiac, diuretic, or emmenagogic in action when consumed with a sympathetic plant. Their corrective is vinegar (♄). For earaches, cook a small onion in the embers of a fire, then place it in a thin cloth with some fresh unsalted butter and apply it to the ear for one minute while it is still warm.[2]

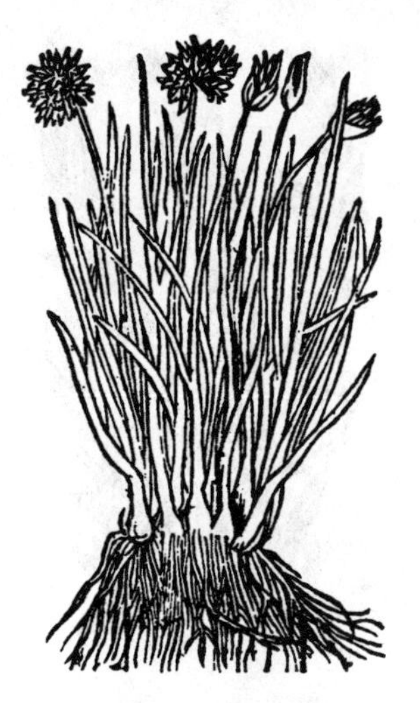

Fig. O.7. Fuchs, *Plantarum effigies,* 365.

ONION, SPRING

Latin binomial: *Allium fistulosum* L.

French names: ciboule, ail fistuleux, chiboule, oignon d'Espagne

Elemental qualities: hot and dry (△)

Ruling planet: ♂

Zodiacal signature: ♏

Occult properties: Spring onion, otherwise known as green onion and scallion, has many of the same occult properties as *onion but is somewhat less active. Its decoction breaks down thick or viscous phlegm and is particularly effective against epilepsy in children.

[OPIUM POPPY ⚕ *See* POPPY, OPIUM]

Fig. O.8. Mattioli, *Discorsi,* 171.

ORANGE

Latin binomial: *Citrus sinensis* (L.) Osbeck

French names: oranger, irangè (Languedocian), orange douce, oranger doux

Elemental qualities: hot and dry (△)

Ruling planet: ☉

Zodiacal signatures: ♌, ♈, and ♐

Occult properties: The orange tree possesses aperitive, carminative, depurative, and stimulant properties. The orange fruit is signed by ♀.[3] Oranges can cure the effects of overeating. To heal metrorrhagia, take seven oranges and cook the rinds in three pints of water, reduce by one-third, and sweeten with twelve spoonfuls of sugar, and then administer the remedy three to four times a day.[4] In the Hermetic language of flowers, orange flowers are emblematic of temperance.

ORCHID

Fig. O.9. Fuchs, *Plantarum effigies,* 316.

Latin binomial (type sp.): *Orchis militaris* L.

French names: satyrion, rognon de prêtre, sabot de Vénus, testicule de prêtre*

Elemental qualities: cold and wet (∇)

Ruling planet: ♀

Zodiacal signature: ♏. Harvest when ♀ is in ♌.

Occult properties: Orchid possesses aphrodisiac, demulcent, and stimulant properties. It has long been celebrated for its aphrodisiac virtue and for its ability to cure male impotence.[5] Its extract also increases fertility. These characteristics prompted the ancient Greeks to classify orchids under the name σατύριον, a diminutive of σάτυρος, "satyr," after the class of lecherous woodland deities. According to Pliny, the aphrodisiac plant described but never identified by Theophrastus was a species of orchid. This plant's aphrodisiac virtue was said to be so great that merely rubbing an ointment made from it on the genitals could enable a person to have sex and climax anywhere from twelve to seventy times in a row.[6] Athanasius Kircher similarly recounts the story of a young man who suffered satyriasis, an uncontrollable sexual desire, every time he walked through a garden filled with orchids.[7]

*[The variant French name Sédir provides, *rognon de prêtre,* meaning "priest's kidney" or "priest's testicle," is quite rare and appears otherwise unattested. Surely this must be a provincial variant of the French common name for the early purple orchid (*Orchis mascula* L.)—namely, *testicule de prêtre.* The common name priest's pintle is also attested in English (see Dodoens, *A New Herball, or Historie of Plants,* 249), and one wonders whether this may be the name behind Shakespeare's "long purples, / That liberal shepherds give a grosser name" (*Hamlet,* act IV, scene vii, lines 168–69). The genus name *Orchis,* in fact, derives from the Greek word ὄρχις, meaning "testicle." —*Trans.*]

Fig. O.10. Fuchs, *Plantarum effigies,* 325.

OREGANO

Latin binomial: *Origanum vulgare* L.

French names: marjolaine bâtarde, majourano salbajo (Languedocian), marjolaine sauvage, origan

Elemental qualities: hot and dry (△)

Ruling planets: ☿ and ☉ (*apud* Mavéric)

Zodiacal signature: ♐ (*apud* Mavéric)

Occult properties: Oregano has antiseptic, carminative, cholagogic, stimulant, sudorific, and stomachic properties. When stripped and reduced to a powder, it drives ants away from wherever it is placed. When worn as an amulet, it is effective against curses, hexes, and bewitchment.[8]

[OX-EYE DAISY ❦ *See* DAISY, OX-EYE]

P

Palm–Purslane

PALM, DATE

Fig. P.1. Mattioli, *Discorsi,* 156.

Latin binomial: *Phoenix dactylifera* L.

French names: palmier, dattier, palmier phoenix*

Elemental qualities: hot and dry (△)

Ruling planet: ☉†

Zodiacal signature: ♐ (*apud* Papus and Jollivet-Castelot)

Occult properties: Date palm possesses aperient, antidiarrheal, demulcent, expectorant, and febrifugal properties. It is sacred to Jupiter. In the Hermetic language of flowers, the palm tree is emblematic of victory and especially

*[In Sédir's original entry, the bizarre variant *Pourkss* (*sic*) appears alongside the names *Tadmor* and *tamar.* See *Les plantes magiques,* 162. This is most likely a typographical error for *Phoenix,* the modern genus name. Antoine Joseph Pernety, Sédir's go-to source for alchemical and Hermetic plant names, notes that "*Phoenix* is also one of the names for the palm tree that bears dates." In another entry, Pernety defines the alchemical name *spatha* as "the bark of the fruit of the palm tree." See Pernety, *Dictionnaire mytho-hermétique,* 382 s.v. *phœnix,* 470 s.v. *spatha.* —*Trans.*]

†[Sédir's placement of the date palm under the exclusive dominion of the ☉ diverges somewhat from the long-standing herbal tradition that assigns it dual planetary rulers—namely the ☉ and the ☾. See, for example, Agrippa, *De occulta philosophia libri tres,* liber I, capita XXIII–XXIIII, xxxi (*palma*), xxxii (*palma arbor*) (English edition: *Three Books of Occult Philosophy,* bk. 1, chaps. 23–24, 93, 98); Lenain, *La science cabalistique,* 136–37; Jollivet-Castelot, *La médecine spagyrique,* 7, 9. Perhaps this is an oversight, but it is interesting to note in support of Sédir's revision that the gematric value of *tamar* (תמר) is 640, the same as that of *shemesh* (שמש), the Hebrew word for the ☉. —*Trans.*]

of mystic triumph. It develops like the latter, from within and without.[1] Tadmor, Palmyra, or "City of Palms," the city built by King Solomon in the wilderness, derives from *tamar* (תמר), the Hebrew name for the date palm.[2]

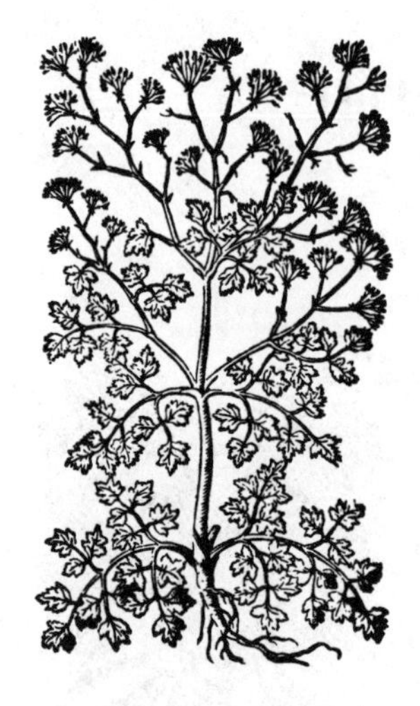

Fig. P.2. Mattioli, *Discorsi,* 455.

PARSLEY

Latin binomial: *Petroselinum crispum* (Mill.) Fuss

French names: persil, ache-persil, jalbert (Languedocian), persin, pétrosine, sersin

Elemental qualities: hot and wet (🜁)

Ruling planets: ♄ and ☉

Zodiacal signature: ♊

Occult properties: Parsley has aperitive, carminative, depurative, diuretic, emmenagogic, stomachic, and tonic properties. Its seeds are subject to the influence of ♋. Rubbing its oleaginous extract on the navel relieves pains caused by bladder and kidney stones. If harvested when the ☉ is in ♉ and administered in lemonade while the ☾ is waning, it becomes cicatrizant, antigout, and purgative in action. If harvested when ♄ is in ♌ and the ☾ is below the horizon, it becomes antiedemic in action. Its infusion restores the menstrual cycle and cures iron-deficiency anemia.

PEACH

Fig. P.3. Mattioli, *Discorsi,* 167.

Latin binomial: *Prunus persica* (L.) Batsch

French names: pêcher, pessiè (Languedocian), reine des vergers

Elemental qualities: cold and dry (🜃)

Ruling planet: ♃

Zodiacal signatures: ♉ and ♎

Occult properties: The peach tree has alterative, aperient, astringent, demulcent, and vermifugal properties. Peach almonds taken on an empty stomach prevent hangovers. A glass of peach-leaf juice has the same effect. When preserved in vinegar with *mint and alum and applied topically on the navel in the form of a poultice, its leaves serve as an infallible vermifuge for children.[3] The peach tree is sacred to Harpocrates, the god of silence. Its fruit has long been thought to resemble the human heart and its leaf the human tongue. In the Hermetic language of flowers, peach blossoms are emblematic of initiatic silence.[4]

PELLITORY-OF-THE-WALL

Fig. P.4. Fuchs, *Plantarum effigies,* 157.

Latin binomial: *Parietaria officinalis* L.

French names: pariétaire, amouroche, casse-pierre, épinard des murailles, herbe à la pierre, herbe aux nonnes, herbe de Sainte Anne, paritoire, perce-mureilles, perce-pierre

Elemental qualities: hot and wet (🜁)

Ruling planets: ☿ and ♄

Zodiacal signatures: ♎ and ♒

Occult properties: Pellitory-of-the-wall, otherwise known as eastern pellitory-of-the-wall, upright pellitory, and lichwort, has aperient, diuretic, and refrigerant properties. Its juice at a dosage of thirty to sixty grams per day is good for inflammatory diseases, edema, and bladder or kidney stones. When administered in the form of a poultice, it mollifies painful tumors and soothes infantile colic. It is also a good herb for consecrating altars and other ritual implements.[5] In the Hermetic language of flowers, it is emblematic of poverty.

Fig. P.5. Fuchs, *Plantarum effigies,* 110.

PENNYROYAL

Latin binomial: *Mentha pulegium* L.

French names: pouliot sauvage, chasse-puces, fénérotet, frétillet, herbe aux puces, herbe de Saint Laurent, menthe pouliot, pouliot, pouliot royal

Elemental qualities: hot and dry (△)

Ruling planet: ♂

Zodiacal signature: ♌ (*apud* Mario)

Occult properties: Pennyroyal has antiseptic, carminative, diaphoretic, emmenagogic, spasmolytic, and stimulant properties. Historically, it is perhaps best known as an abortifacient. It is sacred to the goddess Demeter, who instead of red wine preferred drinking *kykeon,* the sacred brew of the Eleusinian mysteries made of a mixture of water, *barley, and pennyroyal.[6]

Fig. P.6. Fuchs, *Plantarum effigies,* 113.

PEONY

Latin binomial: *Paeonia mascula* (L.) Mill.

French names: pivoine, boule de feu, herbe aux sourciers, penolho (Languedocian), pione, rose bénite, rose de Notre Dame

Elemental qualities: hot, dry, and moderately wet (△)

Ruling planets: ♃ and ☉

Zodiacal signatures: ♈ and ♋. Harvest when the ☉ and ☾ are in ♋. The flower and especially the calyx are subject to the influences of ♈.

Occult properties: Peony has bechic, spasmolytic, and tonic properties. Its distilled water, which should be administered when the ☾, ♂, and ♃ are in ♋, is effective against epilepsy and menstrual irregularity. To cure epilepsy in small children, it suffices to collect the first seeds borne by a young plant and suspend them from a child's neck and then to administer the seeds in the form of a decoction. The same remedy relieves headaches as well as pains in childbirth. When worn as an amulet, it protects the bearer from evil spells and panic attacks. The Greek plant name παιωνία derives from Paean (Παιών in Greek), the physician of the gods.

PEPPER, BLACK

Fig. P.7. Mattioli, *Discorsi*, 351.

Latin binomial: *Piper nigrum* L.

French names: poivrier, pebre (Languedocian), poivier noire, poivre noir

Elemental qualities: hot and dry (△)

Ruling planets: ♂ and ☉

Zodiacal signature: ♌

Occult properties: Black pepper has diaphoretic, calefacient, expectorant, febrifugal, rubefacient, and stimulant properties (see Table 6.1 on p. 58). Its spicy, pungent aroma serves as a potent Martian perfume.[7]

PEPPERMINT

Fig. P.8. Fuchs, *Plantarum effigies*, 164.

Latin binomial: *Mentha piperita* L.

French names: menthe noire, menthe anglaise, menthe poivrée, mento (Languedocian)

Elemental qualities: hot and wet (🜁)

Ruling planet: ☉ (*apud* Mavéric)

Zodiacal signature: ♎

Occult properties: Peppermint has carminative, stomachic, spasmolytic, and stimulant properties. This aromatic perennial herb of the mint family (Lamiaceae or Labiatae) has a strong and sweet odor, a warm and pungent flavor, and a cooling aftertaste that energizes the nervous and digestive systems, making it an ideal ingredient for herbal cordials.[8]

Fig. P.9. Fuchs, *Plantarum effigies,* 203.

PERIWINKLE

Latin binomial (type sp.): *Vinca minor* L.

French names: pervenche, bergère, herbe au pucelage, petite pucelage, proubenco (Languedocian), provence, violette des sorciers

Elemental qualities: cold and dry (🜃)

Ruling planet: ♀ (*apud* Piobb)

Zodiacal signature: ♉

Occult properties: Periwinkle has astringent, bechic, stomachic, and vulnerary properties. Its extract is highly effective as a mouthwash in the treatment of gingivitis, sore throats, and mouth ulcers.[9] When properly magnetized, its distilled water can prove the faithfulness of one spouse to the other. This "herb of fidelity" was sacred to medieval magicians, who gave it the Chaldean name *herisi.** According to the *Grand Albert,* when reduced to a powder combined with earthworms, it incites lust in whoever consumes it with meat. If mixed with sulfur and thrown into a pond, it will kill all the fish. The same mixture, if fed to a buffalo, will cause its immediate demise and, if thrown into a fire, will turn the flames blue.[10]

*[The variant Sédir provides here for Chaldean herb no. 5 appears otherwise unattested. Most Latin editions of the *Grand Albert* read *iterisi,* whereas most French editions offer two variants, *vetisi* and *iterisi.* Cf. Table 0.1 on p. xxiii. —*Trans.*]

PEYOTE

Fig. P.10. Hooker, "Echinocactus Williamsii," Tab. 4296.

Latin binomial: *Lophophora williamsii* (Lem.) J.M.Coult.

French names: mescal, peyote

Elemental qualities: hot and dry (△)

Ruling planet: ☿

Zodiacal signature: ♈

Occult properties: Peyote, from *peyotl,* the Aztec name for the small cactus formerly classified as *Anhalonium lewinii* Henn., has narcotic, stimulant, and tonic properties. Some Native Americans in Texas and New Mexico masticate the desiccated, disk-shaped buttons of the cactus crown to induce mystical, hallucinatory, and psychic visions.

PIMPERNEL

Fig. P.11. Fuchs, *Plantarum effigies,* 9.

Latin binomial (type sp.): *Anagallis arvensis* L.

French names: pimprenelle, mouron des champs, mourrelou (Languedocian)

Elemental qualities: hot and wet (🜁)

Ruling planet: ☿ (*apud* Lenain and Jollivet-Castelot)

Zodiacal signatures: ♊ and ♍

Occult properties: Pimpernel possesses alexitary, astringent, diaphoretic, styptic, and vulnerary properties. When worn as an amulet attached to the collar, it protects the bearer from psychic attacks and malicious spells. A decoction of its juice kills flies and other noxious insects. Pimpernel leaves can be chewed raw during flu season to avoid contracting infectious diseases.[11]

Fig. P.12. Mattioli, *Discorsi,* 81.

PINE

Latin binomial (type sp.): *Pinus sylvestris* L.

French names: pin, daille, dèle, pin de mâture, pin du Nord, pin pectiné, pin silvestre, pinasse

Elemental qualities: cold and wet (🜄)

Ruling planet: ♄

Zodiacal signature: ♋

Occult properties: The pine tree possesses antiseptic, antirheumatic, balsamic, diuretic, and expectorant properties. It is sacred to Cybele and Pan. Pines are one of the oldest species of tree on Earth, and a pine wood is considered by many to be the liveliest cathedral of all for nature worship.[12] Pinecones, which are signed by ♈, can be used to reveal a person's mystical number. To do this, rise early in the morning, and after performing purification rituals, enter a wood of pines. As soon as the disk of the ☉ appears on the horizon, start walking in as large a circle as possible so that you return to your point of departure the moment the entire sphere of the ☉ is visible in the sky. The number of pinecones seen on the ground during this circular perambulation will reveal the walker's mystical number or, if this number is already known, the number governing the particular question or event for which the operation was performed. Alchemists and Hermeticists refer to pines by the names *peuce,* from the Greek name πεύκα, and *pitys,* from the Greek name πίτυς.[13]

Fig. P.13. Linocier, *L'histoire des plantes,* 54.

PLANE TREE

Latin binomial (type sp.): *Platanus orientalis* L.

French names: platane, plane, platane d'Orient, plataniè (Languedocian)

Elemental qualities: moderately hot and wet (🜁)

Ruling planet: ♃

Zodiacal signatures: ♐, ♓, and ♈

Occult properties: The plane tree possesses anti-inflammatory, astringent, ophthalmic, and vulnerary properties. In decoctions of wine, its leaves are effective against inflammations of the eyes and edema, and its roots against dysentery.[14] These trees devote themselves to the Higher Genius of whoever plants them.

PLANTAIN

Fig. P.14. Fuchs, *Plantarum effigies,* 21.

Latin binomial (type sp.): *Plantago major* L.

French names: plantain, grand plantain, plantage (Languedocian), plantain à bouquet, plantain à larges feuilles, plantain des oiseaux*

Elemental qualities: hot, dry, and moderately wet (△)

Ruling planet: ☉

Zodiacal signatures: ♈ and ♌. Harvest when the ☉ and ☾ are in ♓ or when the ☉ is in ♓ and the ☾ is in ♋.

Occult properties: Plantain possesses anodyne, anti-inflammatory, aperient, astringent, bechic, diuretic, febrifugal, styptic, and vulnerary properties. The pharmacological applications of this versatile herb are many and various. Its roots are good for treating migraines, ulcers, and metrorrhagia.[15] When worn as an amulet or hung as a talisman, the whole plant may be used to counteract evil spells.† Its leaves, when crushed and administered in poultices, mollify ulcers. Its seeds, when crushed in wine, and its leaves, when preserved in vinegar, eliminate dysentery. When eaten raw after a period of abstention from alcohol, it cures edema. Its root infused in wine also serves as a valuable antidote to opium.[16] Among the ancient

*[The French common name *plantain des oiseaux* (bird plantain), used for the type species of the genus *Plantago,* could possibly explain the origin of the enigmatic names *ornoglosse* and *langue d'oiseau* (bird-tongue) that Papus gives to ★spurge. (see Papus, *Traité élémentaire de magie pratique,* 236, 547 s.v. *ornoglosse*). If ornoglosse is not simply a typographical error derived from a French version of the *Grand Albert,* Papus may have taken the grecized form *orno-* (Greek ὄρνις or ὄρνεον) from *oiseau* (bird) in the French common name for plantain and combined it with the gallicized form *-glosse* (Greek *γλῶσσα*) from its ancient Greek common name ἀρνόγλωσσον (or its gallicized rendering *arnoglosse*). Why exactly Papus uses this as a secret name for spurge, however, remains something of a mystery—likely this is due to the fact that doves and pigeons are very fond of consuming its poisonous seeds, which nearly all other animals refuse to eat. See further Table 0.2 on p. xxvi and the commentary in the foreword. *—Trans.*]

†[According to Dioscorides, when worn as an amulet, plantain can also dissolve scrofulous tumors (see Dioscorides, *On Medical Materials* 2.126). *—Trans.*]

Greeks, it was known by the names ἀρνόγλωσσον, or "lamb's-tongue," and πολύνευρον, or "many-ribbed herb."[17]

Fig. P.15. Fuchs, *Plantarum effigies,* 228.

PLUM

Latin binomial: *Prunus domestica* L.

French names: prunier, pruniè (Languedocian), prunier domestique

Elemental qualities: moderately cold and dry (🜃)

Ruling planet: ♃

Zodiacal signature: ♍

Occult properties: The plum tree, and especially prunes, its dried fruit, have aperient, stomachic, and febrifugal properties. Its wood is signed by ♏.

Fig. P.16. Fuchs, *Plantarum effigies,* 338.

POLYPODY

Latin binomial (type sp.): *Polypodium vulgare* L.*

French names: polypode, alencados (Languedocian), herbe de gagne, millepieds, réglisse des bois, réglisse sauvage

Elemental qualities: cold and dry (🜃)

Ruling planets: ♄, ♀, and ☾

Zodiacal signature: ♉ (*apud* Mario)

Occult properties: Polypody possesses aperient, cholagogic, pectoral, purgative, and tonic properties. Its powdered roots are effective against polyps of the nose and quartan fever. When burned in suffumigations, it prevents nightmares.

---

*[Sédir's entry appears under the genus name *Polypodium,* from the ancient Greek name πολυπόδιον, meaning "herb with many tiny feet" (from πολύ, "many," + πόδιον, "little feet," a diminutive of πούς), so called because of the foot-like appearance of its rhizome and branches. See *Les plantes magiques,* 164. —*Trans.*]

POMEGRANATE

Fig. P.17. Mattioli, *Discorsi,* 159.

Latin binomial (type sp.): *Punica granatum* L.

French names: grenadier, balaustier, grenadier sauvage, milgraniè (Languedocian)

Elemental qualities: cold and wet (∇)

Ruling planet: ♃

Zodiacal signature: ♓

Occult properties: Pomegranate possesses astringent, cardiotonic, demulcent, emmenagogic, stomachic, and refrigerant properties. The fruit is governed by the ☉ and signed by ♈.[18] Its juice is a natural blood purifier and is both antioxidant and depurative in action.

POPLAR

Fig. P.18. Mattioli, *Discorsi,* 177.

Latin binomial (type sp.): *Populus alba* L.

French names: peuplier, peuplier blanc, pìboul (Languedocian), ypréau

Elemental qualities: cold and wet (∇)

Ruling planet: ♃

Zodiacal signatures: ♓ and ♐ (*apud* Jollivet-Castelot)

Occult properties: Poplar possesses anodyne, anti-inflammatory, astringent, emmenagogic, diuretic, and tonic properties. Homer describes a sacred grove of poplars in the underworld along the shores of the river Acheron (Ἀχέρων in Greek), for which reason Homer names the tree ἀχερωίς, or "Acheron-tree."[19] The poplar is sacred to Hercules, who is credited with bringing the tree up from the underworld and killing the fire-breathing giant Cacus in a cavern on Aventine Hill, which was once forested with poplars. The soporific poplar ointment known as *unguentum populeum* or *populeon,* whose base ingredient is the juice of the leaves and young shoots of various poplar species, procures pleasant dreams and prevents nightmares.[20]

Fig. P.19. Fuchs, *Plantarum effigies,* 293.

POPPY, COMMON

Latin binomial: *Papaver rhoeas* L.

French names: coquelicot, chaudière d'enfer, coprose, coqueliquot, mahon, ponceau, pavot-coq, pavot des champs, pavot rouge, rousèlo (Languedocian)

Elemental qualities: cold and wet (∇)

Ruling planet: ☾

Zodiacal signature: ♓

Occult properties: The common poppy has anodyne, emollient, emmenagogic, and hypnotic properties. Since this herb is highly phlegmatic, it is good to correct it with liquids of the ☉ or ☿. Once corrected, it becomes refreshing and anesthetic in action. Its juice and powdered flower cure pleurisy, and its distilled water heals erysipelas of the head.

Fig. P.20. Fuchs, *Plantarum effigies,* 296.

POPPY, HORNED

Latin binomial: *Glaucium flavum* Crantz

French names: pavot cornu, corblet, glaucie, glaucier jaune, pavot jaune

Elemental qualities: cold and wet (∇)

Ruling planet: ♄

Zodiacal signature: ♊

Occult properties: The horned poppy has bechic, cholagogic, hepatic, and pectoral properties. According to Pliny, when administered with honeyed wine at a dose of one lot, its seeds act as a good purgative. Its main alkaloid component is glaucine, which is anti-inflammatory, hallucinogenic, and sedative in action.[21]

POPPY, OPIUM

Fig. P.21. Fuchs, *Plantarum effigies,* 295.

Latin binomial: *Papaver somniferum* L.

French names: pavot, mecon, œilette, pabot (Languedocian), pavot à opium, pavot somnifère

Elemental qualities: hot and dry (△)

Ruling planets: ♄ and ☾

Zodiacal signatures: ♈, ♌, and ♑ (*apud* Haatan and Jollivet-Castelot)

Occult properties: The opium poppy has anodyne, astringent, narcotic, sedative, and soporific properties (see Table 6.1 on p. 58). Its flowers are governed by ♄ and signed by ♈. In the Hermetic language of flowers, the opium poppy is emblematic of laziness. Sprinkling its juice around your house will rid your home of flies and other noxious insects.[22]

PRIVET

Fig. P.22. Fuchs, *Plantarum effigies,* 272.

Latin binomial (type sp.): *Ligustrum vulgare* L.

French names: troène, bouis salbage (Languedocian), frezillon, liguistre, trougne, trouille

Elemental qualities: hot and dry (△)

Ruling planet: ☿ (*apud* Piobb)

Zodiacal signature: ♈. Harvest under ♌.

Occult properties: Privet possesses astringent, detersive, and vulnerary properties. In decoctions, it is particularly effective against sore throats, mouth ulcers, stomatitis, and chronic engorgement of the tonsils.[23]

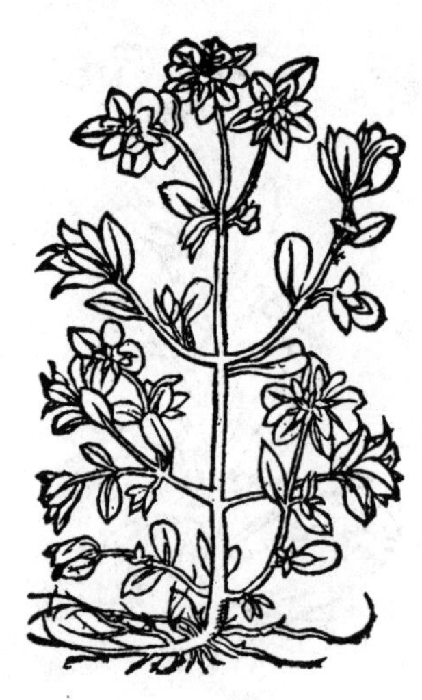

Fig. P.23. Fuchs, *Plantarum effigies,* 62.

PURSLANE

Latin binomial (type sp.): *Portulaca oleracea* L.

French names: pourpier, bourdoulaigo (Languedocian), porcelane, pourcelaine, pourcelane, pourpier maraîcher

Elemental qualities: cold and wet (∇)

Ruling planet: ☾ (*apud* Piobb and Jollivet-Castelot)

Zodiacal signatures: ♋, ♎, and ♓

Occult properties: Purslane possesses antiscorbutic, depurative, diuretic, febrifugal, tonic, and vermifugal properties. This annual succulent not only cures but also prevents hangovers. Suffumigations of its seeds have the same virtue as decoctions of ⋆lily pollen.[24] When placed in someone's bed, purslane will prevent whoever sleeps in it from dreaming. Its juice, when mixed with wine, serves as an antidote to ⋆henbane. Its seeds, when crushed and mixed with good honey, relieve asthma symptoms.[25]

Q

Quince

QUINCE

Fig. Q.1. Fuchs, *Plantarum effigies,* 213.

Latin binomial (type sp.): *Cydonia oblonga* Mill.

French names: cognassier, coignacier, coignier, coudonnier, coudougney

Elemental qualities: cold and dry (🜃)

Ruling planet: ♃

Zodiacal signature: ♑

Occult properties: Quinces possess aperient, anti-inflammatory, astringent, diuretic, emollient, and refrigerant properties. Their juice is particularly effective in the treatment of chronic diarrhea, dysentery, bleeding hemorroids, and irritable bowel syndrome.[1] The quince tree is sacred to Juno, the goddess of marriage, who is frequently depicted wearing a crown of quince leaves.

R

Raisin Tree–Rue

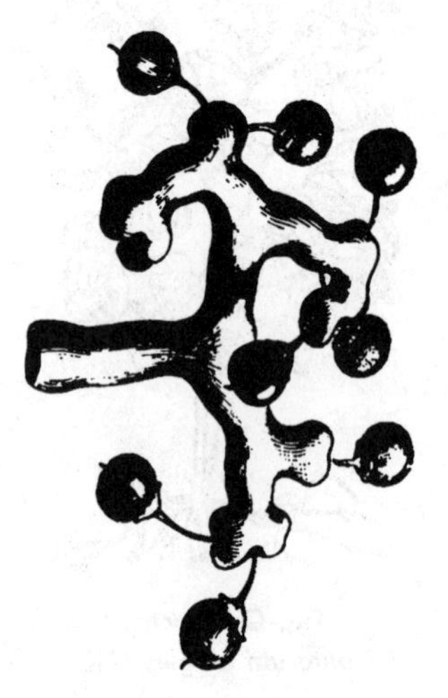

Fig. R.1. Baillon, *Cours élémentaire*, 272.

RAISIN TREE, JAPANESE

Latin binomial: *Hovenia dulcis* Thunb.

French name: raisin de Chine, raisinier de Chine

Elemental qualities: cold and dry (🜃)

Ruling planet: ♄

Zodiacal signature: ♍

Occult properties: The Japanese raisin tree has aperient, febrifugal, diuretic, and spasmolytic properties. When dried and powdered, its fruit is especially good for treating hemorrhages and dysentery.

RAMPION

Fig. R.2. Mattioli, *Discorsi*, 656.

Latin binomial (type sp.): *Phyteuma spicatum* L.

French names: raiponce, aripounxou (Languedocian), raiponce en épi

Elemental qualities: cold and wet (∇)

Ruling planet: ☾ (*apud* Piobb)

Zodiacal signatures: ♋ and, if harvested at the end of October, ♏

Occult properties: Rampion possesses aperient, anti-inflammatory, nutritive, and refrigerant properties. Its leaves are diuretic in action and good for opening obstructions in the liver and gallbladder.[1] The genus name *Phyteuma* derives from the Greek noun φύτευμα, meaning "that which is planted." Rampion is easily cultivated, and its leaves and roots are especially good in salads. Devotees of the ancient temple of Apollo at Delphi not only esteemed rampion as nature's most nourishing plant food, but they also carried it on golden plates in funeral processions as a symbol for the daily "death," or setting, of the ☉.[2]

[RED CABBAGE ⚕ *See* CABBAGE, RED]

[RED SANDALWOOD ⚕ *See* SANDALWOOD, RED]

Fig. R.3. Mattioli, *Discorsi,* 116.

REED

Latin binomial (type sp.): *Arundo donax* L.

French names: roseau, canne blanche, canne de Provence, cannevelle, carabeno (Languedocian), grand roseau, roseau à quenouille

Elemental qualities: cold and wet (🜄)

Ruling planet: ☾

Zodiacal signature: ♓ (*apud* Mario)

Occult properties: Reed roots possess diaphoretic, diuretic, emollient, emmenagogic, and galactofugal properties. To heal dislocations, take two reeds, fit them one into the other, and carry them around as an amulet—this remedy requires only that the bearer of the amulet possess a steadfast will. Alchemists and Hermeticists refer to the reed by the name *kanech,* a medieval transliteration of the ancient Hebrew name קנה, or *qaneh.*[3]

Fig. R.4. Fuchs, *Plantarum effigies,* 35.

RESTHARROW

Latin binomial: *Ononis repens* L.

French names: arrête-bœuf, agabousses (Languedocian), bougrande, bougrane, bougrane rampante, chaupoint, herbe aux ânes, tabouret du diable, tandon, tenon*

Elemental qualities: hot and dry (🜂)

Ruling planets: ♂ and ♃. Harvest under the conjunction of these two planets in the patient's tenth house.

Zodiacal signature: ♌

Occult properties: Restharrow has aperient,

*[Sédir's entry also includes the old pharmaceutical name *remora aratri,* literally "plow-hinderer," which is essentially a Latin equivalent of the English common name (*rest* + *harrow*). See *Les plantes magiques,* 144. Both of these names, much like the French common name *arrête-bœuf* (literally "stop-ox"), originate from the plant's hard, woody roots, which would frequently hinder the progress of horse- or ox-drawn plows and harrows. —*Trans.*]

bechic, detersive, and lithontriptic properties. It is also particularly effective in the treatment of pleurisy and chest pains. When worn as an amulet or hung as a talisman, it becomes apotropaic in action and provides protection against threats of war, theft, and internecine strife.

RHUBARB

Fig. R.5. Mattioli, *Discorsi*, 386.

Latin binomial (type sp.): *Rheum rhaponticum* L.

French names: rhubarbe, rhapontic, rhubarbe sauvage

Elemental qualities: hot and wet (🜁)

Ruling planets: ♃ and ♄

Zodiacal signature: ♊

Occult properties: Rhubarb possesses aperient, astringent, digestive, purgative, and tonic properties. Its roots are particularly effective in the treatment of jaundice. Alchemists and Hermeticists refer to the rhubarb by the secret names *ramed, raved,* and *zipar.*[4]

ROSE

Fig. R.6. Fuchs, *Plantarum effigies,* 377.

Latin binomial (type sp.): *Rosa majalis* Herrm.

French names: rosier, rose d'amour, rose de mai, rose de Pâques, rose du Saint Sacrement, rosier cannelle, rosier de mai, rousiè (Languedocian)

Elemental qualities: cold and dry (🜃)

Ruling planets: ♀ and ♃

Zodiacal signature: ♉

Occult properties: The rosebush possesses anodyne, anti-inflammatory, calmative, cardiotonic, refrigerant, and spasmolytic properties. Rose syrups and infusions, which the alchemists call *mucarum* or *mucharum,* help facilitate conception.[5] The distilled water of rose petals is good for all venereal diseases. It is also possible to compose a rose perfume or liquor that will ready the intellective soul for celestial revelations. This "herb of initiation" was sacred to medieval magicians, who gave it the Chaldean name *eglerisa.*[6] According to the *Grand Albert,* if hung from a tree, a talisman made from a rose hip or seed, a ⋆mustard seed, and the foot of a weasel will cause the tree to become sterile and fruitless. The same talisman placed at the foot of a dying ⋆cabbage will regreen its leaves within half a day. When burned in a lamp, roses can induce visions and reveal the presence of evil spirits.[7] In the Hermetic language of flowers, roses are emblematic of love, patience, initiation, and the Virgin Mother. The rose is one of the twelve sacred plants of the Rosicrucians.[8]

Fig. R.7. Gerard, *The Herball,* 1087.

ROSE, DOG

Latin binomial: *Rosa canina* L.

French names: églantier, agulancier, cynorrhodon, garrabiè (Languedocian), gratte-cul, rose des haies, rose sauvage, rosier des chiens

Elemental qualities: cold and dry (🜃)

Ruling planet: ☿ (*apud* Piobb)

Zodiacal signature: ♍

Occult properties: The petals and hips of the dog rose have aperient, astringent, carminative, diuretic, and tonic properties. In antiquity, its decoction was used to treat bites from rabid dogs, for which reason the ancient Greeks named it κυνόροδον, a compound of the nouns κύνος, "dog," and ῥόδον, "rose."[9]

ROSEMARY

Fig. R.8. Fuchs, *Plantarum effigies,* 271.

Latin binomial: *Rosmarinus officinalis* L.

French names: romarin, encensier, herbe aux couronnes, herbe aux troubadours, rose marine, rosmarin, roumanì (Languedocian)

Elemental qualities: hot and dry (△)

Ruling planets: ☉ and ♃

Zodiacal signature: ♈. Harvest at the beginning of April or under ♏.

Occult properties: Rosemary has astringent, carminative, diaphoretic, nervine, stimulant, and stomachic properties. This versatile plant is most commonly used to treat headaches, head colds, and nervous tension, but it also serves as an effective detersive against leprosy, syphilis, and sores. In Roman antiquity, it was the sacred herb of the ancestral Lares, astral spirits who, when propitiated, watched over the house or community to which they belonged. Alchemists and Hermeticists refer to it by the name *libanotis,* a latinization of the Greek plant name λιβανωτίς, and Paracelsus calls its flower *anthos,* from the Greek word ἄνθος, meaning "blossom," "flower," or "brilliance."[10] The oil of its flowers is white, transparent, aromatic, and vulnerary in action. The water of its flowers is the principal ingredient in the famous panacea known as Queen of Hungary's Water.* When boiled in white wine and made into lotions, its flowers rejuvenate the complexion and, when gargled, perfume the breath.

*[Queen of Hungary's Water, or simply Hungary Water, is a distilled liquor consisting of a spirit of wine infused with the most essential parts of rosemary flowers. It is often considered to be the first alcohol-based perfume in Europe. It was not until the seventeenth century that Hungary Waters began to be used medicinally. For two good recipes, one including a mixture of other herbs such as ⋆thyme, ⋆summer savory, ⋆lavender, ⋆costmary, ⋆sage, and ⋆marjoram, see especially Lémery, *Cours de chymie,* 541–43 (English edition: *A Course of Chymistry,* 403–4). Lémery recommends administering Hungary Water internally for palpitations, fainting, palsy, lethargy, apoplexy, and hysteria and externally for burns, contusions, tumors, and general skin health. —*Trans.*]

Fig. R.9. Bailey, *Standard Cyclopedia,* 1:282.

ROSEMARY, BOG

Latin binomial: *Andromeda polifolia* L.

French names: andromède, andromède à feuilles de polium

Elemental qualities: hot and dry (△)

Ruling planet: ♄

Zodiacal signature: ♈

Occult properties: Bog rosemary is a small bush with pinkish flowers and evergreen, acrid-tasting leaves. It contains a virulent poison that causes vertigo, arrest of blood circulation, and pulmonary paralysis. This narcotic toxin is extremely dangerous for livestock, and sheep especially. Its leaves and stems, which are rich in tannin, are sometimes employed as a mordant in the place of oak galls to give silks a brilliant black sheen.[11]

Fig. R.10. Pomet, *Histoire générale des drogues,* 39.

ROSE OF JERICHO

Latin binomial: *Anastatica hierochuntica* L.

French names: rose de Jéricho, anastatique, main de Marie, plante de la résurrection

Occult properties: The rose of Jericho has the same elemental, astral, and occult properties as the ⋆rose but with an action particular to ♄ in ♋. If a pregnant woman places it in water and it reopens perfectly, this means she will experience a painless labor.[12]

RUE, WILD

Fig. R.11. Fuchs, *Plantarum effigies,* 354.

Latin binomial: *Peganum harmala* L.

French names: rue sauvage, pégane, rue de Syrie, rue verte

Elemental qualities: hot and moderately dry (△)

Ruling planets: ♄, ♂, and ☉

Zodiacal signatures: ♊, ♎, and ♐

Occult properties: Wild rue, otherwise known as Syrian rue, is an extremely versatile herb, having abortifacient, alterative, aphrodisiac, diuretic, emmenagogic, hallucinogenic, and narcotic properties. When ground with ★sage in vinegar, it can be put to use as a cure for quartan fever or even as a parasiticide for chlorotic plants.* According to Pliny, sprinkling a room with a decoction of wild rue mixed with mare's urine exterminates fleas and other noxious insects.[13] It is believed by many to be the magical herb with white flowers and black roots called moly (μῶλυ in Greek), which Hermes gave to Odysseus to protect him against the magical potion of Circe.[14] If gathered when the influence of ♄ is weak and the ☉ is in the tenth house and then worn as an amulet, it will protect the bearer against magical spells. A blade of wild rue tied under the wing of a hen will similarly protect her from foxes and cats. Alchemists often refer to its seed by the Arabic name *harmel* or *harmal.*[15]

*[Chlorosis is the yellowing or whitening of normally green plant tissues due to a decreased production of chlorophyll. Feeding insects are only one of many potential causes that can lead to plant chlorosis. —*Trans.*]

S
Saffron–Strawberry

Fig. S.1. Fuchs, *Plantarum effigies,* 249.

SAFFRON

Latin binomial: *Crocus sativus* L.

French names: safran, crocus, safrà (Languedocian), safran d'automne, safran d'Orient, safran médicinal

Elemental qualities: hot and dry (△)

Ruling planet: ☉

Zodiacal signatures: ♌ and ♐. Harvest when the ☉ is in ♓ and the ☾ is in ♋.

Occult properties: Saffron, the spice derived from the stigmas and styles of saffron flowers, has anodyne, carminative, diaphoretic, emmenagogic, and stimulant properties. The magi gave it the name *sanguis Herculis,* meaning "blood of Hercules."[1]

SAFFRON, MEADOW

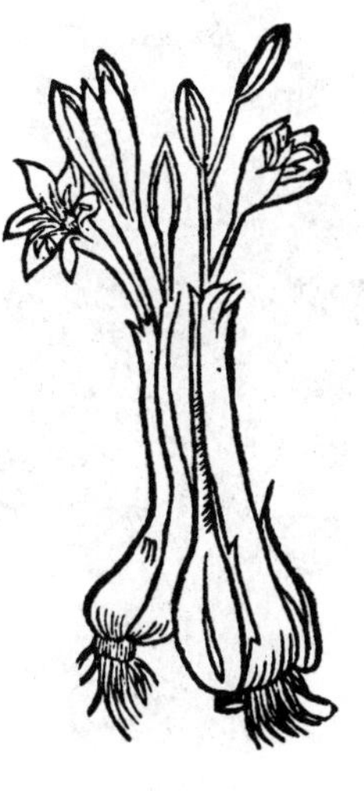

Fig. S.2. Fuchs, *Plantarum effigies,* 201.

Latin binomial: *Colchicum autumnale* L.

French names: colchique d'automne, dame nue, mort chien, safran bâtard, safran des pays, tue-chien, veilleuse*

Elemental qualities: moderately hot and wet (🜁)

Ruling planet: ♃

Zodiacal signature: ♓

Occult properties: Meadow saffron, otherwise known as autumn crocus, has antirheumatic, cathartic, and emetic properties. Its bulb is high in colchicine and strongly diuretic in action, which makes it particularly effective in the treatment of gout.[2] It was the active ingredient in the French regimental officer Nicolas Husson's celebrated *l'Eau médicinale,* or "Medicinal Water." This nostrum, or "secret remedy," for gout, which Husson discovered sometime around the year 1780, was later recognized to be nothing more than a meadow saffron extract. The eminent Alexandrian physicians Oribasius and Aëtius of Amida refer to this herb by the name ἑρμοδάκτυλον, or "Hermes's-finger."[3]

*[Sédir's entry also includes the old pharmaceutical name *diacentauréon de Cælius*—that is, the *diacentaureon* of the Roman physician Caelius Aurelianus. See *Les plantes magiques,* 149. However, as the name of this medicament indicates, the *diacentaureon* was a "[remedy] via centaury" (Greek διά + Latin *centaureum*) and had nothing to do with the meadow saffron and, by extension, nothing to do with the Duke of Portland's remedy for gout. This error appears to be a misreading of Blavatsky, *Isis Unveiled,* 89, where she rehashes in quick succession a variety of the nostrums and specifics described in J. S. Forsyth's *Demonologia.* The information concerning *diacentaureon* and the Duke of Portland's antigout has been moved to its proper place under ★centaury. —*Trans.*]

Fig. S.3. Fuchs, *Plantarum effigies,* 138.

SAGE

Latin binomial: *Salvia officinalis* L.

French names: sauge officinale, herbe sacrée, salbio (Languedocian), sale, sauge, serve, thé de Grèce

Elemental qualities: hot and dry (△)

Ruling planet: ☉

Zodiacal signature: ♈

Occult properties: Sage has aperitive, astringent, cholagogic, stimulant, and tonic properties. The etymology of this plant name is related to the German words *Sol Heil,* meaning "Hail, Sun."[4] Its leaves possess vulnerary properties. Its arcanum and spagyric extract are revitalizing and regenerative in action. Its seeds, which the alchemists call *ebel,* help facilitate conception. This "herb of life" was sacred to medieval magicians, who gave it the Chaldean name *coloricon.** Sage also has extremely powerful apotropaic and exorcistic properties. According to the *Grand Albert,* suffumigations of this Solar herb guard against thunder and lightning and drive away evil spirits.[5]

*[Latin and French editions of the *Grand Albert* both give the variants *colorio* and *coloricon* as names for Chaldean herb no. 12. Cf. Table 0.1 on p. xxiii. —*Trans.*]

SAINT JOHN'S WORT

Fig. S.4. Fuchs, *Plantarum effigies,* 42.

Latin binomial: *Hypericum perforatum* L.

French names: herbe de Saint Jean, chasse diable, herbe à mille trous, herbe aux piqûres, herbe percée, millepertuis, trescalan (Languedocian), trucheron jaune*

Elemental qualities: hot, dry, and moderately wet (△)

Ruling planet: ☉

Zodiacal signatures: ♈ and ♋. Harvest when the ☉ is in ♋ or ♌ and in a good aspect to ♃ or, more simply, on a Friday (the day of ♀) before the rising of the ☉.

Occult properties: Saint John's wort is a remarkably versatile herb with anodyne, aromatic, astringent, cholagogic, diuretic, expectorant, nervine, and stimulant properties. Its juice serves as an excellent vulnerary for healing wounds. Its water is sudorific and vermifugal in action. It is named after Saint John the Baptist because it blooms around the time of his midsummer feast day on June 24. On the morning of the festival of Saint John, celebrants dance around a fire wearing crowns of Saint John's wort leaves and stems, which they bring home afterward to erect as talismanic devices.† If harvested under a new ☾ the morning after the feast day of Saint John and suspended in a field from a stake of ⋆oak before sunrise, it will make the soil exceedingly fertile. If tossed haphazardly in a field at the time of sowing, it will protect the field from hail. In Germany, many believe its apotropaic virtue is much greater when it has been harvested at night. According to Pseudo-Llull, magical perfumes made from its leaves and flowers are useful against demonic hauntings and spirits who guard over

*[Sédir also supplies the name *hypericon,* a transliteration of the ancient Greek plant name ὑπερικόν, from which the modern genus name derives. See *Les plantes magiques,* 155. —*Trans.*]

†[Sédir's source here is Tuchmann, "La fascination: 4. Les fascinés," 253 s.v. *hypéricon,* but these festal crowns were more likely made of mugwort, which, in addition to Saint John's belt (French: *ceinture de Saint Jean*) and Saint John's flower (French: *fleur de Saint Jean*), was also known colloquially as Saint John's crown (French: *couronne de Saint Jean*). On Sédir's confusion of mugwort and Saint John's wort, see the entry ⋆mugwort and the corresponding footnote. —*Trans.*]

hidden treasures.[6] A sprig of Saint John's wort placed inside a shoe also keeps evil spirits at bay. Alchemists and Hermeticists sometimes refer to it by the secret name *porrosa*.[7] It is also one of the twelve magical plants of the Rosicrucians.[8]

Fig. S.5. Pomet, *Histoire générale des drogues*, 107.

SANDALWOOD, RED

Latin binomial: *Pterocarpus santalinus* L.f.
French name: santal rouge
Elemental qualities: hot and dry (△)
Ruling planet: ☉ (*apud* Piobb)
Zodiacal signature: ♌
Occult properties: Red sandalwood has anti-inflammatory, diaphoretic, febrifugal, and tonic properties. It emits a powerful Solar perfume. Its heartwood is especially good for treating hemorrhages.

Fig. S.6. Oliver, *First Book of Indian Botany*, 288.

SANDALWOOD, WHITE

Latin binomial: *Santalum album* L.
French names: santal blanc, santal d'Inde
Elemental qualities: cold and dry (🜃)
Ruling planet: ☾
Zodiacal signature: ♍ (*apud* Jollivet-Castelot)
Occult properties: White sandalwood has anti-inflammatory, astringent, digestive, and febrifugal properties. It emits a powerful Lunar perfume. Its oil is depurative in action and cleanses the blood of toxic viruses.

[SAND LEEK ❦ *See* LEEK, SAND]

[SAND SPURRY ❦ *See* SPURRY, SAND]

SARSAPARILLA

Fig. S.7. Mattioli, *Discorsi,* 665.

Latin binomial: *Smilax ornata* Lem.

French names: salsepareille, salsepareille rouge

Elemental qualities: cold and wet (∇)

Ruling planet: ☿ (root)

Zodiacal signature: ♋ (root)

Occult properties: Sarsaparilla has alterative, demulcent, febrifugal, stimulant, and tonic properties. Its infusion is depurative in action and counters Venusian maladies and obesity.

SAVORY, SUMMER

Fig. S.8. Fuchs, *Plantarum effigies,* 173.

Latin binomial: *Satureja hortensis* L.

French names: sarriette, sarriette annuelle, sarriette des jardins, sarriette vivace

Elemental qualities: hot and dry (∆)

Ruling planet: ☿

Zodiacal signature: ♌

Occult properties: Summer savory has aromatic, carminative, digestive, and stomachic properties. Sprinkle the water of its leaves around your house to get rid of flies and other noxious insects.[9]

Fig. S.9. Fuchs, *Plantarum effigies,* 424.

SAXIFRAGE

Latin binomial (type sp.): *Saxifraga granulata* L.

French names: saxifrage (granulée), casse-pierre, herbe à gravelle, sauluevie

Elemental qualities: cold and wet (🜄)

Ruling planet: ♄

Zodiacal signatures: ♋, ♎, and ♒

Occult properties: Saxifrage possesses astringent, carminative, hepatic, pectoral, and resolvent properties. Its kidney-shaped leaves, which are signed by ♎, are most commonly used in the treatment of kidney and urinary diseases.[10] Its seeds, especially when mixed with the juice of the plant, are likewise adept at dissolving kidney or bladder stones.

Fig. S.10. Fuchs, *Plantarum effigies,* 414.

SCABIOUS, DEVIL'S BIT

Latin binomial: *Succisa pratensis* Moench

French names: scabieuse, herbe aux sabotiers, knautie des champs, langue de vache, mors de diable, scabieuse des champs

Elemental qualities: cold and dry (🜃)

Ruling planet: ☿

Zodiacal signatures: ♉ and ♎. The flower is signed by ♈.

Occult properties: Devil's bit scabious has bechic, demulcent, depurative, febrifugal, stomachic, and vermifugal properties. It is particularly useful in the treatment of scabies, skin conditions, and leprosy.[11] According to herbal lore, the plant's short black root was originally much longer, but the devil became so enraged by its miraculous ability to cure cutaneous ailments that he bit off a portion of the root from below ground. This folktale is the origin of the modern genus name *Succisa,* meaning "cut from below." The devil's bit scabious also has a variety of magical uses. When worn as an amulet, it protects the bearer against all forms of devilishness and wickedness. If reduced to a powder and placed secretly under a dinner table, however, it will cause internecine strife among the diners.[12]

SENNA

Latin binomial (type sp.): *Senna alexandrina* L.

French names: séné, séné d'Alexandrie, séné d'Égypte*

Elemental qualities: cold and dry (🜃)

Ruling planets: ☉, ☾, and ♄

Zodiacal signatures: ♍ and ♉

Occult properties: Senna has aperient, cathartic, stomachic, and stimulant properties. It is most commonly administered as a laxative in herbal teas made from senna pods and leaves.[13] Its decoction has powerful purgative properties.

Fig. S.11. Fuchs, *Plantarum effigies*, 252.

*[Sédir provides separate entries for *casse* and *séné,* which are variant names for senna. See *Les plantes magiques,* 147, 168. These have been merged here to form a single entry. Sédir's *casse* refers not to the cassia otherwise known as Chinese cinnamon (*Cinnamomum cassia* [L.] J.Presl.), whose French common names include *casse aromatique, casse d'Inde, casse de Chine,* and *cassier,* but rather to the genus *Cassia,* which includes the purging cassia (*Cassia fistula* L.), or *casse des boutiques.* This is clear not only from the signatures Sédir provides but also from his brief descriptions of both plants as purgatives. Moreover, the type species *Senna alexandrina* L. was formerly classed under the Latin binomial *Cassia senna* L. —*Trans.*]

Fig. S.12. Fuchs, *Plantarum effigies,* 251.

SENNA, BLADDER

Latin binomial: *Colutea arborescens* L.

French names: baguenaudier, arbre à vessies, baguenaude, balandier, glouglou, panpan, séné du pays

Elemental qualities: hot and wet (🜁)

Ruling planet: ☿ (*apud* Culpeper)

Zodiacal signature: ♎ (especially its leaves)

Occult properties: The flowers of the bladder senna produce inflated, bladder-like seed pods that change in color from pale green to red or copper. Its leaves possess strong purgative properties. An infusion of its seeds is a powerful emetic.[14]

Fig. S.13. Mattioli, *Discorsi,* 275.

SESAME

Latin binomial: *Sesamum indicum* L.

French name: sésame

Elemental qualities: hot and dry (🜂)

Ruling planet: ♃

Zodiacal signature: ♐ (*apud* Jollivet-Castelot)

Occult properties: Sesame has aperient, astringent, diuretic, emollient, hepatic, and tonic properties. Hindu devotees offer the black unhulled seeds of the sesame, or *tila* (तिल) in Sanskrit, in domestic sacrifices to the *pitṛs* or *pitaras*—that is, to the spirits of their departed ancestors, who are comparable to the ancient Roman *manes.*

SHEPHERD'S PURSE

Fig. S.14. Fuchs, *Plantarum effigies,* 350.

Latin binomial: *Capsella bursa-pastoris* (L.) Medik.

French names: bourse de pasteur, bourse à pasteur, bourse de capucin, bourse de Judas, boursette, capselle, tabouret*

Elemental qualities: cold and dry (🜃)

Ruling planet: ♄

Zodiacal signature: ♑ (*apud* Mario)

Occult properties: Shepherd's purse has antiscorbutic, astringent, diuretic, hepatic, and vulnerary properties. It is most commonly used in the treatment of hemorrhages and diarrhea. When held as a talisman in the palm of a man's hand or worn as an amulet around a woman's neck, it cures dysentery. When macerated in vinegar and administered in a compress with the palms of the hands, it is effective against gonorrhea and other Venusian maladies. This "herb of fecundity" was sacred to medieval magicians, who gave it the Chaldean name *lorumborat.* According to the *Grand Albert,* when tempered with ★mandrake juice and administered to animals, it juice helps facilitate conception.[15]

*[In Sédir's original entry, the variant *onagollis* appears alongside the French common names *bourse de pasteur* and *tabouret.* See *Les plantes magiques,* 146. This name derives from Sédir's source, which he does not cite but which can only be Papus's "Sorcerer's Grimoire," which includes the following folk remedy: "*To Stop Bleeding:* Hold on his hand the herb called *bursa pastoris* [i.e., shepherd's purse] or *onagollis* for women, then hang it around the neck over the skin. True oriental turquoise [i.e., calaite] produces the same effect" (Papus, *Traité élémentaire de magie pratique,* 527). The word *onagollis* is an alternate spelling of the Latin plant name *anagallis,* a name in use among old herbalists and apothecaries for the scarlet pimpernel (*Anagallis arvensis* L.). Sédir apparently understands the folk remedy preserved by Papus to say, "the herb called *bursa pastoris* or *onagollis,*" but this is unlikely for the reason already stated. Rather, the original remedy appears to say: "Take the herb called shepherd's purse, for men, or the herb called scarlet pimpernel, for women, hold it in a compress over the wound, then wear it around the neck over the skin as an amulet." Sédir also presents his own interpretation of this folk remedy as cure for dysentery. Cf. Sédir's entry ★pimpernel, where a similar amuletic procedure is said to be apotropaic in action. —*Trans.*]

Fig. S.15. Fuchs, *Plantarum effigies,* 452.

SOAPWORT

Latin binomial: *Saponaria officinalis* L.

French names: saponaire, sabouneto (Languedocian), savonnière

Elemental qualities: cold and wet (▽)

Ruling planet: ♄

Zodiacal signature: ♏ (*apud* Mario)

Occult properties: Soapwort has cholagogic, expectorant, depurative, sternutatory, and tonic properties. It is an especially good herb for consecrating altars and other ritual implements.[16]

Fig. S.16. Fuchs, *Plantarum effigies,* 336.

SOLOMON'S SEAL

Latin binomial (type sp.): *Polygonatum odoratum* (Mill.) Druce

French names: sceau de Salomon, herbe à la rupture, herbe aux panaris, hèrbo de la roumpaduro (Languedocian), faux muguet, genouillet

Elemental qualities: cold and dry (🜃)

Ruling planet: ☿ (*apud* Piobb)

Zodiacal signatures: ♉, ♍, and ♑

Occult properties: Solomon's seal has bechic, cardiotonic, demulcent, diuretic, ophthalmic, tonic, and vulnerary properties. It is commonly used in the treatment of wounds, bruises, and skin inflammation. It also has powerful apotropaic and exorcistic properties. Its root has the same virtue as ⋆Saint John's wort against spirits who guard over buried or hidden treasure.[17] Alchemists and Hermeticists refer to it by the name *secacul,* which derives from the Arabic word for parsnip.[18]

SORB TREE

Fig. S.17. Mattioli, *Discorsi*, 178.

Latin binomial (type sp.): *Sorbus aucuparia* L.
French names: cormier, serbiè (Languedocian), sorbier
Elemental qualities: cold and wet (▽)
Ruling planet: ♃
Zodiacal signature: ♏
Occult properties: The sorb tree, and especially the species known as rowan and mountain ash (*S. aucuparia* L.), possesses antirheumatic, antiscorbutic, astringent, diuretic, and laxative properties. Decoctions of the bark or fruit are particularly effective in the treatment of chronic diarrhea, dysentery, and hemorroids.[19] When worn as amulets or erected as talismans, sorb branches and branchlets provide protection from lightning, hail, evil spirits, and black magic.[20]*

*[Sédir's original entry asserts that the javelin of Romulus was made of sorb wood, but here *cormier* (sorb tree) is an error for *cornier,* a variant name for *cornouiller* (cornelian cherry). See *Les plantes magiques,* 150. The elemental qualities Sédir provides—namely, hot and wet (🜁)—also apply to the cornelian cherry, which, unlike the sorb tree, is cogoverned by ♂ in his domicile of ♏, which increases the tree's quality of hotness. This information has been moved to its proper place under *cornelian cherry. —*Trans.*]

Fig. S.18. Fuchs, *Plantarum effigies,* 260.

SORREL

Latin binomial: *Rumex acetosa* L.

French names: oseille, agreto (Languedocian), aigrette, oseille de brebis, oseille des prés, patience acide, surelle, surette, vinette

Elemental qualities: hot and wet (🜁)

Ruling planet: ☿ (*apud* Piobb)

Zodiacal signatures: ♊ and ♍

Occult properties: Sorrel, an Old French word meaning "sour," has aperient, astringent, diuretic, and refrigerant properties. The root, when cut into small slices and soaked for forty-eight hours in a strong white vinegar, works well in lotions for healing scabs. According to Dioscorides, wearing sorrel root around the neck as an amulet cures goiter and scrofula.[21] Its seeds, when gathered by a young virgin boy, prevent the onset of nocturnal emissions. To cleanse the body of infectious diseases, Papus recommends drinking a warm mixture of one-half ounce of sorrel water and one dram of theriac, which is diaphoretic in action.[22]

SORREL, WOOD

Fig. S.19. Fuchs, *Plantarum effigies,* 324.

Latin binomial: *Oxalis acetosella* L.

French names: alléluia, oseille des Pâques, pain de cocu, pain d'oiseau, petite oseille, petit trèfle, trèfle acide, trèfle aigre*

Elemental qualities: cold and wet (∇)

Ruling planet: ☿

Zodiacal signature: ♓ (*apud* Mavéric)

Occult properties: Wood sorrel, otherwise known as Hallelujah because it blossoms between Easter and Pentecost, when the Hallelujah Psalms are traditionally sung (Psalms 146–150), has anodyne, antiscorbutic, diuretic, emmenagogic, and febrifugal properties. The essence extracted from the leaves and flowers is good for internal use against tonic-clonic, or grand mal, seizures. Alchemists and Hermeticists refer to it by the name *oxus,* a transliteration of the Greek word ὀξύς, meaning "sharp," from which both the ancient Greek name ὀξαλίς and the modern genus name *Oxalis* derive.[23] In mysticism, it is emblematic of the Trinity. Whereas the more common three-leaf variety presages bad weather whenever it bows toward the ground, which gives it a more fragrant scent than usual, the less common four-leaf variety (*Oxalis tetraphylla* Cav.) brings good luck.

*[Sédir's original entry appears under the imprecise French common name *trèfle,* meaning "clover" or "trefoil," along with the variants *alléluia* (hallelujah) and *pain de cocu* (cuckoo-bread). See *Les plantes magiques,* 170. Although the name *trèfle* by itself is not commonly used to refer to ⋆wood sorrel, the names *petit trèfle* (small clover) and *trèfle acide* or *trèfle aigre* (sour clover) are well attested. —*Trans.*]

Fig. S.20. Fuchs, *Plantarum effigies*, 4.

SOUTHERNWOOD

Latin binomial: *Artemisia abrotanum* L.

French names: abrotone, armoise mâle, aurone, aurone mâle, citronelle, garderobe, herbe royale, ivrogne*

Elemental qualities: hot and dry (△)

Ruling planet: ☿ (*apud* Culpeper and Mavéric)

Zodiacal signature: ♈. Harvest at the beginning of April or under ♏.

Occult properties: Southernwood has natural diuretic, deobstruent, emmenagogic, and tonic properties. It is an especially good herb for pregnant women, who may take it as a uterine tonic to ease the process of parturition.

[SPEARMINT ⚕ *See* MINT]

*[Sédir's entry appears under the species name *abrotanum,* which derives from the ancient Greek plant name ἀβρότονον. See *Les plantes magiques,* 141. —*Trans.*]

SPEEDWELL

Fig. S.21. Fuchs, *Plantarum effigies*, 93.

Latin binomial (type sp.): *Veronica officinalis* L.

French names: véronique, berounica (Languedocian), herbe aux ladres, herbe de Sainte Véronique, véronique mâle

Elemental qualities: hot and dry (△)

Ruling planet: ☿ (*apud* Gessmann and Jollivet-Castelot)

Zodiacal signature: ♈. Harvest after the full ☾ terminating the dog days of summer.

Occult properties: Speedwell, otherwise known as bird's-eye and gypsyweed, has alterative, antipruritic, astringent, diaphoretic, and tonic properties. The genus *Veronica* was named after Saint Veronica, who, according to legend, offered Jesus her veil as he carried the cross to Golgotha to wipe the blood and sweat from his face.[24]

[SPHAGNUM MOSS ⚕ *See* MOSS, SPHAGNUM]

[SPRING ONION ⚕ *See* ONION, SPRING]

Fig. S.22. Parkinson, *Theatrum botanicum,* 224.

SPURGE

Latin binomial (type sp.): *Euphorbia antiquorum* L.

French names: euphorbe, laxusclo (Languedocian)*

Elemental qualities: hot and dry (△)

Ruling planet: ♂

Zodiacal signature: ♏

Occult properties: Spurge has antiseptic, cathartic, diuretic, emetic, purgative, and rubefacient properties. Several species of the genus *Euphorbia,* most notably the petty spurge (*E. peplus* L.) and the Mediterranean spurge (*E. characias* L.), are effective in the treatment of skin conditions like warts, fistulae, carbuncles, and boils, as well as Venusian diseases like gonorrhea.[25] The magi gave spurge the secret name *gonos Areos,* from the Greek γόνος Ἄρεως, meaning "semen of Ares."[26] The powdered stems of spurge serve as excellent perfumes for attracting the influences of ♄ and his domicile of ♒.†

*[Sédir's entry for *euphorbe* (spurge) includes French common names for a variety of spurges: *réveille-matin* for sun spurge (*Euphorbia helioscopia* L.), *omblette* for petty spurge (*Euphorbia peplus* L.), and *lait de couleuvre* and *tithymale* for cypress spurge (*Euphorbia cyparissias* L.), for which Sédir provides a separate entry. See *Les plantes magiques,* 152. As stated in the foreword, in cases where Sédir uses either the genus name alone or a collection of names for several different species, this means that the information provided applies to multiple species. For this reason, only the Latin binomial for the type species of the *Euphorbia* genus, the antique spurge, is listed here. The Languedocian name *laxusclo* or *lachusclo,* literally "milk-burner" (from Latin *lac,* "milk," + *ustulo,* "I burn"), also applies to all spurges (so Barthés, *Glossaire botanique languedocien, français, latin,* 125). —*Trans.*

†As the etymology of the provincial name *luxusclo* would suggest, the elemental qualities Sédir provides—namely, cold and wet (▽)—apply solely to the powdered stems (♄ in his domicile of ♒ bears these same elemental qualites) and are incorrect for spurges in their natural state. From the standpoint of *magical* herbal actions, an herb in its powdered form can have the opposite magical effect of the same herb in its natural state (see, for example, Sédir's entries ⋆periwinkle, ⋆vervain, and ⋆devil's bit scabious). For this reason, the elemental qualities for spurge have been changed to hot and dry (△), as Sédir's entries for ⋆caper spurge and ⋆cypress spurge would suggest and, indeed, as is to be expected from the vast majority of Martian herbs. See further Sédir's comments on Martian herbs in chapter 3 and my own comments concerning the discrepencies between Sédir's sources—namely, the *Grand Albert* and Papus—in the foreword. —*Trans.*]

SPURGE, CAPER

Fig. S.23. Fuchs, *Plantarum effigies,* 256.

Latin binomial: *Euphorbia lathyris* L.

French names: catapuce, catapusso (Languedocian), euphorbe catapuce, euphorbe épurge*

Elemental qualities: hot and dry (△)

Ruling planet: ♂

Zodiacal signature: ♈. Harvest under ♌.

Occult properties: All parts of the caper spurge are violently purgative in action. Like other spurges, such as the resin spurge (*E. resinifera* A.Berger), it exudes a highly poisonous milky latex upon cutting or bruising called euphorbium, which has extremely powerful emetic, purgative, and vesicant properties. Its leaves are also rubefacient in action.[27]

SPURGE, CYPRESS

Fig. S.24. Fuchs, *Plantarum effigies,* 469.

Latin binomial: *Euphorbia cyparissias* L.

French names: tithymale, euphorbe faux cyprès, euphorbe petit cyprès, herbe au lait, lait de couleuvre, rhubarbe des pauvres, rhubarbe du paysan

Elemental qualities: hot and dry (△)

Ruling planet: ♂

Zodiacal signature: ♌

Occult properties: The cypress spurge has violent emetic, irritant, and purgative properties. Its root, however, when infused for three days in vinegar, is effective against edema.

*[Sédir's entry appears under the Latin name *catapultia,* an uncommon but nonetheless attested variant of the plant name *catapulta,* meaning "catapult," a pun highlighting the plant's emetic properties. See *Les plantes magiques,* 147. Both of these argotic variants appear to derive from the obsolete genus name *Cataputia* (in which the ★castor bean was also formerly classed), a latinized diminutive of Greek καταποτον, meaning "pill," "bolus," or "something swallowed." —*Trans.*]

Fig. S.25. Parkinson, *Theatrum botanicum,* 562.

SPURRY, SAND

Latin binomial: *Spergularia rubra* (L.) C. Presl.*

French names: sabline rouge, spergulaire rouge

Elemental qualities: hot and wet (🜁)

Ruling planet: ♂

Zodiacal signature: ♎

Occult properties: Sand spurry has aromatic, diuretic, and lithontriptic properties. This ferruginous herb, which grows in thick tufts low to the ground, has small red flowers with five sepals, five petals, ten stamens, and three styles. It produces capsules from April to September. In an infusion of forty grams per liter of liquid, it helps pass bladder and kidney stones and alleviates renal colic.

Fig. S.26. Fuchs, *Plantarum effigies,* 494.

STRAWBERRY

Latin binomial (type sp.): *Fragaria vesca* L.

French names: fraisier, fraisier de montagne, fraisier des bois, maxoufo (Languedocian)

Elemental qualities: cold and wet (∇)

Ruling planet: ♃

Zodiacal signature: ♓

Occult properties: Strawberry possesses aperient, astringent, diuretic, refrigerant, and tonic properties. It is a superb emollient and is especially good for treating jaundice. Wearing a belt made of strawberry stems and leaves keeps evil spirits at bay.[28]

*[The Latin binomial Sédir provides, *Arenaria rubra* L., is synonymous with *Spergularia rubra* (L.) C. Presl. —*Trans.*]

[SUMMER SAVORY ❦ ***See*** **SAVORY, SUMMER]**

[SWEET ALYSSUM ❦ ***See*** **ALYSSUM, SWEET]**

[SWEET CLOVER ❦ ***See*** **CLOVER, SWEET]**

[SWEET FLAG ❦ ***See*** **FLAG, SWEET]**

[SWINE CRESS ❦ ***See*** **CRESS, SWINE]**

T

Tamarind–Turnip

Fig. T.1. Pomet, *Histoire générale des drogues,* 220.

TAMARIND

Latin binomial (type sp.): *Tamarindus indica* L.

French name: tamarinier

Elemental qualities: cold and wet (∇)

Ruling planets: ♄ and ☾. The fruit is governed by the ☉

Zodiacal signature: ♋

Occult properties: Tamarind possesses aperient, astringent, digestive, febrifugal, and hepatic properties. Wines infused with tamarind wood are helpful in the treatment of splenomegaly, leprosy, and toothaches. The pods of Asian tamarinds, which are filled with six to twelve seeds and have an acidulous and sweet, reddish-black pulp, are not only effective medicinally, but their seeds are also frequently used in divination, most notably the geomantic method of the Malagasy known as *sikidy.*[1]

TANSY

Fig. T.2. Fuchs, *Plantarum effigies,* 27.

Latin binomial: *Tanacetum vulgare* L.

French names: tanaisie, barbotine, herbe aux vers, herbe de Saint Marc, tanacée, tanarido (Languedocian)

Elemental qualities: hot and wet (🜁)

Ruling planet: ☉

Zodiacal signature: ♊ (*apud* Haatan)

Occult properties: Tansy has aromatic, diaphoretic, spasmolytic, stimulant, and vermifugal properties. Its leaves have a strong and bitter but not unpleasant taste, and its yellow blooms and flowers have a robust, balsamic, and camphorous aroma that is parasiticidic in action.[2] It is particularly effective in the treatment of nervous disorders.

TEA

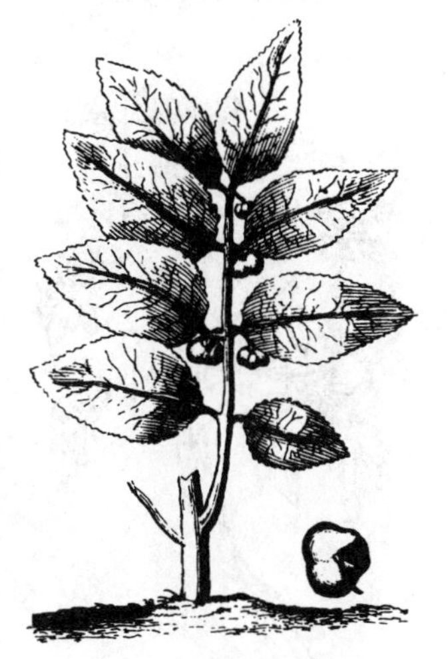

Fig. T.3. Pomet, *Histoire générale des drogues,* 143.

Latin binomial: *Camellia sinensis* (L.) Kuntze

French names: thé, té (Languedocian), théier, thé vert

Elemental qualities: hot, dry, and moderately wet (△)

Ruling planet: ☿

Zodiacal signatures: ♐, ♉, ♊, and ♍ (*apud* Mavéric)

Occult properties: The tea plant has astringent, cardiotonic, diuretic, odontalgic, and stimulant properties. Its infusion was formerly used by Japanese Buddhists to magically influence and strengthen their monastic communities. Drinking green tea infused with fresh *parsley leaves is an effective remedy for lowering blood pressure.[3]

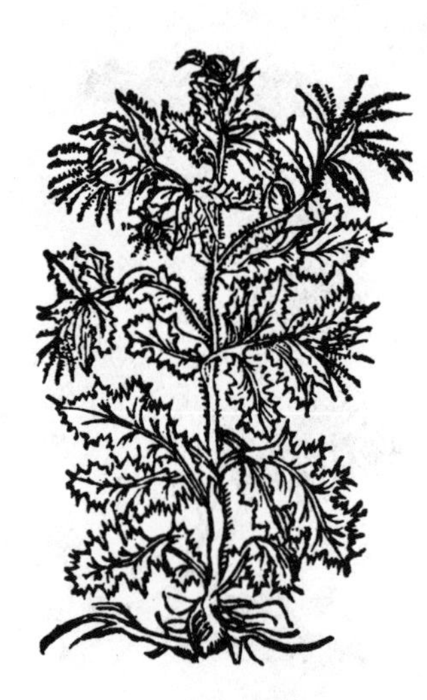

Fig. T.4. Fuchs, *Plantarum effigies,* 67.

THISTLE, BLESSED

Latin binomial: *Cnicus benedictus* L.

French names: chardon bénit, centaurée bénite, chardon béni, chardon marbré, safran sauvage

Elemental qualities: hot and dry (△)

Ruling planet: ♂

Zodiacal signature: ♌. Harvest in June before its yellow flowers blossom.

Occult properties: Blessed thistle, also known as holy thistle, has astringent, cholagogic, depurative, detersive, diuretic, stimulant, and stomachic properties. To make a febrifuge more powerful than quinine,* consume a mixture of its leaves and flowering tops macerated in a small glass of wine. The dew that collects in its capsules is good for scrofulous and catarrhal ophthalmia. Its infusion heals pulmonary ulcers.[4]

Fig. T.5. Fuchs, *Plantarum effigies,* 508.

THISTLE, CARLINE

Latin binomial: *Carlina vulgaris* L.

French names: chardon carline, cardounilho (Languedocian), carline commune, herbe à la pluie

Elemental qualities: hot and dry (△)

Ruling planet: ♂

Zodiacal signature: ♌

Occult properties: Carline thistle has diaphoretic and purgative properties. When harvested at the end of October, its aromatic and pungent roots submit to the influences of ♏ and take on aphrodisiac properties. Some alchemists and Hermeticists refer to it by the name *ixia,* a Latin transliteration of ἰξία, the ancient Greek name for the pine thistle (*Carlina gummifera* L.).[5]

*[Quinine, the medicament obtained from the bark of the cinchona tree (*Cinchona officinalis* L.), is best known for its effectiveness in treating malaria. —*Trans.*]

THYME

Fig. T.6. Fuchs, *Plantarum effigies,* 477.

Latin binomial: *Thymus vulgaris* L.

French names: thym, frigoule, frigoulo (Languedocian), pote, thym des jardins

Elemental qualities: hot and dry (△)

Ruling planet: ☉

Zodiacal signature: ♌. Harvest when the ☉ is in ♓ and the ☾ is in ♋.

Occult properties: Thyme has antiseptic, carminative, diaphoretic, spasmolytic, tonic, and vermifugal properties. In addition to its myriad medicinal and therapeutic uses, it also serves as a natural apotropaic, not only for the property on which it grows but also for whoever wears it as an amulet.[6] In the Hermetic language of flowers, thyme is emblematic of action.

THYME, BASIL

Fig. T.7. Fuchs, *Plantarum effigies,* 314.

Latin binomial: *Acinos arvensis* (Lam.) Dandy

French names: basilic sauvage, basèli salbage (Languedocian), petit thym sauvage, pouliot des champs, thym basilic, thym calament

Elemental qualities: hot and dry (△)

Ruling planet: ☉

Zodiacal signature: ♈

Occult properties: Basil thyme has odontalgic, stimulant, and rubefacient properties. If a whole plant is placed under a dish of meat, it will prevent anyone from touching it. When worn as an amulet, it protects the bearer from infernal apparitions.[7]

Fig. T.8. Fuchs, *Plantarum effigies,* 140.

THYME, WILD

Latin binomial: *Thymus serpyllum* L.

French names: serpolet, menudet (Languedocian), poleur, pouliet, poulieu, thym sauvage, thym serpolet

Elemental qualities: hot and dry (△)

Ruling planet: ♀

Zodiacal signature: ♐ (*apud* Mario)

Occult properties: Wild thyme, otherwise known as creeping thyme and mother-of-thyme, has many of the same properties as *thyme. Its flowering tops are astringent, nervine, and tonic in action. It serves as a particularly efficacious remedy against poisonous bites and stings. When burned as a suffumigant, it drives away all venomous creatures.[8]

Fig. T.9. Parkinson, *Theatrum botanicum,* 712.

TOBACCO

Latin binomial: *Nicotiana tabacum* L.

French names: tabac, herbe à la reine, petun, tabac de Virginie, tabat (Languedocian)

Elemental qualities: cold and dry (🜃)

Ruling planet: ☾

Zodiacal signature: ♍ (*apud* Mario)

Occult properties: Tobacco has discutient, diuretic, demulcent, sedative, and sialagogic properties. When distilled, it produces a powerful emetic and a good astringent liquor for ringworm. When smoked in a pipe, it has calmative properties that make it useful for contemplation. According to Papus, decoctions of tobacco leaves administered in the form of a lotion eliminate pimples and reduce redness of the face.[9]

TORMENTIL

Fig. T.10. Fuchs, *Plantarum effigies,* 145.

Latin binomial: *Potentilla erecta* (L.) Raeusch.

French names: tormentille, herbe de Sainte Catherine, potentille tormentille

Elemental qualities: cold and dry (🜃)

Ruling planet: ♂

Zodiacal signatures: ♉ and ♍

Occult properties: Tormentil has powerful astringent, odontalgic, styptic, and tonic properties. Its decoction is particularly effective as a gargle against sore throat, mouth ulcers, and infected gums. Since antiquity, the distilled water of its leaves and roots has served as the main ingredient in numerous antidotes and counterpoisons.[10]

TURNIP

Fig. T.11. Fuchs, *Plantarum effigies,* 98.

Latin binomial: *Brassica rapa* L.

French names: rave, navet, rabo (Languedocian)

Elemental qualities: cold and wet (🜄)

Ruling planet: ☾

Zodiacal signature: ♓. Harvest under ♏ or when the ☾ is in ♓.

Occult properties: Turnip possesses anti-inflammatory, antirheumatic, emollient, and cardiotonic properties. It is particularly effective in the treatment of gout, arthritis, and chilblains. Turnip juice is mildly laxative and sedative in action.[11] Its seeds are often used in love philters, diuretic counterpoisons, and remedies for smallpox.

U
Usnea

Fig. U.1. Gerard, *The Herball*, 1374.

USNEA

Latin binomial (type sp.): *Usnea florida* (L.) F.H.Wigg.

French names: usnée, moufo d'albre (Languedocian)

Elemental qualities: cold and wet (▽)

Ruling planets: ♄ and ☾

Zodiacal signature: ♓

Occult properties: Usnea has antimicrobial, cosmetic, demulcent, febrifugal, and vulnerary properties. This pale grayish-green fruticose lichen grows on the barks of trees and on the bones of decayed corpses. Paracelsus is said to have harvested one from the skull of a hanged man and used it to compose the vulnerary ointment known as the Weapon-Salve.[1]

V
Valerian–Violet

VALERIAN

Latin binomial (type sp.): *Valeriana officinalis* L.

French names: valériane, baleriano (Languedocian), herbe à la meurtrie, herbe aux chats, herbe de Saint Georges, valériane sauvage

Elemental qualities: cold and dry (🜃)

Ruling planet: ☿

Zodiacal signature: ♉

Fig. V.1. Fuchs, *Plantarum effigies,* 496.

Occult properties: Valerian, otherwise known as garden heliotrope, has carminative, diuretic, nervine, sedative, and spasmolytic properties. Its root, which is governed by ♄ and is the most potent part of the plant, can be used to counter asthma and edema and stave off infections. In infusions, it helps facilitate conception. Like *catnip, valerian has a hypnotizing and euphorizing effect on felines, and it is well known that whenever the root is presented to hypnotic subjects, it makes them crawl on the ground on all fours, meowing and clawing like cats.[1] Offerings of this upright perennial with white flowers and hollow stems are typically made to Saint George in conjunction with medicinal prayers to cure nervous disorders.[2] Alchemists and Hermeticists refer to it by the names *leucolachanum*—a compound of Greek λευκός, meaning "white," "bright," or "light," and λάχανον, meaning "garden-herb"—and *phu* or *phy.**

Fig. V.2. Fuchs, *Plantarum effigies,* 407.

VERVAIN

Latin binomial (type sp.): *Verbena officinalis* L.

French names: verveine, berbeno (Languedocian), guérit-tout, herbe à tous les maux, herbe aux enchantements, herbe aux sorcières, herbe du foie, herbe du sang, herbe sacrée, verveine mâle, verveine sauvage

Elemental qualities: hot and moderately wet (🜂)

Ruling planets: ♀ and ☉

Zodiacal signature: ♎. The *Grand Albert* and Papus recommend harvesting when the ☉ passes into ♈.[3]

Occult properties: Vervain has astringent, depurative, diaphoretic, emmenagogic, galactagogic, spasmolytic, tonic, and vulnerary properties. The distilled water of this extremely versatile herb is effective against anemia of the optic nerve. By prolonging its distillation, Jan Baptiste van Helmont asserts, practitioners can obtain an excellent liquor, to be prescribed in small homeopathic doses, for curing tuberculosis and

*[Sédir provides these alchemical names in a separate entry under *valériane sauvage* (wild valerian), but this is simply another French common name for *Valeriana officinalis* L., as is now indicated in the French names category. See *Les plantes magiques,* 170. This peculiarity appears to stem from Sédir's source, Antoine-Joseph Pernety, who equates *leucolachanum* with *valériane sauvage* and *phu* or *phy* with *valériane* (Pernety, *Dictionnaire mytho-hermétique,* 248, 382). Sédir's original entry, however, supplies the names *leucophagum* (*sic*), *phu,* and *phy* as variants for *valériane sauvage.* As for Sédir's *leucophagum,* which Pernety identifies as the so-called doctor's blancmange, an old stew remedy for consumption made with capon, partridge, and almonds (*Dictionnaire mytho-hermétique,* 248), this is surely an error for *leucolachanum* (Pernety's entry for the former immediately precedes his entry for the latter). The plant name *phu* is best known from Dioscorides, *On Medical Materials* 1.11, where it is spelled φοῦ, but ultimately it derives from فو, the Arabic plant name for valerian, as Agrippa correctly points out (see Agrippa, *De occulta philosophia libri tres,* liber I, caput XXVIII, xxiiii; English edition: *Three Books of Occult Philosophy,* bk. 1, chap. 28, 102). Carl Linnaeus even named one of the Valerianaceae after it (Linnaeus, *Species plantarum,* 1:32, no. 6), but it is hard to know whether *Valeriana phu* L., now known in English by the common name Turkey valerian, is one and the same as the ancient nervine and antihysteric described by Dioscorides. —*Trans.*]

dissolving blood clots. When administered in the form of a poultice, its roots heal scrofula, ulcers, and abrasions. When administered in infusions or in cataplasms, its leaves cure rabies. Its seeds, when mixed with year-old ★peony seed, are effective against caducity. If planted with the appropriate rites in a field or next to a home, it will increase the prosperity of that field or home. If the hosts of a gathering put four vervain leaves in some wine and then sprinkle the mixture around their dining room, all their guests will be jovial and the atmosphere of the gathering convivial. If practitioners secretly hold vervain in the palm of their hand while asking patients to report on their health, patients will recover if their answer is positive but die if it is negative. Since antiquity, its juice has been used to make aphrodisiacs and irresistible love potions. This "herb of love" was sacred to medieval magicians, who gave it the Chaldean name *olphanas.** According to the *Grand Albert,* when worn as an amulet, it takes on aphrodisiac properties and makes the bearer vigorous in bed, but if it is placed in powdered form between two lovers, it will drive a wedge between them.[4] Alchemists and Hermeticists call it *vena Veneris,* or "vein of Venus," and *peristerion,* a transliteration of the ancient Greek plant name περιστέριον, meaning "pigeon's-herb," because whenever it is placed in a dovecote, all the pigeons invariably assemble around it.[5] Vervain is one of the twelve magical plants of the Rosicrucians.†

*[In most French editions of the *Grand Albert,* the name of Chaldean herb no. 13 appears as *olphanas,* but some Latin editions read *olphauas.* Cf. Table 0.1 on p. xxiii. —*Trans.*]

†[Franz Hartmann's *With the Adepts,* Sédir's primary source for the "twelve Rosicrucian plants," makes no explicit mention of vervain. Here Sédir's source must be Stanislas de Guaïta's discussion of Jan Baptiste van Helmont's so-called *plante attractive.* After quoting the relevant passage from Helmont's *De magnetica vulnerum curatione,* Guaïta states the following: "This famous plant, whose identity is well known among the Brothers of the Rosy Cross, is none other than *Verbena rustica*" (*Le serpent de la Genèse,* 1:371). Guaïta's pseudo-scientific nomenclature is a latinization of the French common name *verveine sauvage,* which is none other than *Verbena officinalis* L. In the original entry, Sédir cites both Helmont's text and Guaïta's commentary, but he does not discuss the matter himself (*Les plantes magiques,* 170–71). Helmont describes the so-called love-procuring plant, without revealing its identity, as follows: "I know an herb, commonly obvious, which, if you rub and cherish it in your hand until it becomes warm, then hold fast the hand of another person until it, too, becomes warm, will make that person continually burn with an ardent love for you for several days" (Helmont, *De magnetica vulnerum curatione,* 29–30; Helmont, *A Ternary of Paradoxes,* 14–15 § 27). In any case, it is reasonable to assume that Sédir's citation is an endorsement of Guaïta's identification of Helmont's love-procuring plant as vervain. —*Trans.*]

Fig. V.3. Fuchs, *Plantarum effigies,* 96.

VETCH

Latin binomial (type sp.): *Vicia sativa* L.

French names: vesce, dravière, jarosse, poisette, vesce des champs

Elemental qualities: hot and wet (🜁)

Ruling planet: ☾ (*apud* Culpeper)

Zodiacal signatures: ♊ and ♍

Occult properties: Vetch possesses anti-inflammatory, astringent, and emollient properties. It is most commonly used in the treatment of eczema and other skin conditions. The ancient Latin name *vicia,* which is the same as the modern genus name, derives from the verb *vincire,* meaning "to bind." Many of the common names for the vetch in various languages are etymologically related. Some believe Latin *vicia,* Greek βικίον, Lithuanian *wikke,* German *Wichen,* Flemish *vitse,* French *vesce,* English *vetch,* and so on, all stem from the Celtic radical *gwig.*[6]

VIOLET

Fig. V.4. Fuchs, *Plantarum effigies,* 177.

Latin binomial (type sp.): *Viola odorata* L.

French names: violette, biuleto (Languedocian), fleur de mars, violette de carême, violette de mars, violette des haies, violette odorante

Elemental qualities: cold and dry (🜃)

Ruling planets: ♀ and ♃

Zodiacal signature: ♉

Occult properties: Violet possesses anti-inflammatory, bechic, cordial, emollient, laxative, and pectoral properties. Infusions of violet flowers are most commonly used in the treatment of coughs, bronchitis, respiratory catarrh, and strep throat.[7] Alchemists and Hermeticists refer to it by the name *matronalis flos,* or "matron's-flower."[8]

W

Wallflower–Wormwood

Fig. W.1. Fuchs, *Plantarum effigies,* 258.

WALLFLOWER

Latin binomial: *Erysimum cheiri* (L.) Crantz

French names: giroflée (des murailles), giroflée jaune, girouflado (Languedocian), violier jaune

Elemental qualities: moderately hot and wet (🜁)

Ruling planet: ♀

Zodiacal signatures: ♊ (*apud* Mario)

Occult properties: Wallflower has cardiotonic, antirheumatic, emmenagogic, nervine, and resolvent properties. The indefatigable botanist Carl Linneus derived the species name *cheiri* from *khayrī* (sometimes spelled *keiri, kheyri,* or *kheyry*), a Hispano-Arabic variant of the Arabic plant name *khīrī* (a derivative of Persian یریخ), which was used to describe various sweet-scented flowers such as wallflowers and gillyflowers like the carnation, or clove pink (*Dianthus caryophyllus* L.). Linneus then cleverly adapted this term into a Greek homonym and embellished it with the termination *-anthus* to form the wallflower's original genus name *Cheiranthus,* from Greek χείρ, "hand," and ἄνθος, "flower"—that is, a flower to be carried constantly in the hand on account of its pleasant aroma.[1]

WALNUT

Fig. W.2. Mattioli, *Discorsi,* 186.

Latin binomial (type sp.): *Juglans regia* L.

French names: noyer, noix royale, nougo (Languedocian)

Elemental qualities: hot and dry (△)

Ruling planet: ☾

Zodiacal signature: ♐

Occult properties: Walnuts possess anti-inflammatory, astringent, depurative, laxative, and vermifugal properties. The bark of the root serves as a useful emetic and counter-poison and is particularly effective in the treatment of stomatitis. A decoction of the walnut tree's leaves, taken at a dose of one cup in the morning and one cup in the evening, works extremely well against scrofula, cutaneous eruptions, and tumefactions, but the patient will need to continue this regimen for a lengthy period of time before seeing any meaningful results. This same decoction also serves as the basis of a method for curing syphilis, but for this cure to be effective, the patient must be endowed with an exceptionally energetic vitality. Alchemists and Hermeticists call the walnut tree *lignum Heraclei,* or "Hercules's-wood," and refer to its oil by the name *oleum ligni Heraclei,* or "oil of Hercules's-wood."[2]

WATERCRESS

Fig. W.3. Fuchs, *Plantarum effigies,* 419.

Latin binomial: *Nasturtium officinale* W.T. Aiton

French names: cresson des fontaines, cresson d'eau, crinson, crussoun (Languedocian)

Elemental qualities: cold and dry (🜃)

Ruling planet: ☾ (*apud* Mavéric)

Zodiacal signature: ♉

Occult properties: Watercress has antiscorbutic, expectorant, purgative, stimulant, and resolutive properties. This aquatic plant species is most commonly used in the treatment of glandular tumors, lymphatic swellings, and inflammations of the skin.[3] Alchemists and Hermeticists refer to it by the secret name *saure.*[4]

[WATER LILY ❦ *See* LILY, WATER]

[WATER MINT ❦ *See* MINT, WATER]

WATER SOLDIER

Latin binomial (type sp.): *Stratiotes aloides* L.

French names: joubarbe aquatique (faux aloès), aloès d'eau, stratiote aquatique*

Properties: Water soldier has the same elemental, astral, and occult properties as ⋆yarrow.

Fig. W.4. Gerard, *The Herball,* 587.

*[Sédir introduces the rare plant name *joubarbe aquatique* (water soldier) as a variant of *mille-feuilles* in his entry ⋆yarrow. See *Les plantes magiques,* 160. Once again, Sédir is following his go-to source, Pernety, who equates the same two plant names (Pernety, *Dictionnaire mytho-hermétique,* 307). Since these are two different plant species, albeit plants with very similar properties, a separate entry has been created here. On the proper identification of the French common name *joubarbe aquatique,* see Linnaeus, *Des Ritters Carl von Linné vollständiges Pflanzensystem,* 7:278. —*Trans.*]

WHEAT

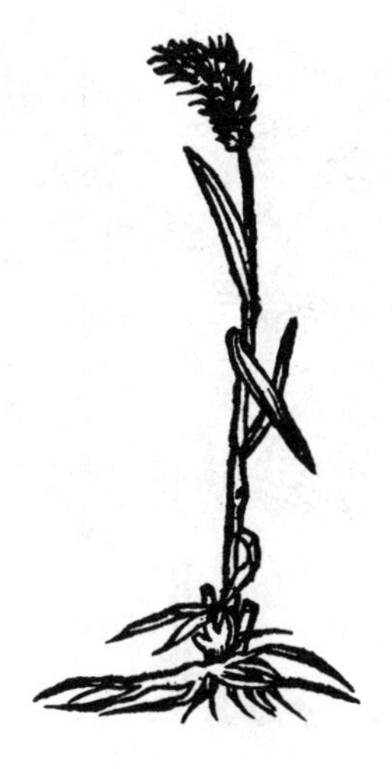

Fig. W.5. Fuchs, *Plantarum effigies*, 372.

Latin binomial (type sp.): *Triticum aestivum* L.

French names: blé, blat (Languedocian), blé tendre

Elemental qualities: cold and dry (🜃)

Ruling planet: ☉

Zodiacal signature: ♍

Occult properties: Wheat has astringent, antivinous, cholagogic, febrifugal, and stomachic properties. Ears of wheat grain roasted over fires on the feast day of Saint John the Baptist (June 24) can be used to alleviate toothaches and prevent boils.

[**WHITE BRYONY** ⚕ *See* BRYONY, WHITE]

[**WHITE HELLEBORE** ⚕ *See* HELLEBORE, WHITE]

[**WHITE SANDALWOOD** ⚕ *See* SANDALWOOD, WHITE]

[**WILD ANGELICA** ⚕ *See* ANGELICA, WILD]

[**WILD RUE** ⚕ *See* RUE, WILD]

[**WILD THYME** ⚕ *See* THYME, WILD]

Fig. W.6. Fuchs, *Plantarum effigies,* 192.

WILLOW

Latin binomial (type sp.): *Salix alba* L.

French names: saule, osier blanc, saule blanc, sause (Languedocian), sausse*

Elemental qualities: cold and wet (∇)

Ruling planet: ♄

Zodiacal signatures: ♋ (tree) and ♐ (leaves)

Occult properties: The willow tree possesses anodyne, anti-inflammatory, astringent, febrifugal, and sedative properties. Its bark, which is high in salicin, is often referred to as "nature's aspirin."[5] Its seeds and oleaginous extract are anaphrodisiac in action. Apollo fashioned handcuffs from the pliant twigs of a willow (ἰτέα in Greek) and tried to bind Hermes for sacrificing two of his cattle, but the bands fell from Hermes's wrists and immediately began growing from beneath the cattle's feet.[6] The ancient Germans used willow (*Wida* in Old High German) to perform rhabdomancy, and medieval magicians employed its branches as dowsing rods for discovering metal ores and buried treasure. When worn as an amulet, a willow branchlet prevents nightmares and infernal visions.

---

*[Sédir's original list of variant names includes the following: "*Fitea,* for *Fitegæ* in Aeolian Greek, *Wida* in Old High German" (*Les plantes magiques,* 168). Here Sédir is following Lenglet-Mortier and Vandamme, who make the similar claim: "In Aeolian Greek, FΙτεα, ἡ, for FΙτεσα (waitasa) = willow" (*Nouvelles et véritables étymologies,* 179–80). The purpose of the authors' capitalized letters is to draw connections between the initial syllables in names for the willow in various languages, but Sédir appears to mistake the archaic Greek letter Ϝ, the *waw* or *diagamma* (lowercase ϝ), which fell out of use in the fifth century BCE, for the more familiar Roman letter *F.* At any rate, Greek ἰτέα indeed derives from ϝιτέα (which would be better transliterated as *witea*), and ϝιτέα from Aeolian ϝίτυς, and ϝίτυς in turn from Proto-Indo-European **weh$_1$y-,* meaning "to twist" or "to twine." As for Sédir's *Fitegæ,* this must be a typographical error for ϝιτέσα—that is, for Lenglet-Mortier and Vandamme's hypothetical Greek form between ϝίτυς and ϝιτέα. Most likely Sédir's intention in reproducing these variants here was to connect Lenglet-Mortier and Vandamme's "FΙτεσα (waitasa)" with Sanskrit वेतस (*vetasa*) or वैतस (*vaitasa*), meaning "rod," "stick," or "cane," and in so doing to linguistically link willow branches with the practice of rhabdomancy (from Greek ῥάβδος, "rod" or "wand," + μαντεία, "divination"). See further Lenglet-Mortier and Vandamme, *Nouvelles et véritables étymologies,* 84; cf. Monier-Williams, *A Sanskrit-English Dictionary,* 963 s.v. वेत and 968 s.v. वैतस. —*Trans.*]

[WINTER WOLFSBANE ☙ *See* WOLFSBANE, WINTER]

WOLFSBANE

Fig. W.7. Fuchs, *Plantarum effigies,* 49.

Latin binomial: *Aconitum napellus* L.

French names: aconit (napel), capuchon (de moine), capuze de moine, casque bleu, casque de Jupiter, char de Vénus, coqueluchon, napel bleu, madriette, tue-loup*

Elemental qualities: cold and dry (🜃)

Ruling planet: ♄

Zodiacal signature: ♑

Occult properties: Wolfsbane, or monkshood, has powerful anodyne, febrifugal, irritant, narcotic, and sudorific properties. According to the ancient Greeks, the λυκοκτόνον, or "wolfsbane," originated from the slavering mouth of three-headed dog Cerberus when Hercules dragged him from the underworld.[7] Its leaves heal old ulcers and inflammatory swellings of the lymphatic glands in the cellular tissue of the groin or armpit. Its roots, if harvested when ♄ conjuncts the ☉ and infused in wine, possess similar curative properties. Although sudorific and poisonous, the wolfsbane is good for paralysis, kidney and bladder stones, jaundice, and asthma. It also stops epistaxis, helps hair grow back, and serves as an antidote for venomous bites. It is one of the twelve sacred plants of the Rosicrucians.[8]

*[Sédir provides separate entries for *aconit* and *capuchon de moine,* which are variant names for the same species of *Aconitum.* See *Les plantes magiques,* 142, 146. These have been merged here to form a single entry. Sédir's inclusion of *pardalianches* as a variant for ★wolfsbane is likely due to the sixteenth-century binomial *Aconitum pardalianches* Fuchs, which is now commonly identified as herb Paris (*Paris quadrifolia* L.) and should not be confused with great false leopard's-bane (*Doronicum pardalianches* L.). In any case, the species name *pardalianches* does not belong here. —*Trans.*]

Fig. W.8. Parkinson, *Theatrum botanicum,* 318.

WOLFSBANE, WINTER

Latin binomial: *Eranthis hyemalis* Salisb.

French names: ellébore jaune, hellébore d'hiver, hellébore jaune

Elemental qualities: hot and dry (△)

Ruling planet: ♄

Zodiacal signature: ♈

Occult properties: Winter wolfsbane, or winter aconite, is one of the first flowers to bloom in the early spring. All parts of this plant are highly caustic, toxic, and dangerous for both humans and animals. In the Hermetic language of flowers, it is emblematic of calamity.

[WOODLAND GERMANDER ⚕ *See* GERMANDER, WOODLAND]

[WOOD SORREL ⚕ *See* SORREL, WOOD]

Fig. W.9. Fuchs, *Plantarum effigies,* 2.

WORMWOOD

Latin binomial: *Artemisia absinthium* L.

French names: absinthe, absince, absynthe, aloyne, aluine, aluyne, alvine, armoise amère, gìusses (Languedocian), herbe-sainte

Elemental qualities: hot and dry (△)

Ruling planet: ♂

Zodiacal signature: ♑

Occult properties: Wormwood has febrifugal, nervine, stomachic, tonic, and vermifugal properties. This herb is a receptacle of the lower astral realm. In a certain sense, it could justly be described as "the hashish of the West." Its flowering tops, when prepared spagyrically, can be used to attain certain altered states of consciousness and psychic awareness.

WORMWOOD, FIELD

Fig. W.10. Mattioli, *Discorsi*, 413.

Latin binomial: *Artemisia campestris* L.

French names: armoise rouge, armoise des champs, aurone des champs, aurone sauvage, blanchette

Elemental qualities: hot and dry (△)

Ruling planet: ♂

Zodiacal signature: ♈. Harvest after the full ☾ terminating the dog days of summer.

Occult properties: Field wormwood has antiseptic, cholagogic, emmenagogic, and vermifugal properties. When hung in the home as a talisman, it provides protection from malicious magical spells, lightning strikes, and evil spirits. When worn as an amulet, it dispels epilepsy and Saint Vitus's dance, otherwise known as Sydenham's chorea.

Y

Yarrow

Fig. Y.1. Fuchs, *Plantarum effigies,* 422.

YARROW

Latin binomial (type sp.): *Achillea millefolium* L.

French names: mille-feuilles, fenoulheto (Languedocian), herbe aux charpentiers, herbe aux cochers, herbe aux militaires, herbe de Saint Joseph, millefeuille, saigne-nez, sourcil de Vénus

Elemental qualities: hot, dry, and moderately wet (△)

Ruling planet: ☿ (*apud* Haatan, Piobb, and Bélus)

Zodiacal signature: ♈. Harvest when the ☉ and ☾ are in ♋.

Occult properties: Yarrow has antiseptic, astringent, odontalgic, febrifugal, and vulnerary properties. It is probably the best of all vulnerary herbs. Styptic powders made from its dried leaves not only stop bleeding instantaneously but also disinfect wounds.[1] Alchemists and Hermeticists refer to it by the ancient names *militaris,* meaning "soldier's-herb," and *stratiotes,* a transliteration of its Greek equivalent, στρατιώτης.[2]

Appendix 1
Occult Medicine

There are many different methods of curing diseases: allopathy, homeopathy, isopathy, surgery, chromokinetics, and electro-psychotherapy, for example, not to mention the many different disciplines of ordinary medicine.[1] The keen observer will notice among the more recent curative methods our doctors put into practice on a daily basis a host of procedures that, unbeknown to their inventors or practitioners, have their origins in the Hermetic sciences.

We define occult medicine as any therapeutic system that, when confronted with the pathological symptoms of the physical body, bases its diagnoses on an astral examination of the patient and treats the patient's life force in its invisible form. There are several divisions in this esoteric temple of Aesculapius—which is to say that occult therapists are aware of a certain number of truths, that each of their souls is endowed with specific magical powers, and that, consequently, each of the theories formulated in their respective disciplines comprises one aspect of the same universal truth. We make the following general distinctions among initiates of this esoteric temple:

1. The first group of initiates makes use of drugs appropriated from physical nature and energizes them by means of learned chemical preparations: these are the *spagyrists.*
2. The second group of initiates invokes and summons to their aid the healing agents and denizens of spiritual nature: these are the *magicians.*
3. The third group of initiates relies solely on the action of the forces of their inner being, which they energize by means of the will, and abandons all cures to the care of the divine goodness: these are the *mystics* and *theurgists.*

SPAGYRIC MEDICINE

In spagyric theory, human beings are microcosms, and as such, their bodies must therefore contain in some abridged form all of the material substances of the physical planet upon which they live. In other words, their physiological life is the result of the reciprocal actions of the four elements and three constituent principles (Salt, Sulfur, and Mercury). When an organic disorder arises, it is due to some disruption in the harmony among these seven forces. Spagyric healing consists in restoring this natural harmony either by dissolving congestions or by strengthening weaknesses. To this end, the spagyrist supplies the human organism with an external substance, not in its natural state but in a purified and exalted form.

All bodies of nature can serve as medicines—which is to say that all bodies, even and especially poisons, contain a balm whose extraction constitutes one of the most important operations of practical alchemy. Although Paracelsus is often considered the father of alchemical pharmacopoeia, the spagyric art was also held in high regard in antiquity, chiefly in India and Iran, and more recently the Taoist doctors have become its expert practitioners.

Among the spagyrists there are several different schools of thought. Some believe that chemical manipulation alone is enough to give any medicine the requisite therapeutic strength. Others believe that the spagyrist's will must plant a seed of active dynamism within the medical material. A third school of thought takes into account the astrological correspondences of the months, days, and hours. Finally, the last school of thought, which is that of the Western alchemists, believes that the light of nature—which is to say, the theoretical knowledge of matter and the alchemical process—must receive the collaboration of the divine light of grace for it to be effective. We shall see at the conclusion of this essay what effects these different methods produce in the invisible world.

MAGICAL MEDICINE

Magical medicine is quite different. It utilizes material objects as fulcrums for the operator's will and as capacitors for the forces of invisible agents. One can get a very crude idea of this process by recalling the cures of the mediumistic healers who would lend their vital forces to "spirits." But whereas spiritists abandon themselves to their instinct, pursuant to their level of knowledge and internal evolution, magicians operate according to rigorous and traditional rules. The will of the adept is capable of moving a band of more or less elevated invisible agents. The magician's powers may be confined to the cure of a single disease or only a few. These powers, acting either in the presence of the patient or at a distance, can cure the merest malaises and the most serious maladies by driving out or destroying the ailment. Circles, pentacles, and altars and correspondences of colors, aromas, regalia, talismans, and sacred names—such are the tools the magician employs. In this fundamentally occult art, inner revelation is the magician's only true master, and this is why Agrippa and Paracelsus sketch out its rudiments in deliberately obscure terms.

Magical healing consists in developing the astral sensibility within oneself; studying the changes of the astral forces in the invisible atmosphere of the planet; preparing a cleansed and purified material object such that its spirit—which is to say, its astral archetype—is inversely analogous to the astral force one wishes to evoke; and causing this force to descend to the planet, thus imitating the operation of the mystery of the real presence.*

Practical magic so conceived is a true and exact science, a hyperphysical mechanics with its own instrumentation and methods of calculation, and this is why it often exerts a special charm over artistic minds, whereas mysticism more often fascinates scientific minds.

Broadly speaking, there are two categories of magic in terms of morality: white magic, which is benevolent, and black magic, which is malevolent. There are also two categories of magic in terms of the operator's attitude: *admonitive magic* and *supplicative magic.* Magical admonitions can be given only to beings whose condition of existence is inferior to the human condition, such as elementals, animal spirits, and larvae or lemures, provided, of course, they are not giants among their species. In other words, to command a particular invisible agent, it is essential that none of the invisible beings who live in the orb of the magician's astral spirit be weaker than that invisible agent. This kind of magic can be very dangerous.

*[The mystery of the real presence refers to the theological doctrine(s) that Jesus Christ is substantially or *really,* rather than just symbolically, present at performances of the Christian ritual known as the Eucharist. —*Trans.*]

Magical petitions, on the other hand, are made by abandoning oneself to the egregore of a chain through prayer.* The biblical psalms are some of the most beautiful and enduring testaments of this type of magic. Many initiates of the Muslim brotherhoods use the verses of the Quran in this manner, and the Hindu mantrikas do the same with the verses of their sacred books, so, too, the Catholic faithful with their litanies and monastic liturgies. In supplicative magic, the practitioner's mind—taking this word in the astral sense—vibrates simultaneously and synchronously with the vibrations of others making the same petition with the very same words, and the egregore assumes the form of the sentiments expressed in these petitionary formulae, whether they be sentiments of healing, fortune, revenge, or death.

THEURGIC MEDICINE

Some initiatory schools teach that the human being or, more pointedly, the essential principle of the human being—namely, the will—is the only true power in the macrocosm. Sensitive beings, spirits, gods, demigods, demons, mountains, and seas—all are mere appearances that dissolve before the eye of the intelligence when it comes to understanding the universal interplay of the subtle forces, of which these higher objects in some way constitute their points of intersection. Any organic disorder capable of being perceived by such cerebrality must also be capable of being dissolved by it. This theoretical point of view is found in esoteric Buddhism and in rāja yoga. Exoterically, it takes shape in the schools of the magnetizers, who heal by operating on patients via the pure will. Each occult therapeutic system has its own secular copy and substitute: for the spagyrist, it is the pharmacist; for the magician, it is the psychic healer; for the yogi, it is the magnetizer, hypnotist, or suggestioner. But any absolutely pure mystical system can have no vulgar counterpart.

*[The French term *égrégore* derives from the ancient Greek adjective ἐγρήγορος, meaning "wakeful." The plural ἐγρήγοροι is employed as a substantive in the Book of Enoch to refer to the Watchers (literally, the "Wakeful Ones"). French *égrégore,* although originally a neologism of Victor Hugo, came to refer to "an artificial psychic entity created by a group of humans." See especially Bosc, "Égrégore," 151–53, *contra* Éliphas Lévi, *Le grand arcane ou l'occultisme dévoilé,* 255–59. —*Trans.*]

We admit that our division of therapeutics into these three main groups is somewhat artificial. In traditional history, as in social life and even in the initiatic schools, we find a variety of transitional forms. For example, there are some forms of spagyrics, in addition to the aforementioned schools of thought, that verge on magical medicine, such as the various methods by which knowledge of the virtues of plants, stones, and animals is acquired through the revelations of spiritual beings. The iatric procedures of healing via the *mumia*—that is, through magical action on the image of a diseased organ[2]—and healing via telepathy represent transitional forms between magic and magnetism. Furthermore, one may equally characterize the prayers of the Christian Scientists as a transitional form between magnetism and supplicative magic.

Theurgic therapeutics, however, is set apart from all the others. What follows is the general framework of its theory and practice.

What is a disease? For the spagyrist, it is a constitutional alteration. For the magician, it is the work of a malevolent invisible agent. For the magnetist, it is a rupture in the fluid equilibrium. Whether the intellectual horizon of the healer is limited to the physical or astral plane will determine whether he or she seeks for the cause of a disease in the one or the other of these planes: in environmental influences, microbes, heredity, or diet, for example, or in astrological influences, enchantments, spiritual vampirism, or astral parasitism. The human principles susceptible to these diverse intrusions are obviously the physical body and the astral body. The human will, however, remains outside both spheres of influence. There must therefore be another, primordial cause of diseases: this is what the mystic calls sin.

The mystic questions the orthodox tradition, which asserts that the original state of *Homo sapiens* was beautiful, perfect, and healthy, that the whole of nature worked in this species favor, that pain descended upon the world at the first wrongdoing, and that one of its forms was disease. The masters of occultism, with Fabre d'Olivet in the lead,[3] have taught us how the effects of the Fall of the universal Adam came to be mitigated by their extension in space and their multiplication over time, but however low the human race may have fallen, it has never lost that magnificent power it possessed in its majesty—namely, the power to create beings through action. Each human action generates a form in the invisible world. All seers know this to be true from experience. Depending on their capabilities, the seer will perceive this creation either as a geometrical form or as an organic being. In either case, however, the created being radiates from and reacts to the source of its emanation. Now, because every action is either good or evil, it will bear fruits either of health, intelligence, and happiness or of sickness, ignorance, and misfortune. The rich man who receives the beggar by beating him with a stick is responsible for his action, with respect not only to his will but also to the organ that performed the action. If in a subsequent incarnation of his soul the rich man should receive a body with a paralyzed arm, a spagyrist, magician, or magnetist might be able to cure him of his paralysis, but in reality they will merely remove it from the man's arm and transfer it to some other place. Only the theurgist can obliterate the disease by eradicating from the rich man's soul the darkness that was the first cause of his deficiency.

We could relay a host of other details on the subject of occult epidemiology, but we must limit ourselves to what we have written here for the time being.[4] We hope that the few ideas we have expressed above are sufficient to incite readers to study the subject of occult therapeutics in greater detail.

Appendix 2
Paracelsian Physiology

In Paracelsian physiology, the microcosmic human being is divided into three principal parts:[1]

1. The *soul,* which has divine origins, comprises the free will and the will. The soul of the sage is entirely capable of governing the other two principal parts.
2. The *spirit,* which has astral origins, comes from the firmament, or "astral wheels"—that is, the ethereal world comprised of the zone of the fixed stars and the planets.
3. The *body,* which has terrestrial origins.

Paracelsus was a physiologist, a naturalist, a physician, and a magician. He was neither a theurgist nor a mystic, however, and so his investigations concentrate not on the soul but rather on the body and the interactions between its dynamic functions and on the spirit and its relationship to the body.

The spirit originates in an aerial, stellar chaos, and it carries within it the summation of all astral forces. It is otherwise known as the *mens,* or "mind," and its faculties encompass all facets of physiological life and what the Paracelsian school would call "psycheology." When a soul is called to descend to Earth, it selects a *spiritus* from the astral sea that accords with its own nature, with the nature of its earthly parents, and with its future incarnation. The spirits of the parents participate in the generation of the spiritual seed of the child. This spiritual seed, which contains the vital spirit in a potential state, will develop in the fluidic atmosphere of Earth and give birth to its organic life. Paracelsus understands its development as an unstable equilibrium or polarization. In the static state, this dualism is known as mind and body; in the physiological state, absorption and elimination; and in the biological state, intellectual life and vegetative life. Thus, the visible being and the invisible being exist in a relationship of reciprocity. The former acts upon the latter by the perceptions; the latter acts upon the former by the imagination. The imagination, therefore, should be regarded as a *living faculty,* much like Éliphas Lévi's "plastic mediator."[2]

The spiritual being descends into matter through the five senses; the material being ascends into the astral by purifications and digestions. When a being is conceived, the mind of the fetus receives the impressions of the stars through the mind of its mother, hence the importance of the mother's psychic well-being during gestation. When the child is born, it acts first according to the inclinations of the stars and then according to its atavism, but all of its inclinations remain dirigible and are corrigible by the will. The will, whose instrument is faith, therefore governs and directs the motor skills.

Just as the material seed produces a vital humor that is physiological in nature, the radical humor of the spirit, which is determined by the collaboration of the spirits of the parents, produces a vital spirit. The vital spirit dies as soon as life is extinguished—this happens at a predetermined moment, which can be ascertained astrologically. Time exerts upon the vital humor of organic life a corruptive action that can be slowed but never stopped. Thus, according to Paracelsus, terrestrial immortality is impossible.

The vital spirit is the source of strength, power, life, and balm. The balm, whose seat is in the heart, is a force of conservation that specializes in the various anatomical organs. The body is thus the battlefield of two forces: the one astral, consisting of the three constituent principles Salt, Sulfur, and Mercury, which seeks to return to its cosmic matrix; the other electromagnetic, existing in the blood, the membranes, and the flesh, which fights against the former by means of the preserving agent of the balm.

The vital spirit, or liquor, is the source of all mental and psychic qualities. This invisible and impalpable power can dominate the body via the *mens* through thought. Such work is the province of the will, at least with respect to its development. The vital spirit specializes in directing each of the organic functions, which are seven in number, corresponding to the seven planets:

1. The heart has a Solar spirit (☉).
2. The brain has a Lunar spirit (☾).
3. The spleen has a Saturnian spirit (♄).
4. The lungs have a Mercurial spirit (☿).
5. The kidneys have a Venusian spirit (♀).
6. The gallbladder has a Martian spirit (♂).
7. The liver has a Jupiterian spirit (♃).

Each of these spirits sets out from the heart to its particular organ and returns back to the heart. The matter of the body itself represents Earth.

The planets exercise powerful influences over the body, and the organs—the planets of our body—exert powerful influences over each other. The combined motion of the planets emanates a spiritual exhalation or radiation that mingles with the igneous atmosphere of Earth, and this in turn produces an astral essence that transmits the planetary influx to our spirit. It is not the astral essence that furnishes the substance of our spirit but rather the very medium through which the planets move. In other words, the astral medium, being put into vigorous action by the atavistic force of the parents, forms the essence of the spiritual seed.

The astral medium is vitalized and organized under the direction of the "universal M," or *magnale magnum*—that is, the magnetic principle of worlds. On the microcosmic scale, this same magnetic principle is the tie that binds life to our body. It fights against the action of a caustic agent known as the *destructive archaeus,* the primary cause of decrepitude, which seeks to reduce the body to its last matter through heat and organic combustions.

Thus, each child is born with its own unique celestial firmament and planetary configuration: this is the *ens naturale,* or natural *ens.* And just as the seven planets are found within the human body, so, too, are the four elements:

1. FIRE (🜂) burns in and exits through the eyes.
2. WATER (🜄) flows in and around all the vessels.
3. AIR (🜁) is the means and medium of bodily motion.
4. EARTH (🜃) enters the body in the form of food.

The four humoral temperaments, which should not be mistaken for mental or psychic qualities, are in turn determined by the "flavor" of the natural *ens:*

1. Bitterness produces a *choleric temperament.*
2. Acidity produces a *melancholic temperament.*
3. Sweetness produces a *phlegmatic temperament.*
4. Salinity produces a *sanguine temperament.*

All things exist within the microcosmic human being: the movements of the planets, the properties of the elements, the substances of the three lower kingdoms of nature, the atmospheric fluids, and so on. But each of these things exists within the human organism in a *virtual* rather than a substantial state. For example, gold exists within every human body; however, this gold does not have the same appearance as the gold we find in nature.

The physical substances of the human body are divisible into four main groups:

1. Blood
2. Fat and muscles
3. Water of the marrow and bones
4. Resins and gums of the viscera and tendons

All mixed bodies can be divided according to the same schema and so used to create invaluable medicines. It is expedient to recall the following in the administration of such medicines:

- The life of *Homo sapiens* is an astral balm or celestial fire.
- The life of the blood is a saline spirit (*spiritus salis*).
- The life of wood is the resin.
- The life of the plant is the terrestrial liquor (*liquidum terrae*).
- The life of the metal is a hidden fat buried in the Sulfur.

Lastly, it is important to note that all beings in the three lower kingdoms of nature have Salt for a body, Mercury for a spirit, and Sulfur for a soul.

Everything in nature, moreover, is a mixture of good and evil, of pure and impure. Every food therefore contains both a balm and a venom, a force of conservation and a force of destruction.[3] The human stomach, in fact, is the greatest of alchemists. When it functions properly, it transmutes foods into nutritive elements and eliminates poisons absorbed through the openings of the body. In a similar manner, mercury is eliminated through the skin, white sulfur through the nostrils, arsenic through the ears, salt through the gallbladder, and sulfur through the anus. Furthermore, every function of organic life is governed by a vital spirit, or *archaeus.* Thus, there is an *archaeus* of digestion, an *archaeus* of respiration, an *archaeus* of excretion, and so on.

Just as the stomach protects the body from internal corruptions brought on by the ingestion of foods, the skin acts as a shield and protects the body from external corruptions brought on by the environment. Diseases enter or, more precisely, *develop* in the body when the balm is caught by surprise, for we carry the germs of all diseases within our *spiritus,* which is imperfect as a result of the Fall of our first father. Thus, every sickness is a form of expiation.

Diseases enter the body either through the *ens,* our physical being, or through the *mens,* our psychic being. In the latter case, they can originate in the pernicious influxes of planets, in the enchantments of black magicians, or even in our own imaginations. In cases of succubation and incubation, however, it is more often than not the patients who have harmed themselves by opening a door, wittingly or unwittingly, to the actions of evil spirits. In general, all diseases deriving from the natural *ens* may be divided into four types:

1. Chronic diseases (*morbi chronici*) come from the planets.
2. Acute diseases (*morbi peracuti*) come from the elements.
3. Natural diseases (*morbi naturales*) come from the temperaments.
4. Diseases of color (*morbi tingentes*) come from the humors.*

The death of the human organism consists in the loss of its vital air, the depletion of its spirituous balm, the extermination of its natural light, and,

*[According to E. Wolfram, the Paracelsian diseases of color, or *morbi tingentes,* comprise the various infectious diseases (see Wolfram, *The Occult Causes of Disease,* 210). —*Trans.*]

finally, the separation of its body, soul, and spirit. The physical body returns to its terrestrial matrix. The celestial or spiritual body continues to live on for some time—it is this body that sometimes appears in the form of ghosts, specters, or phantoms—until it dissolves back into the aerial chaos. And lastly, the soul returns to its divine source.

It is to be noted as well that when human beings die, their constituent parts continue to act for some time. Their particular vital spirits do not exit the body immediately. These effluvia constitute the *mumia,* which, when properly extracted, has a magnetic force capable of effectuating the most remarkable cures and panaceas.[4]

Appendix 3
On Opium Use

Much has already been written both for and against the use of excitants.[1] Our purpose here is not to provide an exhaustive bibliographic overview, since most of our readers are probably already familiar with the principal works on the subject. Many celebrated artists and writers, such as Honoré de Balzac and Charles Baudelaire, have occupied themselves with this question, and most, with the exception of Thomas de Quincey, have spoken out against the use of such drugs. On the other hand, the journal *L'Initiation* has published copious studies on the use of opium and hashish, the most authoritative of which are those by Matgioï, Numa Pandorac, and Gabriel de Lautrec, all of whom ring the bell of optimism.[2] Here, we would like to counter these views and toll a bell of pessimism.

There are two kinds of opium smokers: thrill seekers or lovers of sensation and spiritual seekers or lovers of the secret science. Only the former are known in the West, whereas the latter are found throughout the East, primarily in India, Indo-China, China, and Malaysia. Since plenty has already been written about the morality of the former, we would like to focus on the latter, who have not hitherto been discussed in any meaningful way in print. Readers, of course, are welcome to take the contents of these pages for pure fantasy or true reality, for it is the duty of every reader to examine, weigh, and judge what he or she reads.

Where does opium, or rather the plant that supplies it, come from? We know that human beings are masters of the universe and that the inferior creatures came into being for their benefit. Some of the inferior creatures emerged alongside them, while others owe their existence to the mannerisms, general behaviors, and collective habits and customs of these dethroned monarchs. Thus, a cruel people incites the invisible attraction and progressive materialization of cruelty, and a few centuries of uninterrupted cruelty will give rise to broods of ferocious beasts. Likewise, an unsanitary people will summon unsanitary creatures, and their great-grandchildren will one day suffer an indestructible plague of vermin.

Millions of years ago a fragment from a neighboring planet collided with our Earth and filled it with its flora, fauna, and humanity. The adaptation of all these beings to the biological conditions of Earth occurred slowly, but their internal character remained largely unchanged. These humans preserved a subtle form of egoism that removes them from practical life, drives them toward escapism, and deprives them of any scruple over the means by which they free themselves from action. They actively sought to bring about what they desired—for desire is an active force—and gradually concretized their ideal by removing it from the domain of the spirit and bringing it progressively closer to Earth, and because of the mysterious ties that bind the spiritual faculties of humans to those of plants, a new being came to make contact with the "place down here" in the form of a plant.

Such is the legend we heard about the origin of the *opium poppy, an etiology that accounts for the pleasures all opium users seek—namely, to forget life, work, and worry, to forget the rain, the cold, and the mud, to forget for a moment the neverending desire for progress, to find oneself as one is, to destroy effort, negate progress, and lethargize all that exists outside individual tranquillity, for such is the effect of the infamous brown powder.

It is important to realize that the vast majority of young people who smoke hashish do so only to "impress the bourgeois," and by "bourgeois" we mean the materialist hidden in the depths of each one of us. All users, and nervous personalities especially, experience fear during that moment of emptiness that precedes the narcotic reverie, unless, of course, they have used the drug at so low a dose as to produce little or no effect, as though they were smoking boiled tobacco. When the narcotic experiment is a success, however, the hashish itself does not create anything, but rather, by exalting the individuality, it develops the present disposition of the individual to its final limit under the guise of a particular sensation. In this state, users can experience nightmarish hallucinations, and if their use of the drug becomes habitual, physiological habituation can lead to a veritable poisoning and a wholescale depletion of users' energy levels, both mental and physical. All these effects can be summed up in the following symptom: loss of will. Later, we shall discuss by what sophistry some have tried to turn this exhaustion of energy into the absolute of science.

But let's leave these dilettantes behind for the moment. We would like to discuss a category of experimentalists whom we fear are more numerous than is generally believed. Suppose a sincere spiritual seeker learns that the astral realm is the key to a great many occult conceptions and that opium can help the user project into the astral, and suppose this incites the resolute seeker to experiments *in anima vili.** Let me stress that I neither have nor desire the right to influence the freedom of any individual. Our purpose here is to reveal the flip side of the narcotic coin and expose what inconveniences may arise from such practices.

Most opium users, as we have said, pursue the intoxication of "artificial paradises" to escape daily platitudes or general fatigue brought on by constant labor. Users seek to avoid boredom like soldiers seek relief from barrack duties. But users and potential users would be wise to recall two axioms. The first is that it is we ourselves who, in the past, have determined or chosen our present destiny, either because our past faults require correction in the form of labor and toil or because our soul has willingly accepted a future rich in merit that necessitates difficulties in the present. Any refusal to move forward or delay in this progression is therefore unreasonable.

The second axiom follows logically from the first. We must, with our utmost energy and to the greatest extent possible, fulfill our daily duty and perform all the actions everyday life imposes on us. To do this requires constant effort and self-supervision. Such an uninterrupted succession of personal effort and endeavor may well be more difficult than any isolated act of heroism. The Belgian playwright Maurice Maeterlinck beautifully illustrates this point in *The Treasure of the Humble.*[3] Many in our modern era seem to believe that a single act of brilliance outweighs an obscure life devoid of

*[The expression *in anima vili* comes from the famous Latin dictum: *Fiat experimentum in anima* (or *corpori*) *vili,* or "Let the experiment be performed on a worthless soul (or body)," which Immanuel Kant and others used to justify the use of animals in experimental research. In Sédir's view, opium users essentially become their own guinea pigs. —*Trans.*]

glory, but whoever ascribes to this view is sorely mistaken. From a moral, intellectual, and spiritual point of view, we all have within ourselves *exactly what we need.* To quest for something more or other may be nothing more than avarice and may well attract even greater responsibilities and hardships. If we desire knowledge and science, the heavens will fulfill our desire as soon as we begin doing our daily duty with all our heart. If, on the other hand, we do not make every effort to live according to spiritual law, we shall add yet another breach to our soul by trying to acquire powers or insights we have neither earned nor deserve.

All paths must be followed to their ultimate destinations, the narrow, obscure, and disagreeable as much as the broad, well lit, and convenient. There is therefore no reason to envy the pathways of our neighbors, who may possess admirable and seemingly innate qualities, for whom life may be easy, or whom fortune may favor. Just as we must force ourselves to welcome someone who has been unkind to us as we would a friend, so we must smile at all our trials, tribulations, hardships, and sufferings. We must be able to have sympathy for all things.

These arguments are intended for those who would use opium only to experience relief, oblivion, or pleasure. But there are other opium users, the vast majority residing in Eastern countries, who ask something different from the bamboo pipe. Their goal is the tao, nirvana, and *paramātman*—which is to say, serenity, immutability, and impassibility.* Prior to reaching this goal, however, they pass through many invisible, enchanting, and even infernal kingdoms, and they experience many pitfalls and charms from which they must escape. Such astral trials and tribulations help the initiate develop certain spiritual faculties, for blunders are the ferment of the soil of the soul.

To be sure, it is an egregious error to assume that everything in nature is free. Those who desire to make themselves strong must compel themselves for years to incessant labor on the visible plane. To be successful, however, the appropriate conditions must be met on one of the invisible planes. As for those who desire powers or wisdom, how much longer and

*[Elsewhere Sédir translates *paramātman* (परमात्मन् in Sanskrit), which is usually defined as the primordial self or the self beyond, as the supreme soul. See Sédir, "L'esotérisme indou: Les avatars," 288. —*Trans.*]

harder must they work to acquire and develop them? Suppose some laborers wish to enter the stock exchange. Not only must they become brokers themselves, but they must first have made a meaningful amount of money to enter the market. The situation is no different for spiritual seekers: to acquire any real occult power or knowledge, one must not only work hard but also pay a price.

This is not because there is some malicious, all-powerful genius who demands that we work hard for such things but because any real power or living science is an *active force,* a force a thousand times more active than a grain of radium or a few grams of dynamite. Even the simplest intellectual conception implies a veritable encyclopedia of preliminary data that one must comprehend first before grasping it. How many of us truly understand or can explain the force of gravity or, for that matter, any phenomenon whose quotidian nature has stripped away our interest in it? And how many of us truly understand or can explain the modus operandi of even the simplest magnetic cures? To effect such a cure, how many thousands of minute beings must be set into motion, not only in patients, so that they feel confident and become receptive, so that their nervous currents change and their physical fluids accelerate or delay their transmission, and so that, sometimes, even their flesh itself changes form, but also in healers, so that they accurately perceive the symptoms of a disease and establish a proper diagnosis, so that they set their own vital forces in motion and bring them into contact with those of the patient, and, finally, so that they triturate and manipulate the disease until it is diminished, destroyed, removed, or transformed? All of these actions are performed outside the realm of matter, and yet they may all be understood, weighed, and analyzed by the mind and perceived in their effects by the nervous apparatus. This briefest of outlines highlights only the principal gestures of such cures. Irrespective of the method employed, however, they always progress in the same fashion, even when performed solely by the imposition of hands.

Psychic abilities are a lot like plants: they must be sown on favorable ground and require time to grow and develop, time to flower, and time to bear fruit. If we want to hasten their growth, we must nurture them as a gardener would a hothouse flower, with a great deal of artistic finesse and care, and even then we may still obtain only a delicate and feeble product that withers and fades at the first wind. If the faculties acquired as a result of magical, volitional, or mental training are liable to perish as soon as they leave their greenhouse shelter—which is to say, as soon as the material body is delivered unto death—then there is all the more reason to believe that any faculties acquired from the use of an excitant like opium will last no longer than a single existence. Why, then, waste time on such a precarious practice of so little profit?

Let's now examine initiated opium users. By "initiated" we simply mean persons who are preoccupied with intellectual advancement or with developing a strong will and a sound judgment, who are accustomed to mental examination, and who are capable of deep contemplation for extended periods of time. Of course, when they use opium, they will enter a realm that was hitherto unknown to them. But how will they discern what laws prevail in this realm, what probable dangers may lie in wait, or which fruits are healthy and which are poisonous? By deduction? And how will these users know whether this unknown realm will look anything like the known? By intuition? Can they be certain that the crystal mirrors of their private lives will not be tarnished or laden with deceptive images by the inhabitants of this realm or by the very atmosphere in which they desire to bathe? Will their will still be all powerful? In theory, yes. In practice, maybe not. Will their brain remain lucid? Will their senses remain calm? There are temptations even more subtle than those of this world.

Whether or not any or all of these dangers manifest, there are some chains that mortals simply cannot break. Here's what we mean. Suppose a young man in a school of arts and crafts is taken out and placed in an elementary workshop where he plies his trade reasonably well and manufactures devices to perform his job with greater efficiency and less fatigue. Opium is comparable to one of these devices. Now, if he wants to leave the working class and become an engineer—which is to say, if he desires to ascend to a higher plane—he has no choice but to go back to school. Thus, each one of us, in accordance with the logical consequences of our invisible pasts, belongs to a spiritual "caste," for lack of a better term. Each one of us is capable of becoming first in our own respective caste, but to be permitted to pass into another, higher caste requires radical purification. In other words, we must be capable of suppressing whatever causes may have brought us into our present caste by exhausting their consequences. This payment cannot be made with the mind, through magnetism, in the astral, or by means of magic. It can only be made by the spiritual heart, and its currency is spiritual purification. Spiritual purification alone can furnish the soul with the light that will remain beyond the torrent of generations.

If you want to go from Paris to Brest, it is much easier to take the train than to walk on foot. A truism, to be sure, provided you have enough money to purchase a ticket. But, more pointedly, why do you desire to go elsewhere? Our duty is to the place in which we ourselves chose to be born; the thirst for someplace other is just another form of escapism, if not fruitless puerility.

We all have within ourselves particular physical faculties, muscles, passions, preconceived notions, and so on. It is not by extinguishing our innate nature that we improve ourselves but by combatting our bad qualities and cultivating our good qualities. Opium use, on the other hand, encumbers physical activity, passion, and even the desire for knowledge itself, leading our thousands of invisible energies down the road to apathy and bringing the legions of cells traversing our brain into a state of drowsiness as they wait to be transported to the invisible realm and there build that dismal and dead palace pseudo-intellectuals call the "temple of deliverance."

Of course, we do not mean to suggest that investigations into the mysteries are forbidden, only that investigations into the mysteries should not be carried out by illicit means. The invisible realm is intimately connected to all aspects of our daily life; every minute of every day, it offers us ample proof of its existence and assistance, such that we do not lack a subject of study. As for the most common of its manifestations—namely, dreams—do they not furnish us on a daily basis with reflections, analogies, correspondences, and teachings? We must closely guard what reflections our dreams provoke in us and not generalize them, and the same goes for anything else we receive from nature. We walk down wicked paths only when we desire to go our own way and follow our whims and fancies. A sad commentary, to be sure, but one we can prove to ourselves on a daily basis.

It is time now to close our sermon on opium use. Artistic types will probably laugh it away because of its defects in form or its disorderly arrangement. We accept all of their criticisms in advance, so long as they subscribe to our conclusion that pride is the perdition of humanity and that it is counterproductive to search for artificial foods when the self already provides us with so many natural foods. It is better to advance one step together with all of humanity than to go one mile alone in an uncharted forest, where our footprints will quickly be erased and all our courage and presumptuous efforts reduced to naught.

Glossary of Herbal Actions

ABORTIFACIENT plants (or **abortifacients**) contain substances that induce abortion. From Latin *abortus,* "abortion" or "miscarriage," + *faciens* (genitive: *facientis*), the present participle of *facere,* "to cause."

ALEXITARY plants (or **alexitaries**) ward off contagions and infectious diseases. The old pharmaceutical term *alexitary* has no exact modern equivalent. Unlike an ANTIBACTERIAL or ANTIMICROBIAL, an alexitary is a form of *preventive medicine.* From the Greek adjective ἀλεξητήριος, meaning "able to keep off, defend, or help," but as a neuter substantive ἀλεξητήριον takes on the meaning of "remedy," "medicine," or "protection," or even "charm against (a specific disease)."

ALTERATIVE plants (or **alteratives**) promote health and healing by changing or altering the processes of the body (rather than by evacuating something from the body). From the Medieval Latin *alterativum,* "productive of change" or "causing alteration."

ANAPHRODISIAC plants (or **anaphrodisiacs**) suppress the libido and reduce sexual desire. From Greek ἀναφρόδιτος, "insensibility to love" or "without sexual desire," a compound made up of the Greek alpha privative (ἀ- before consonants, ἀν- before vowels), expressing negation or absence, + ἀφροδισιακός, "inducing sexual desire." Synonymous with *antaphrodisiac* and *anti-aphrodisiac,* both terms being formations that utilize the Greek preposition ἀντί, "against," in place of the α privative, "without." Cf. APHRODISIAC.

ANESTHETIC plants (or **anesthetics**) induce insensitivity to pain, most often in combination with a loss of consciousness. From Greek ἀναίσθητος, "without sense or feeling," a compound made up of the Greek α privative (ἀν- before vowels), "without," + αἰσθητικός, "of or relating to sense-perception (αἴσθησις)." Cf. ANODYNE.

ANODYNE plants (or **anodynes**) relieve or allay pain, but without a loss of consciousness. From Greek α privative (ἀν- before vowels), "without," + ὀδύνη, "pain of body." Roughly synonymous with *analgesic,* from Greek ἀναλγησία, "insensibility." NB: Sédir's dictionary uses "anodyne" instead of "analgesic" because the latter is so often confused with "anesthetic." Cf. ANESTHETIC.

ANTHELMINTIC plants (or **anthelmintics**) prevent, destroy, or expel intestinal worms. From Greek ἀντί, "against," + ἕλμινς, "intestinal worm." Synonymous with VERMIFUGAL.

ANTIBACTERIAL plants (or **antibacterials**) destroy bacteria, inhibit their growth, and prevent their reproduction. From Greek ἀντί, "against," + βακτήριον, a neuter diminutive of βακτήρια, meaning "staff" or "cane." Antony Leeuwenhoek, who first observed bacteria in 1676, originally called them *animalcules,* meaning "tiny animals," but the term *bacteria* (plural of *bacterium*) later came into favor because the first bacteria discovered were shaped like tiny rods or canes. Cf. ANTIMICROBIAL.

ANTIBILIOUS plants prevent or cure bilious disorders caused by an excessive secretion of bile, such as indigestion, stomach pains, constipation, or excessive flatulence. From Greek ἀντί, "against," + Latin *biliosus,* "full of bile (*bilis*)" or "biliousness." Cf. CHOLAGOGIC.

ANTIEDEMIC plants (or **antiedemics**) counter edema or swelling caused by the accumulation of excess fluids in the tissues of the body. From Greek ἀντί, "against," + οἴδημα, "swelling" or "tumor," the latter being a derivative of the verb οἰδεῖν, "to swell."

ANTIEMETIC plants (or **antiemetics**) stop or reduce nausea and vomiting. From Greek ἀντί, "against," + ἐμετικός, "causing vomiting." Cf. EMETIC.

ANTIGOUT plants relieve or prevent the onset of gout, a form of inflammatory arthritis characterized by severe pain, tenderness, and redness of the joints, chiefly those in the feet and hands. From Greek ἀντί, "against," + Latin *gutta,* "drop (of fluid)"—in antiquity, the disease was believed to be caused by the "dropping" of viscous humors from the blood into the joints, and hence its archaic name, dropsy.

ANTIHYSTERIC plants (or **antihysterics**) counter or calm hysteria, a nervous disorder characterized by excessive or uncontrollable emotion. From Greek ἀντί, "against," + ὑστερικός, "suffering of the womb (ὑστέρα)" or "hysterical." Hysteria, a Neolatin medical term no longer in clinical use, was formerly thought to be caused by a dysfunction of the uterus and hence to be peculiar to women.

ANTI-INFLAMMATORY plants (or **anti-inflammatories**) reduce inflammation, excessive redness, or swelling in a part or parts of the body. From Greek ἀντί, "against," + Latin *inflammatus,* the past participle of *inflammare,* "to set on fire."

ANTIMICROBIAL plants (or **antimicrobials**) destroy microbial life and counter affections caused by microorganisms. From Greek ἀντί, "against," + French *microbe,* a portmanteau of the Greek words μικρός, "small," + βίος, "life." Synonymous with *antibiotic,* from Greek ἀντί, "against," + βιωτικός, "of or relating to life (βίος)," specifically the life of invasive microorganisms. Cf. ANTIBACTERIAL.

ANTIOXIDANT plants (or **antioxidants**) inhibit the process of oxidation in the body and the production of unstable atoms or "free radicals," which damage cells and cause illness and aging. From Greek ἀντί, "against," + French *oxidant* (modern French: *oxydant*), "an oxidizing agent," the present participle of *oxider,* "to oxidize."

ANTIPRURITIC plants (or **antipruritics**) relieve itching. From Greek ἀντί, "against," + Latin *pruritus,* "itching." Cf. IRRITANT.

ANTIRHEUMATIC plants (or **antirheumatics**) counter rheumatism and prevent or reduce joint damage caused by inflammatory disorders. From Greek ἀντί, "against," + ῥευματικός, "subject to a humor, flux, or discharge from the body (ῥεῦμα)."

ANTISCORBUTIC plants (or **antiscorbutics**) prevent or cure scurvy, a disease caused by a deficiency of vitamin C that is characterized by bleeding gums and the opening of previously healed wounds. From Greek ἀντί, "against," + Medieval Latin *scorbutus,* "scurvy."

ANTISEPTIC plants (or **antiseptics**) destroy microorganisms that cause sepsis and septic diseases. From Greek ἀντί, "against," + σηπτικός, "of or relating to putrefaction (σῆψις)."

ANTIVINOUS plants prevent or dispel the effects of drinking excess alcohol and are useful in the treatment of alcohol addiction. From Greek ἀντί, "against," + Latin *vinosus,* "fondness of wine."

APERIENT plants (or **aperients**) relieve constipation. From Latin *aperiens* (genitive: *aperientis*), the present participle of *aperire,* "to open." Aperients are mildly LAXATIVE in action. Not to be confused with APERITIVE, which stems from the same Latin verb.

APERITIVE plants (or **aperitives**) stimulate the appetite. From Medieval Latin *aperitivus,* a derivative of the verb *aperire,* "to open." Not to be confused with APERIENT, which stems from the same Latin verb.

APHRODISIAC plants (or **aphrodisiacs**) increase sexual desire. From Greek ἀφροδισιακός, "inducing sexual desire," a derivative of ἀφροδίσιος, "belonging to the goddess of love"—that is, to Aphrodite (Ἀφροδίτη), whose name means "sexual love" or "pleasure." Cf. ANAPHRODISIAC.

APOTROPAIC plants (or **apotropaics**) ward off or expel evil, especially when worn as amulets or installed as talismanic devices. From Greek ἀποτροπαῖος, "averting evil," a derivative of the verb ἀποτρέπειν, "to turn away."

AROMATIC plants (or **aromatics**) possess strong and often pleasant odors that stimulate the body or parts of the body and have a variety of therapeutic and medicinal applications. From Greek ἀρωματικός, "aromatic," a derivative of ἄρωμα, "aroma."

ASTRINGENT plants (or **astringents**) contract the tissues of the body and stop the flow of discharges of blood or mucus. From Latin *astringens* (genitive: *astringentis*), the present participle of *astringere,* "to bind together." Cf. STYPTIC.

BALSAMIC plants (or **balsamics**) are a class of AROMATIC herbs that have a soothing or healing effect on the human organism. From Greek βάλσαμον, "balsam," the ancient plant name for both the balsam tree, *Commiphora gileadensis* (L.) C.Chr., and the herb *costmary, *Tanacetum balsamita* L.

BECHIC plants (or **bechics**) prevent or relieve coughs. From Greek βηχικός, "of or for a cough (βήξ)." Synonymous with *antitussive,* from Greek ἀντί, "against," + Latin *tussis,* "cough."

BOTANOMANTIC plants (or **botanomantics**) fortell future events through their incorporation into a variety of ritualistic, divinatory operations. The term *botanomancy,* from Greek βοτάνη, "plant," + μαντεία, "divination," is often defined as divination by burning tree branches or leaves, but this is merely one form of plant divination. Not all forms of plant divination require the burning of vegetal matter. For example, whereas daphnomancers draw omens from the burning of bay branches and leaves, sycomancers draw omens from the amount of time it takes a plucked fig leaf to dry (see Sédir's entries *bay and *fig in part 3).

CALEFACIENT plants (or **calefacients**) cause a sensation of warmth in the body and dispel chills. From *calefaciens* (genitive: *calefacientis*), the present participle of *calefacere,* "to make warm." Cf. REFRIGERANT.

CALMATIVE plants (or **calmatives**) have a soothing, relaxing, or quieting effect on the mind and body. From French *calmatif,* a derivative of Old French *calmer,* "to make still or quiet." Calmatives are mildly SEDATIVE in action.

CAPILLARY plants (or **capillaries**) stimulate hair growth. From Latin *capillaris,* "an ointment for the hair (*capillus*)."

CARDIOTONIC plants (or **cardiotonics**) strengthen and counter affections of the heart by increasing its efficiency, improving the contraction of its muscles, and increasing blood flow. From Greek καρδία, "heart," + τονικός, "of or relating to stretching or contraction," a derivative of τόνος, "tone." Cf. CORDIAL.

CARMINATIVE plants (or **carminatives**) reduce or expel flatulence and allay pain in the stomach and bowels. From Old French *carminatif* or Latin *carminatus,* "healed (by incantation)," the past participle of *carminare.* Each of these terms stems from Latin noun *carmen,* meaning "song," "incantation," or "magical formula," from which modern English *charm* also derives.

CATHARTIC plants (or **cathartics**) stimulate the evacuation of the bowels and accelerate the process of defecation by increasing the bulk of the feces. From Greek καθαρτικός, "of or fit for cleansing or purifying (κάθαρσις)." Cf. LAXATIVE.

CHOLAGOGIC plants (or **cholagogues**) promote the flow of bile from the body, specifically the discharge of bile from the liver and gallbladder by purging it downward. From Greek χολαγωγός, "carrying off bile," a compound made up of χολή, "bile," + ἄγωγος, "drawing forth or eliciting."

CICATRIZANT plants (or **cicatrizants**) promote healing through the formation of scar tissue. From French *cicatrisant,* a derivative of Latin *cicatrix,* "scar."

CORDIAL plants (or **cordials**) have a stimulating and invigorating effect on the human organism, especially on the heart and circulatory system. From Medieval Latin *cordialis,* "of or relating to the heart," a derivative of Latin *cors* (genitive: *cordis*), meaning "heart," "mind," or "soul." Cf. CARDIOTONIC.

COSMETIC plants (or **cosmetics**) preserve and beautify the body, especially the face. From Greek κοσμητικός, "skilled in ordering or arranging," a derivative of κόσμος, meaning both "cosmos" or "world" and "order," "arrangement," or "ornament."

DEMULCENT plants (or **demulcents**) soothe or allay irritation or pain in inflamed tissues. From Latin *demulcens* (genitive: *demulcentis*), the present participle of *demulcere,* "to stroke or caress."

DEOBSTRUENT plants (or **deobstruents**) remove obstructions by opening and clearing the natural ducts of the fluids and secretions of the body. From Latin *deobstruens* (genitive: *deobstruentis*), the present participle of *deobstruere,* "to unblock."

DEPURATIVE plants (or **depuratives**) purify and detoxify the human organism, especially the blood and humors. From Latin *depuratus,* the past participle of the verb *depurare,* "to depurate, purify, or refine." Depuratives are otherwise known as "blood purifiers."

DESICCANT plants (or **desiccants**) absorb moisture and induce or sustain a state of dryness, or desiccation. From Latin *desiccans* (genitive: *desiccantis*), the present participle of *desiccare,* "to dry up." Desiccants are the opposite of *humectants,* from *humectans* (genitive: *humectantis*), the present participle of *humectare,* "to moisten," which induce or sustain a state of moisture, or humectation.

DETERSIVE plants (or **detersives**), which are usually administered in the form of topical applications, cleanse the surfaces over which they pass. From Latin *detersus,* the past participle of the verb *detergere,* "to cleanse or remove." Synonymous with *detergent,* from Latin *detergens* (genitive: *detergentis*), the present participle of *detergere.*

DIAPHORETIC plants (or **diaphoretics**) induce perspiration. From Greek διαφορητικός, a derivative of the verb διαφορεῖν, meaning "to generate sweat or perspiration." Diaphoretics are mildly SUDORIFIC in action.

DIGESTIVE plants (or **digestives**) promote or aid in the digestion of food. From Latin *digestivus,* "of or relating to digestion (*digestio*)," a derivative of the verb *digerere,* meaning "to dissolve, divide, or carry apart."

DISCUTIENT plants (or **discutients**) discuss or disperse morbid humors, tumors, and coagulated matter. From Latin *discutiens* (genitive: *discutientis*), the present participle of *discutere,* "to dissipate, break up, or disperse." Synonymous with *discussive,* a derivative of *discussus,* the past participle of *discutere.*

DIURETIC plants (or **diuretics**) promote diuresis, an increased or excessive production of urine. From Greek διουρητικός, a compound made up of δία, "though," + οὐρητικός, "promoting urine (οὖρον)," and a derivative of the verb διουρεῖν, "to pass in urine," a compound made up of δία, "through," + οὐρεῖν, "to urinate."

EMETIC plants (or **emetics**) induce nausea and vomiting. From Greek ἐμετικός, "causing vomiting," a derivative of the verb ἐμεῖν, "to vomit." Cf. ANTIEMETIC.

EMMENAGOGIC plants (or **emmenagogues**) stimulate blood flow in the uterus and regulate or induce menstruation. From Greek ἔμμηνα, "the menses" (plural of ἔμμηνος, "monthly"), + ἄγωγος, "drawing forth or eliciting."

EMOLLIENT plants (or **emollients**) soften, lubricate, and soothe the skin. From Latin *emolliens* (genitive: *emollientis*), the present participle of *emollire*, "to soften out or make soft," a compound made up of the preposition *e* (*ex* before vowels and voiceless consonants), "out," + *mollire*, "to soften."

ERRHINE plants (or **errhines**) cause an increase of mucus within the nose and hence promote or induce nasal discharges and secretions. From Greek ἔρρινον, a compound made up of ἐν, "in," + ῥίς, "nose" (errhines being most commonly applied inside or snuffed up the nose). Cf. STERNUTATORY.

EXPECTORANT plants (or **expectorants**) promote or induce the discharge of phlegm and other fluids from the respiratory tract. From Latin *expectorans* (genitive: *expectorantis*), the present participle of *expectorare*, "to expel from the chest (*pectus*)."

FEBRIFUGAL plants (or **febrifuges**) prevent, dispel, or reduce fever. From Latin *febris*, "fever," + *fugare*, "to chase away." Synonymous with *antipyretic*, from Greek ἀντί, "against," + πυρετικός, "of or relating to fever (πυρετός)."

GALACTAGOGIC plants (or **galactagogues**) promote or increase lactation, the secretion of milk from the mammary glands of a nursing mother. From Greek γάλα (genitive: γάλακτος), "milk," + ἄγωγος, "drawing forth or eliciting." Cf. GALACTOFUGAL.

GALACTOFUGAL plants (or **galactofuges**) stop or reduce lactation, the secretion of milk from the mammary glands of a nursing mother. From Greek γάλα (genitive: γάλακτος), "milk," + φεύγειν, "to flee." Cf. GALACTOGOGIC.

HALLUCINOGENIC plants (or **hallucinogens**) induce hallucinations—that is, perceptions having all the qualities of real perceptions but in the absence of external stimuli. The term was first coined by Sir Thomas Browne in 1646 from the Latin verb *alucinari*, "to wander in mind," + the

Greek suffix -γενης, meaning "producing," "generating," or "formed by," a derivate of γίγνεσθαι, "to be born."

HEPATIC plants (or **hepatics**) strengthen and counter affections of the liver. From Greek ἡπατικός, "of or relating to the liver (ἧπαρ)."

HYPNOTIC plants (or **hypnotics**) induce a trance state resembling sleep that is characterized by a heightened susceptibility to suggestion. From Greek ὑπνωτικός, "causing sleep (ὕπνος)." Cf. SOPORIFIC.

HYPOTENSIVE plants (or **hypotensives**) lower the blood pressure. From Greek ὑπό, "under," + Latin *tensus,* the past participle of *tendere,* "to stretch." Synonymous with *antihypertensive,* from Greek ἀντί, "against," + ὑπέρ, "over," + Latin *tensus,* "being stretched."

IRRITANT plants (or **irritants**) contain substances that elicit inflammatory responses when they come in contact with the human body (usually on the skin). From Latin *irritans* (genitive: *irritantis*), the present participle of *irritare,* "to provoke, excite, or aggravate." Cf. ANTIPRURITIC.

LAXATIVE plants (or **laxatives**) soften the feces, loosen the bowels, and relieve constipation. From Medieval Latin *laxativus,* "loosening," a derivative of the verb *laxare,* "to loosen or relax." Cf. CATHARTIC.

LITHONTRIPTIC plants (or **lithontriptics**) dissolve or destroy stones in the bladder or kidneys. From Greek λίθον, "stone," + τρῖψις, "rubbing" or "friction," a derivative of τρίβειν, "to rub, grind, or wear away."

LYMPHATIC plants (or **lymphatics**) strengthen and counter affections of the lymph, lymph vessels, or lymph nodes and help restore the immune system. From Latin *lymphaticus,* a derivative of *lympha,* "water."

MYCTERIC plants (or **mycterics**) clear and cleanse the nasal cavity. A modern adjective meaning "of or relating to the nose," coined from Greek μυκτήρ, "nose."

NARCOTIC plants (or **narcotics**) blunt the senses and induce stupor, coma, or insensibility to pain. From Greek ναρκωτικός, "benumbing," a derivative of νάρκη, "numbness."

NERVINE plants (or **nervines**) soothe and counter affections of the nerves and help treat nervous disorders and anxiety. From Latin *nervinus,* "of or relating to the nerves (*nervi*)."

NUTRITIVE plants (or **nutritives**) are high in vitamins and minerals and promote general health and well-being. From Medieval Latin *nutritivus,* a derivative of the verb *nutrire,* "to nourish."

ODONTALGIC plants (or **odontalgics**) strengthen and counter affections of the teeth. From Greek ὀδονταλγία, "toothache," a compound of ὀδούς (genitive: ὀδόντος), "tooth," + ἄλγος, "pain."

OPHTHALMIC plants (or **ophthalmics**) strengthen and counter affections of the eyes. From Greek ὀφθαλμικός, "of or relating to the eyes (ὀφθαλμοί)."

OXYTOCIC plants (or **oxytocics**) stimulate the contraction of the uterine smooth muscle and hasten the process of parturition. From Greek ὠκυτόκος, "causing quick and easy birth," a compound made up of ὠκύς, "swift" + τόκος, "childbirth."

PARASITICIDIC plants (or **parasiticides**) kill parasites. From Latin *parasitus,* "parasite" (a loanword from Greek παράσιτος), + *caedere,* "to kill."

PECTORAL plants (or **pectorals**) strengthen and counter affections of the respiratory system. From Latin *pectoralis,* "of or relating to the chest (*pectus*)."

PSYCHOTROPIC plants (or **psychotropics**) affect the mind or mental processes in a profound manner. Psychotropics are literally "mind-altering" plants, from Greek ψυχή, "soul" or "mind," + τρέπειν, "to turn or alter."

Synonymous with *psychoactive,* a neologism generated from Greek ψυχή, "soul" or "mind," + Latin *activus,* "active."

PULMONARY plants (or **pulmonaries**) strengthen and counter affections of the lungs. From Latin *pulmonarius,* "of or relating to the lungs (*pulmones*)."

PURGATIVE plants (or **purgatives**) cleanse and purge the body of unwanted waste, especially through the evacuation of the bowels. From Latin *purgativus,* a derivative of the verb *purgare,* "to cleanse." Purgatives are strongly LAXATIVE in action.

REFRIGERANT plants (or **refrigerants**) lower the body temperature and help reduce fevers. From Latin *refrigerans* (genitive: *refrigerantis*), the present participle of *refrigerare,* "to make cool."

RESOLVENT plants (or **resolvents**) promote the resolution and dispersal of abnormal growths, swellings, or inflammations. From Latin *resolvens* (genitive: *resolventis*), the present participle of *resolvere,* "to resolve."

RESTORATIVE plants (or **restoratives**) preserve or restore the equilibrium of the body, especially the humors of the body. From Medieval Latin *restaurativum,* a derivative of the verb *restaurare,* "to restore."

RUBEFACIENT plants (or **rubefacients**) stimulate circulation locally and produce redness of the skin when administered in the form of topical applications. From Latin *rubefaciens* (genitive: *rubefacientis*), the present participle of *rubefacere,* "to make red."

SEDATIVE plants (or **sedatives**) allay irritability or excitement, assuage pain, and induce sleep. From Medieval Latin *sedativus,* a derivative of the verb *sedare,* "to allay, restrain, or calm down." Cf. CALMATIVE and SOPORIFIC.

SIALAGOGIC plants (or **sialagogues**) stimulate or increase the flow of saliva. From Greek σίαλον, "saliva," + ἄγωγος, "drawing forth or eliciting."

SOPORIFIC plants (or **soporifics**) induce drowsiness or sleep. From Latin *sopor,* "deep sleep." Cf. SEDATIVE.

SPASMOLYTIC plants (or **spasmolytics**) relieve or suppress the sudden involuntary contraction of muscles. From Greek σπασμός, "convulsion" or "spasm," + λύσις, "loosening" or "releasing." Synonymous with *antispasmotic,* from Greek ἀντί, "against," + σπασμός, "spasm."

SPLENIC plants (or **splenics**) strengthen and counter affections of the spleen. From Greek σπληνικός, "of or relating to the spleen (σπλῆν)."

STERNUTATORY plants (or **sternutatories**) cause or induce sneezing. From Latin *sternutatorius,* "causing to sneeze," a derivative of *sternuere,* "to sneeze." Cf. ERRHINE.

STIMULANT plants (or **stimulants**) activate and temporarily quicken the vital processes and physiological functions of the body. From Latin *stimulans* (genitive: *stimulantis*), the present participle of *stimulare,* "to stimulate." Roughly synonymous with *excitant,* from Latin *excitans* (genitive: *excitantis*), the present participle of *excitare,* "to excite."

STOMACHIC plants (or **stomachics**) stimulate gastric digestion and strengthen the appetite. From Latin *stomachicus,* "of or relating to the stomach (*stomachus*)." Cf. APERITIVE and DIGESTIVE.

STYPTIC plants (or **styptics**) stop or reduce bleeding and hemorrhaging. From Greek στυπτικός, "astringent," a derivative of στύφειν, "to contract or draw together." Synonymous with *hemostatic,* from Greek αἷμα, "blood," + στάσις, "halting." Cf. ASTRINGENT.

SUDORIFIC plants (or **sudorifics**) induce sweating. From New Latin *sudorificus,* "inducing sweat," a derivative of Latin *sudor,* "sweat." Cf. DIAPHORETIC.

TONIC plants (or **tonics**) preserve, increase, or restore the tone and health of the body or an organ of the body and promote general health and well-being. From Greek τονικός, "of or relating to stretching or contraction," a derivative of τόνος, "tone."

VERMIFUGAL plants (or **vermifuges**) prevent, destroy, or expel intestinal worms. From Latin *vermis,* "worm," + *fugare,* "to expel." Synonymous with ANTHELMINTIC.

VESICANT plants (or **vesicants**) contain chemical compounds that can cause extensive tissue damage and blistering to the skin. From Medieval Latin *vesicans* (genitive: *vesicantis*), "causing blistering," the present participle of *vesicire,* "to blister," a derivative of *vesica,* "bladder" or "blister."

VULNERARY plants (or **vulneraries**) promote the healing of wounds. From Latin *vulnerarius,* "of or relating to wounds," a derivative of *vulnus,* meaning "wound."

Concordance of Elemental and Astral Plant Signatures

The following concordance to Sédir's dictionary provides classifications of plants according to their elemental, planetary, and zodiacal signatures. In the groupings of plants by astral signature (planetary and zodiacal), a plant name that appears without square brackets indicates that the signature is the plant's primary signature, whereas a plant name that appears in square brackets indicates that the signature is a secondary signature.

ELEMENTAL SIGNATURES

Air Plants (🜁): *Hot and Wet*

acacia
alyssum, sweet
amaranth
anise
bindweed
borage
caraway
castor bean
catnip
chamomile
cherry, cornelian
cinquefoil
clover, sweet
coltsfoot
comfrey
cowslip
cuckoopint
cyclamen
daisy
daisy, ox-eye
dill
dittany of Crete
elecampane
fig
fennel
goat's-beard
grapevine
hazel
hyacinth
iris

Air Plants (🜁): *Hot and Wet* (cont.)

knotweed
lemon
linden
mallow, marsh
needleleaf
parsley
pellitory-of-the-wall
peppermint
pimpernel
plane tree
rhubarb
saffron, meadow
senna, bladder
sorrel
spurry, sand
tansy
vervain
vetch
wallflower

Fire Plants (🜂): *Hot and Dry*

agaric
Alexanders
aloeswood
angelica, garden
angelica, wild
arnica, mountain
artichoke
asparagus
balm, lemon
basil
bay
bear's ear
betony
birch
blackthorn
boxwood
broom
bryony, white
buckthorn
bugle
buttercup
cardamom
cedar
celandine, lesser
celery
centaury
chicory
chive
chrysanthemum
cinnamon
clove
coca
colocynth
coriander
costmary
cress, bitter
cress, garden
cress, swine
cypress
edelweiss
elderberry
elderberry, dwarf
eyebright
fern, hart's-tongue
fern, male
frankincense
fuchsia
fumitory
garlic
gentian
grains of paradise
heliotrope
hellebore, black false
hellebore, green
hellebore, white
henbane
holly
horehound
horseradish
hound's tongue
juniper
kusha grass
larkspur
lavender
leek
leek, sand
madder
marjoram
mint
mint, water
mustard
myrrh
nettle
nutmeg
old-man's beard

Fire Plants (△): *Hot and Dry* (cont.)

olive
olive, Niçoise
onion
onion, spring
orange
oregano
palm, date
pennyroyal
peony
pepper, black
peyote
plantain
poppy, opium
privet
restharrow
rosemary
rosemary, bog
rue, wild
saffron
sage
Saint John's wort
sandalwood, red
savory, summer
sesame
southernwood
speedwell
spurge
spurge, caper
spurge, cypress
tea
thistle, blessed
thistle, carline
thyme
thyme, basil
thyme, wild
walnut
wolfsbane, winter
wormwood
wormwood, field
water soldier
yarrow

Earth Plants (🜃): *Cold and Dry*

agrimony
alkanet
almond
apple
ash
asphodel
banyan
barberry
barley
beet greens
birthwort
bistort
blackberry
bryony
burdock, greater
burdock, lesser
camellia, oilseed
celandine, greater
cherry, winter
chinaberry
cleavers
cumin, horned
daffodil
figwort
fern, maidenhair
flag, sweet
flax
forget-me-not
germander
germander, woodland
gladiolus
hellebore, black
hemlock
hemp
honeysuckle
hornbeam
houseleek
ivy
lily
mallow
mandrake
mignonette
mistletoe
mugwort
mulberry
mullein
myrtle
oak
oat
peach
periwinkle
plum
polypody
quince

Earth Plants (🜃): *Cold and Dry* (cont.)

raisin tree
rose
rose, dog
rose of Jericho
sandalwood, white
scabious, devil's bit
senna
shepherd's purse
Solomon's seal
tobacco
tormentil
valerian
violet
wheat
wolfsbane

Water Plants (🜄): *Cold and Wet*

alder
aloe
bachelor's button
bamboo, black
bear's breeches
beetroot
belladonna
cabbage
cabbage, red
camphor
chaste tree
cherry
chickweed
cucumber
fava bean
foxglove
gladiolus, marsh
heather
jimsonweed
lettuce
licebane
lichen, lungwort
lily, water
lotus
lungwort
mercury
moss, sphagnum
orchid
pine
pomegranate
poplar
poppy, common
poppy, horned
purslane
rampion
reed
sarsaparilla
saxifrage
soapwort
sorb tree
sorrel, wood
strawberry
tamarind
turnip
usnea
watercress
willow

PLANETARY RULERS

Saturnian Plants (♄)

asphodel
bamboo, black
belladonna
bistort
blackberry
burdock, greater
burdock, lesser
chaste tree
chinaberry
cleavers
coca
cumin, horned
cypress
fava bean
fern, hart's-tongue
fern, maidenhair
fern, male
[fumitory]
hellebore, black
hellebore, black false
hellebore, green

Saturnian Plants (♄) (cont.)

hellebore, white
hemlock
hemp
henbane
holly
jimsonweed
licebane
lichen, lungwort
[lungwort]
mandrake
moss, sphagnum
parsley
[pellitory-of-the-wall]
pine
polypody
poppy, horned
poppy, opium
raisin tree
[rhubarb]
rosemary, bog
[rose of Jericho]
rue, wild
saxifrage
[senna]
shepherd's purse
soapwort
tamarind
usnea
willow
wolfsbane
wolfsbane, winter

Jupiterian Plants (♃)

agrimony
alkanet
[almond]
amaranth
apple
ash
bachelor's button
[balm, lemon]
banyan
barberry
beetroot
betony
birch
birthwort
borage
bugle
[cabbage, red]
castor bean
cedar
centaury
cherry
cherry, cornelian
[chickweed]
comfrey
edelweiss
elecampane
fig
flax
[fumitory]
germander
[grapevine]
hazel
[henbane]
[hornbeam]
knotweed
lemon
[lily]
madder
mint
mullein
[nutmeg]
oak
olive
olive, Niçoise
peach
peony
plane tree
plum
pomegranate
poplar
quince
[restharrow]
rhubarb
[rose]
[rosemary]
saffron, meadow
sesame
sorb tree
strawberry
[violet]

Martian Plants (♂)

artichoke
asparagus
[barberry]
basil
bear's breeches
bear's ear
blackthorn
broom
bryony
buckthorn
[bugle]
buttercup
[cherry, cornelian]
chive
colocynth
cress, garden
cress, swine
cuckoopint
elderberry, dwarf
[fern, male]
foxglove
[fumitory]
garlic
germander
gladiolus
gladiolus, marsh
[holly]
horehound
horseradish
hound's tongue
leek
leek, sand
[madder]
[mint]
mint, water
mugwort
mustard
nettle
[nutmeg]
onion
onion, spring
pennyroyal
pepper, black
restharrow
[rue, wild]
spurge
spurge, caper
spurge, cypress
spurry, sand
thistle, blessed
thistle, carline
tormentil
wormwood
wormwood, field

Solar Plants (⊙)

aloe
aloeswood
alyssum, sweet
angelica, garden
angelica, wild
arnica, mountain
[ash]
balm, lemon
barley
bay
camellia, oilseed
caraway
cardamom
celandine, greater
celandine, lesser
cinnamon
clove
clover, sweet
[coca]
costmary
cress, bitter
[cyclamen]
dittany of Crete
eyebright
flag, sweet
frankincense
gentian
grains of paradise
grapevine
heliotrope
hornbeam
[hyacinth]
[knotweed]
kusha grass
lavender
lotus
marjoram
[mignonette]
mistletoe
myrrh
nutmeg
[oat]

Solar Plants (☉) (cont.)

[olive]
[olive, Niçoise]
orange
[oregano]
palm, date
[parsley]
[peony]
[pepper, black]
peppermint
plantain
rosemary
[rue, wild]
saffron
sage
Saint John's wort
sandalwood, red
senna
tansy
thyme
thyme, basil
[vervain]
wheat

Venusian Plants (♀)

almond
[apple]
[bamboo, black]
boxwood
cherry, winter
coriander
cyclamen
daffodil
dill
figwort
forget-me-not
fuchsia
germander, woodland
goat's-beard
houseleek
hyacinth
iris
larkspur
[lily]
[lily, water]
mallow
mallow, marsh
mignonette
myrtle
orchid
periwinkle
[polypody]
rose
rose of Jericho
thyme, wild
vervain
violet
wallflower

Mercurial Plants (☿)

acacia
agaric
Alexanders
anise
beet greens
bindweed
[bistort]
bryony, white
[bugle]
catnip
celery
chamomile
chicory
chrysanthemum
cinquefoil
coltsfoot
cowslip
daisy
daisy, ox-eye
elderberry
[fava bean]
fennel
[hazel]
heather
honeysuckle
ivy
juniper
lungwort
[marjoram]
mercury
mulberry
needleleaf
old-man's beard

Mercurial Plants **(☿) (cont.)**

oregano
pellitory-of-the-wall
peyote
pimpernel
privet
rose, dog
sarsaparilla
savory, summer
scabious, devil's bit
senna, bladder
Solomon's seal
sorrel
sorrel, wood
southernwood
speedwell
tea
valerian
water soldier
yarrow

Lunar Plants **(☾)**

alder
cabbage
cabbage, red
camphor
chickweed
cucumber
[jimsonweed]
[leek]
[leek, sand]
lettuce
lily
lily, water
linden
[mandrake]
oat
[polypody]
poppy, common
[poppy, opium]
purslane
rampion
reed
sandalwood, white
[senna]
[tamarind]
tobacco
turnip
[usnea]
vetch
walnut
watercress

ZODIACAL SIGNATURES

Arien Plants **(♈)**

agaric
[arnica, mountain]
asparagus
bear's ear
betony
bugle
cardamom
[castor bean]
chicory
coca
colocynth
[coltsfoot]
cress, garden
cress, swine
elderberry
elderberry, dwarf
eyebright
fern, hart's-tongue
fumitory
[gentian]
grains of paradise
hellebore, black false
hellebore, white
henbane
holly
horehound
hound's tongue
juniper
larkspur
marjoram
mint
mint, water
mustard

Arien Plants (♈) (cont.)

myrrh
nutmeg
old-man's beard
olive
olive, Niçoise
[orange]
peony
peyote
[plane tree]
plantain
poppy, opium
privet
rosemary
rosemary, bog
sage
Saint John's wort
southernwood
speedwell
spurge, caper
thyme, basil
wolfsbane, winter
wormwood, field
water soldier
yarrow

Taurean Plants (♉)

agrimony
alkanet
almond
ash
asphodel
barberry
celandine, greater
cherry, winter
chinaberry
daffodil
figwort
[fern, hart's-tongue]
fern, maidenhair
flax
forget-me-not
germander
gladiolus
hemp
[hyacinth]
ivy
[licebane]
lily
mallow
[mallow, marsh]
mistletoe
[moss, sphagnum]
myrtle
[oak]
peach
periwinkle
polypody
rose
rose of Jericho
scabious, devil's bit
[senna]
Solomon's seal
[tea]
tormentil
valerian
violet
watercress

Geminian Plants (♊)

amaranth
anise
bindweed
[blackthorn]
borage
[burdock, greater]
[burdock, lesser]
caraway
castor bean
[cedar]
chamomile
[chickweed]
coltsfoot
comfrey
cuckoopint
dittany of Crete
elecampane
fennel
[hemp]
knotweed
[leek]
[leek, sand]
[lettuce]
[lily, water]

Geminian Plants **(♊) (cont.)**

mallow, marsh
parsley
pimpernel
poppy, horned
rhubarb
rue, wild
sorrel
tansy
[tea]
vetch
wallflower

Cancerian Plants **(♋)**

aloe
[balm, lemon]
bear's breeches
cabbage
cabbage, red
camphor
chaste tree
chickweed
[cinquefoil]
cucumber
fava bean
[figwort]
[fern, hart's-tongue]
gladiolus, marsh
[hazel]
lettuce
lichen, lungwort
lily, water
lungwort
[old-man's beard]
[peony]
pine
purslane
rampion
[rose of Jericho]
[Saint John's wort]
sarsaparilla
saxifrage
tamarind
willow

Leonian Plants **(♌)**

Alexanders
angelica, garden
celery
arnica, mountain
balm, lemon
basil
bay
boxwood
broom
bryony, white
buttercup
cedar
centaury
chrysanthemum
cinnamon
clove
[coca]
coriander
cress, itter
[cyclamen]
cypress
edelweiss
frankincense
fuchsia
gentian
hazel
heliotrope
[juniper]
kusha grass
lavender
[mullein]
nettle
orange
pennyroyal
pepper, black
[plantain]
[poppy, opium]
restharrow
saffron
sandalwood, red
savory, summer
spurge, cypress
thistle, blessed
thistle, carline
thyme

Virgoan Plants (♍)

acacia
[anise]
apple
[balm, lemon]
banyan
barley
beet greens
birthwort
bistort
camellia, oilseed
[chicory]
cinquefoil
cleavers
cumin, horned
flag, sweet
germander, woodland
hemlock
honeysuckle
houseleek
mugwort
mullein
oak
oat
[pimpernel]
plum
raisin tree
rose, dog
sandalwood, white
senna
[Solomon's seal]
[sorrel]
[tea]
tobacco
[tormentil]
[vetch]
wheat

Libran Plants (♎)

alyssum, sweet
[boxwood]
buckthorn
catnip
[celandine, greater]
[chamomile]
[cherry, winter]
[chickweed]
cowslip
cyclamen
dill
[dittany of Crete]
[figwort]
[fern, hart's-tongue]
goat's-beard
grapevine
hyacinth
iris
lemon
[licebane]
linden
[mallow]
[mint]
[mugwort]
needleleaf
[peach]
pellitory-of-the-wall
peppermint
[purslane]
[rue, wild]
[saxifrage]
[scabious, devil's bit]
senna, bladder
spurry, sand
vervain

Scorpion Plants (♏)

artichoke
[ash]
[bamboo, black]
[barberry]
beetroot
belladonna
blackthorn
[cabbage]
cherry, cornelian
[cuckoopint]
daffodil
[figwort]
heather
jimsonweed
leek

Scorpion Plants (♏) (cont.)

leek, sand
licebane
mercury
[mugwort]
onion, spring
orchid
[rampion]
soapwort
sorb tree
spurge

Sagittarian Plants (♐)

aloeswood
angelica, wild
[beetroot]
birch
celandine, lesser
chive
[coca]
costmary
[cress, garden]
[cucumber]
fern, male
garlic
hellebore, green
[henbane]
horseradish
[ivy]
[juniper]
[knotweed]
[larkspur]
madder
onion
[orange]
oregano
palm, date
plane tree
[poplar]
[rue, wild]
[saffron]
sesame
tea
thyme, wild
walnut
[willow]

Capricornian Plants (♑)

bryony
hellebore, black
[henbane]
mandrake
mignonette
mulberry
[poppy, opium]
quince
shepherd's purse
[Solomon's seal]
wolfsbane
wormwood

Aquarian Plants (♒)

[agrimony]
[angelica, garden]
[arnica, mountain]
[bindweed]
[bistort]
blackberry
[borage]
burdock, greater
burdock, lesser
[cedar]
clover, sweet
[cumin, horned]
daisy
daisy, ox-eye
fig
[fennel]
[germander, woodland]
[hemlock]
[hemp]
[larkspur]
[myrrh]
[pellitory-of-the-wall]
[saxifrage]

Piscean Plants (♓)

alder
bachelor's button
cherry
foxglove
lotus
moss, sphagnum
[plane tree]
pomegranate
poplar
poppy, common
[purslane]
reed
saffron, meadow
sorrel, wood
strawberry
turnip
usnea

Notes

TRANSLATOR'S FOREWORD

1. See further Michelet, *Les compagnons de la hiérophanie,* 95.
2. Papus, "Échos," 293.
3. See Papus, "L'École hermétique," 47.
4. See Barlet, "Université libre des hautes études," 3; and Eléazar, "Faculté des sciences hermétiques," 100.
5. See further Turland et al., *International Code of Nomenclature for Algae, Fungi, and Plants;* and Turland, *The Code Decoded.*
6. Linnaeus, *Species plantarum,* 2:1172.
7. Dioscorides, *On Medical Materials* 3.1.
8. See Lamarck, *Encyclopédie méthodique: Botanique,* 1:111, no. 45.
9. Interested readers may consult either Brummit and Powell's *Authors of Plant Names* or the database of the International Plant Names Index (IPNI) online.
10. See especially the detailed compendium of animal signatures in Mavéric and Monfloride's *La magie rurale,* 14–30.
11. Fulcanelli, *Le mystère des cathédrales,* 42.
12. See Rabelais, *Tiers liure des faictz et dictz heroïques du noble Pantagruel,* 336. This passage is from chap. 46 of the first edition of bk. 3 (cited here), but in subsequent editions and translations of *Gargantua and Pantagruel* it will be found in bk. 3, chap. 50.
13. Revelation 8:11. See further Deuteronomy 29:18; Proverbs 5:4; Jeremiah 9:15 and 23:15; Lamentations 3:15 and 3:19; and Amos 5:7 and 6:12.
14. See, for example, Delvau, *Dictionnaire de la langue verte,* 249 s.v. *herbe-sainte;* and Barrère, *Argot and Slang,* 205 s.v. *herbe.*
15. Pliny, *Natural History* 27.28.
16. Sédir, *Les plantes magiques,* 154.

17. Pliny, *Natural History* 24.116.
18. See Blankaart, *Lexicon medicum renovatum,* 729 s.v. *philadelphus* [English edition: *The Physical Dictionary,* 270 s.v. *philadelphus*]; Haller, *Onomatologia medica completa,* 1123 s.vv. *philadelphus und philanthropos;* and Hooper, *Lexicon medicum or Medical Dictionary,* 641 s.v. *Galium.* It is to be noted, however, that the name *philadelphus* does not appear in French alchemist Leméry's influential *Traité universel des drogues simples,* 47–48 s.vv. *aparine sive aspergo.*
19. Papus, "Les facultés occultes de l'homme," 129 (italics added).
20. See Agrippa, *De occulta philosophia libri tres,* liber I, caput XXXII, xxxvii [English edition: *Three Books of Occult Philosophy or Magic,* bk. 1, chap. 32, 107].
21. See Papus, *Traité élémentaire de magie pratique,* 233–40.
22. See further Pseudo-Albertus, *Liber aggregationis seu secretorum,* 6r (planetary herb no. 4) = *Les admirables secrets d'Albert le Grand,* 77–78; cf. Papus, *Traité élémentaire de magie pratique,* 236, 547 s.v. *ornoglosse.*
23. Mavéric, *Hermetic Herbalism,* 42.

CHAPTER 1. PHYTOGENESIS

1. *Translator's note:* See further Papus, *Traité élémentaire de science occulte,* 273; Papus, *Premiers éléments de lecture de la langue égyptienne,* 37–41.
2. L'Aulnaye, *Histoire générale et particulière des religions et du culte de tous les peuples du monde, tant anciens que modernes,* 1:173.
3. See further Papus, *Traité méthodique de science occulte,* 952–59.
4. *Translator's note:* Sédir derives this outline as well from L'Aulnaye, *Histoire générale,* 1:269.
5. Burgoyne, *The Light of Egypt,* 2:35.
6. Cf. Burgoyne, *The Light of Egypt,* 2:35–36.
7. Trevisan, *Le texte d'alchymie, et le songe-verd,* 16–18.
8. *Translator's note:* Sédir derives this outline from Papus, *Traité méthodique de science occulte,* 583.
9. *Translator's note:* Cf. Paracelsus, *The Hermetic and Alchemical Writings,* 2:363 s.v. *derses;* Ruland, *Lexicon alchemiae, siue, dictionarium alchemisticum,* 182 s.v. *derses;* and Hartmann, *The Life and the Doctrines of Philippus Theophrastus,* 34 s.v. *derses.*
10. *Translator's note:* Cf. Paracelsus, *The Hermetic and Alchemical Writings,* 2:361 s.v. *clissus;* Ruland, *Lexicon alchemiae,* 156–57 s.v. *clissus;* and Hartmann, *The Life and the Doctrines of Philippus Theophrastus,* 33 s.v. *clissus.*
11. *Translator's note:* Cf. Paracelsus, *The Hermetic and Alchemical Writings,* 2:372

s.v. *leffas;* Ruland, *Lexicon alchemiae,* 302 s.v. *leffas;* and Hartmann, *The Life and the Doctrines of Philippus Theophrastus,* 38 s.v. *leffas.*

12. *Translator's note:* See especially the masterful works by the botanist and classicist Dierbach, *Flora mythologica,* and by the renowned philologist Gubernatis, *La mythologie des plantes.*

CHAPTER 2. PLANT PHYSIOLOGY

1. *Translator's note:* Descriptions of the three plant organs have been augmented with recourse to Sédir's sources. See Papus, *Anatomie philosophique et ses divisions,* 105–6; and the German naturalist Oken's *Esquisse du système d'anatomie et d'histoire naturelle,* 13–14.
2. *Translator's note:* Descriptions of the interrelationships between plant organs have been augmented with recourse to Sédir's sources. See Papus, *Traité méthodique de science occulte,* 266–67.
3. *Translator's note:* Descriptions of the three stages have been augmented with recourse to Sédir's sources. See Papus, *Traité méthodique de science occulte,* 267–70.
4. *Translator's note:* Cf. Ruland, *Lexicon alchemiae,* 334 s.v. *Mercurius mineralium.*
5. *Translator's note:* Cf. Genesis 3:17–19.
6. *Translator's note:* See Reichenbach, *Reichenbach's Letters on Od and Magnetism.* For a much more detailed and thorough analysis of the *od* of plants, see especially Reichenbach, *Die Pflanzenwelt in ihren Beziehungen zur Sensitivitat und zum Ode.*
7. *Translator's note:* See further Reichenbach, *Physico-Physiological Researches on the Dynamics of Magnetism, Electricity, Heat, Light, Crystallization, and Chemism,* 251–55 (§§ 248–52). Sédir silently corrects Reichenbach's confusion of the terms *positive* and *negative,* on which see Babbit, *The Principles of Light and Color,* 422–23.
8. *Translator's note:* See further Mattei, *Elettromiopatia* [English edition: *Electro-Homœopathy*].
9. *Translator's note:* Descriptions of the pre-Socratic botanical theories have been supplemented with recourse to Sédir's probable sources. See Pseudo-Aristotle, *On Plants* 1.1.815a–b; and Plutarch, *Causes of Natural Phenomena* 1.911c–d.
10. Boscowitz, *L'âme de la plante,* 1–13.
11. *Translator's note:* See Darwin, *The Botanic Garden,* 227–45; cf. Darwin, *Zoonomia, or the Laws of Organic Life,* 1:101–7.
12. See Martius, *Reise in Brasilien;* Martius, *Die Pflanzen und Thiere des tropischen America, ein Naturgemälde;* and Martius, "Die Unsterblichkeit der Pflanze: Ein

Typus," in *Reden und Vorträge über Gegenstände aus dem Gebiete der Naturforschung,* 263–86.

13. See Fechner, *Nanna oder über das Seelenleben der Pflanzen.*
14. *Translator's note:* Sédir takes this figure from Boscowitz, *L'âme de la plante,* 27. For more precise calculations, see Bradley, *A Philosophical Account of the Works of Nature,* 44–46.
15. *Translator's note:* See further Boscowitz, *L'âme de la plante,* 28–32.
16. Bonnet, *Contemplation de la nature,* 2:76–77.
17. *Translator's note:* Table 2.2 on p. 23 appears to be an extended graphic representation of the chart attributed to Adrian von Mynsicht (alias Madathanus) in Sédir's *Histoire et doctrines des Rose-Croix,* 202.

CHAPTER 3. PLANT PHYSIOGNOMY

1. Saint-Martin, *De l'esprit des choses, ou coup d'œil philosophique sur la nature des natures et sur l'objet de leur existence,* 1:140–42.
2. *Translator's note:* For the myth of Saturn devouring his own children (Saturn being the Roman equivalent of the Greek god Kronos), see especially Hesiod, *Theogony* 453–91.
3. Saint-Martin, *De l'esprit des choses,* 1:152–58.
4. *Translator's note:* Descriptions of the zodiacal signatures have been augmented with additional information concerning their medicinal actions and therapeutic effects. See the following works by Sédir's occult colleagues: Papus, *Traité élémentaire de magie pratique,* 250–51; Bélus, *Traité des recherches,* 35–40; and Mario, "La flore mystérieuse," *La vie mystérieuse* 29; 30; 31; 32; 34; 35; 36; 38; 39; 41; 45; 47; and 50. On the characteristic perfumes of the signs of the zodiac, see especially Agrippa, *De occulta philosophia libri tres,* liber I, caput XLIIII, lii [English edition: *Three Books of Occult Philosophy or Magic,* bk. 1, chap. 44, 136].

CHAPTER 4. ALIMENTATION

1. *Translator's note:* Paragraph slightly augmented with recourse to Sédir's probable sources; see Papus, *Traité élémentaire de magie pratique,* 129–32. The leading advocates of vegetarianism in Sédir's day were Bonnejoy, *Principes d'alimentation rationnelle hygiénique et économique avec des recettes de cuisine végétarienne* and especially *Le végétarisme et le régime végétarien rationnel,* and Pivion, *Étude sur le régime de Pythagore.* There were also many advocates of vegetarianism among Sédir's

occult peers. See, for example, Papus, *Traité élémentaire de magie pratique,* 129–32 (with some reservations); Mavéric, *La médecine hermétique des plantes ou l'extraction des quintessences par art spagyrique,* 7, 75–81 [English edition: *Hermetic Herbalism,* 3, 61–67]; and Bosc, *Traité de la longévité ou l'art de devenir centenaire,* 16–49. For contemporary works by English occultists, see Kingsford, *The Perfect Way in Diet;* Besant, *Vegetarianism in the Light of Theosophy;* and Leadbeater, *Vegetarianism and Occultism.*

CHAPTER 5. PHYTOTHERAPY

1. *Translator's note:* See Schuré, *The Great Initiates,* 1:33; Davidson, *The Mistletoe and Its Philosophy,* 68; cf. Sédir's French translation of Davidson's work under the title *Le gui et sa philosophie,* 86.
2. *Translator's note:* See further Mavéric, *La médecine hermétique des plantes,* 43, 48–49, 79–80, 152 [English edition: *Hermetic Herbalism,* 36, 43, 67–68, 132].
3. *Translator's note:* For a more complete exposition of Paracelsian transplantation, see Vergnes, "De la transplantation des maladies," *Le Voile d'Isis* 26, no. 13: 33–40; no. 14: 118–21; no. 15: 192–96; and no. 16: 263–71; cf. Hartmann, *The Life and the Doctrines of Philippus Theophrastus,* 151–60, 187–90.
4. *Translator's note:* See especially Bourru and Burot, *La suggestion mentale et l'action à distance des substances toxiques et médicamenteuses;* Luys, "Sur l'action des médicaments à distance chez les sujects hypnotisés"; and Durville, *Les actions psychiques à distance.*
5. *Translator's note:* Sédir recommends the following old pharmaceutical works in the bibliography to *Les plantes magiques:* Bauderon, *La pharmacopée de Bauderon;* Charas, *Pharmacopée royale galénique et chymique* [English edition: *The Royal Pharmacopœa, Galenical and Chymical*]; Schröder, *La pharmacopée raisonnée de Schroder;* Lémery, *Pharmacopée universelle* [English edition (abridged): *Pharmacopoeia Lemeriana contracta*]; and Lémery, *Traité universel des drogues simples.*
6. *Translator's note:* See Theophrastus, *Enquiry into Plants* 9.10.
7. *Translator's note:* See Dioscorides, *On Medical Materials* 4.148.
8. Monginot, *Traitté de la conservation et prolongation de la santé,* chap. 9.
9. Berkeley, *Siris,* 2, 5, 170.
10. Berkeley, *Siris,* 172–73.
11. Störck, *Observations nouvelles sur l'usage de la cigüe,* liv.
12. *Translator's note:* Cf. Theophrastus, *Enquiry into Plants* 9.15.8.
13. *Translator's note:* See Hippocrates, *Nature of Women and Barrenness* 71; Galen,

On the Powers and Mixtures of Simple Remedies 4.13, *On Temperaments* 3.4, *On the Composition of Local Remedies* 7.5, and *On Antidotes* 2.13; Mercuriale, *De morbis muliebribus praelectiones,* 235; and Astruc, *Traité des maladies des femmes,* 2:391.

14. *Translator's note:* See further chapter 7. For more detailed information on harvesting plants according to Hermetic principles, see Mavéric, *La médecine hermétique des plantes,* 40–48 [English edition: *Hermetic Herbalism,* 33–41].
15. See Boerhaave, *Elementa chemiae,* 2:149–50; cf. Berkeley, *Siris,* 21–22.
16. *Translator's note:* Cf. Paracelsus, *The Hermetic and Alchemical Writings,* 3:358 s.v. *cagastric* and 2:371 s.v. *iliaster;* Ruland, *Lexicon alchemiae,* 263–64 s.v. *iliaster;* and Hartmann, *The Life and the Doctrines of Philippus Theophrastus,* 43 s.v. *yliaster.*
17. Pseudo-Aquinas, *Secreta alchimiae magnalia,* 17.
18. *Translator's note:* See further appendix 1. For more thorough investigations into the spagyric art of Hermetic herbalism, see especially Mavéric, *La médecine hermétique des plantes* [English edition: *Hermetic Herbalism*] and Jollivet-Castelot, *La médecine spagyrique.*

CHAPTER 6. PLANT MAGIC

1. *Translator's note:* See Augustine, *City of God* 25.23.
2. *Translator's note:* See Paracelsus, *The Hermetic and Alchemical Writings,* 2:255–58.
3. *Translator's note:* See further Bjerregaard, "The Elementals, the Elementary Spirits, and the Relationship between Them and Human Beings," *The Path* 1, no. 10 and no. 11—a work favorably cited in Blavatsky, *The Secret Doctrine,* 1:630.
4. Cf. Bjerregaard, *Lectures on Mysticism and Nature Worship,* 15–16.
5. B., "Haunted Trees and Stones." See further Prel, "Pflanzenmystik."
6. *Translator's note:* On the winged oak (ὑπόπτερος δρῦς) of Pherecydes, see Clement of Alexandria, *Miscellanies* 6.6.53.5.
7. See in particular De Quincey, *Confessions of an English Opium-Eater;* and Baudelaire, *Les paradis artificiels.*
8. *Translator's note:* See further appendix 3.
9. *Translator's note:* See Ragon, *Maçonnerie occulte suivie de l'initiation hermétique,* 82–85; cf. Ragon, *Orthodoxie maçonnique, suivie de la maçonnerie occulte et de l'initiation hermétique,* 498–501.
10. Nynauld, *De la lycanthropie, transformation, et extase des sorciers.* [Cf. Philipon, *Stanislas de Guaita et sa bibliothèque occulte,* 98. —*Trans.*]

11. Nynauld, *De la lycanthropie,* 24–25.
12. *Translator's note:* See Nynauld, *De la lycanthropie,* 27–63; cf. Guaïta, *Le serpent de la Genèse,* 2:172–74 n. 1.
13. Nynauld, *De la lycanthropie,* 37–38.
14. Nynauld, *De la lycanthropie,* 49.
15. *Translator's note:* See Agrippa, *De occulta philosophia libri tres,* liber I, caput XLIII, xlix–li [English edition: *Three Books of Occult Philosophy,* bk. 1, chap. 43, 132–35]. These same recipes also appear in Nynauld, *De la lycanthropie,* 73–74; and Collin de Plancy, *Dictionnaire infernal, ou bibliothèque universelle,* 4:244 s.v. *parfums.*

CHAPTER 7. OCCULT HORTICULTURE

1. See Lenglet-Mortier and Vandamme, *Nouvelles et véritables étymologies médicales tirées du gaulois,* 28.
2. *Translator's note:* Cf. Matthew 13:3–9.
3. *Translator's note:* See Sédir's discussions of the subjects of plant embryogenesis and plant morphology and development in chapter 2; cf. Papus, *Traité méthodique de science occulte,* 269.
4. *Translator's note:* See Sédir's discussion of the subject of allelopathy and polyculture (or companion planting) in chapter 3.
5. Flammarion, "Les radiations solaires et les couleurs."
6. *Translator's note:* On this subject, see especially Mavéric, *La médecine hermétique des plantes,* 50–109 [English edition: *Hermetic Herbalism,* 42–95].
7. *Translator's note:* The brief description of the planetary days and hours has been supplemented from Bélus's *Traité des recherches,* 50; and the English astrologer Joseph Blagrave's magnum opus, *Blagrave's Astrological Practice of Physick,* 9–17. Sédir references the planetary "hours of the night" at the end of this chapter and provides precise planetary hours for plant harvests in his dictionary in part 3 (see the entries ⋆bay, ⋆chicory, and ⋆cyclamen), but he neglects to provide readers with any description or tabulation of the planetary days and hours.
8. Thiers, *Traité des superstitions selon l'écriture sainte, les décrets des conciles et les sentimens des saints pères et des théologiens,* 1:45–46.

CHAPTER 8. VEGETATION MAGIC

1. Pseudo-Clement, *Clementine Recognitions* 2.9.
2. Prel, "Forciertes Pflanzenwachstum."

3. Langhans, *Neue Ost-Indische Reise,* 560–62.
4. Hingston, *The Australian Abroad,* 2:56–57.
5. *Translator's note:* Sédir's examples of vegetation magic have been augmented to include the account from the so-called *Agrouchada-Parikchai,* which is not nearly so well known as the famous example cited by Sédir at the end of this section—namely, Jacolliot's description of a fakir's spontaneous vegetation of a papaya, or pawpaw, seed. See Jacolliot, *Le spiritisme dans le monde,* 41–43 [English edition: *Occult Science in India and among the Ancients,* 37–39].
6. Franck von Franckenau, *De palingenesia sive resuscitatione artificali plantarum, hominum et animalium,* 140–41. [See further Agricola, *Versuch der universal Vermehrung aller Bäume, Stauden und Blumen-Gewächse,* 1:81; cf. Agricola, *The Experimental Husbandman and Gardener,* 86. —*Trans.*]
7. *Translator's note:* See Oxley, "Fac-simile of Plant with Flower (*Ixora crocata*) Produced by a Materialised Spirit Form (Yolande), at Newcastle-on-Tyne, August 4th, 1880." For fuller accounts of these events and others, see D'Espérance, *Shadow Land or Light from the Other Side,* 247–58, 259–73, 322–33.
8. See Tavernier, *Les six voyages en Turquie, en Perse, et aux Indes,* 2:44–45; Maldigny, "Confession spiritualiste: Quatrième et dernier article," 416–17; Mousseaux, *Les hauts phénomènes de la magie, précédés du spiritisme antique,* 230 n. 1; and Görres, *Die christliche Mystik,* 3:554. For further accounts, see Prel, "Forciertes Pflanzenwachstum."
9. *Translator's note:* See Jacolliot, *Le spiritisme dans le monde,* 309–14 [English edition: *Occult Science in India,* 259–64]; cf. Blavatsky, *Isis Unveiled,* 1:139–41.
10. Hartmann, *Der Spiritismus,* 53n**; cf. Hartmann, *Spiritus,* translated by Charles C. Massey, 50n†. —*Trans.*]
11. Prel, "Forciertes Pflanzenwachstum," 150–51; cf. Prel, *Die Philosophie der Mystik,* 160–279. See further Saint-Denys, *Les rêves et les moyens de les diriger.*
12. See Haeckel, *Anthropogenie oder Entwickelungsgeschichte des Menschen,* 177.
13. Papus, *Traité élémentaire de magie pratique,* 20–25.
14. *Translator's note:* See, for example, Agricola, *Versuch der universal Vermehrung aller Bäume, Stauden und Blumen-Gewächse,* 1:11; Agricola, *The Experimental Husbandman and Gardener,* 7.

CHAPTER 9.
THE VEGETABLE PHOENIX

1. *Translator's note:* See Duchesne, *Ad veritatem hermeticae medicinae,* liber I, caput XXIII, 231–32; cf. Duchesne, *The Practise of Chymicall, and Hermeticall Physicke, for the Preservation of Health,* bk. 1, chap. 10 (pages unnumbered).

2. Brosse, *De la nature, vertu, et utilité des plantes,* 44.
3. Duchanteau, *La grand livre de la nature,* 18–19.
4. Guaïta, *Le serpent de la Genèse,* 2:697.

CHAPTER 10. PLANT PALINGENESIS IN HISTORY AND PRACTICE

1. *Translator's note:* For the final chapter of part 2, Sédir reproduces L. Desvignes's French translation of Karl Kiesewetter's "Die Palingenesie in ihrer Geschichte und Praxis geschildert," which was published in Papus's journal *L'Initiation* (see Kiesewetter, "La palingénésie historique et pratique"). This translation is based on both Kiesewetter's original German and Desvignes's French translation, in which a great number of Kiesewetter's footnotes are regrettably displaced. In addition to such corrections, Kiesewetter's article has also been further revised and edited for the sake of greater uniformity and continuity with Sédir's preceding chapters.
2. See Prel, "Der Pflanzenphönix," 197.
3. Ovid, *Metamorphoses* 7.275–84.
4. Eckartshausen, *Aufschlüsse zur Magie,* 2:390.
5. On Albertus Magnus and the homunculus, see especially Campanella, *De sensu rerum et magia,* 215. See also Hollandus, *Opera vegetabilia,* a work that has been reprinted on numerous occasions.
6. *Translator's note:* Cf. Paraclesus, "A Book of Renovation and Restauration" in *Paracelsus, His Archidoxis, Comprised in Ten Books,* 18 (translation slightly revised).
7. Paracelsus, *De natura rerum,* 2v. Our own remarks appear in brackets.
8. See Maurer, *Amphitheatrum magiae universae theoreticae et practicae,* 517.
9. Agrippa, *De occulta philosophia libri tres,* liber I, caput XXXVI, xlii [English edition: *Three Books of Occult Philosophy,* bk. 1, chap. 36, 118].
10. See Duchesne, *Ad veritatem hermeticae medicinae;* Claves, *Philosophia chymica tribus tractatibus comprehensa;* Béguin, *Tyrocinium chymicum* [English edition: *Tyrocinium chymicum, or Chymical Essays*]; Borel, *Historiarum, et observationum medico-physicarum;* Tachenius, *Hippocrates chymicus* [English edition: *Otto Tachenius, His Hippocrates chymicus*]; Sennert, *De chymicorum cum Aristotelicis et Galenis consensu ac dissensu;* Digby, *Dissertatio de plantarum vegetatione;* Becke, *Experimenta et meditationes circa naturalium rerum principia;* Maxwell, *De medicina magnetica libri III;* and Pezoldt, "De palingenesia seu artificiali rerum ex regno vegetabili."
11. See Franck von Franckenau, *De palingenesia, sive resuscitatione artificiali plantarum.*

12. Eckartshausen, *Aufschlüsse zur Magie,* 2:386.
13. Maxwell, *De medicina magnetica,* liber II, caput V, 100–101. Our own remarks appear in brackets.
14. Maxwell, *De medicina magnetica,* liber II, caput XX, 164. Our own remarks appear in brackets.
15. Maxwell, *De medicina magnetica,* 160. Our own remarks appear in brackets.
16. Becke, *Experimenta et meditationes,* 253–55.
17. Becke, *Experimenta et meditationes,* 244.
18. Anonymous, *Künstliche Auferweckung der Pflanzen, Menschen, Thiere aus ihrer Asche,* 51–54.
19. See Becher, *Chymischer Glücks-Hafen,* 784–85 § 90 (*Alius modus*); cf. Oetinger, *Gedanken über die Zeugung und Geburten der Dinge,* 47; and Prel, "Der Pflanzenphönix," 198.
20. Becher, *Chymischer Glücks-Hafen,* 784 § 89 (*Regeneratio plantarum*). Our own remarks appear in brackets.
21. *Translator's note:* See further Kiesewetter, "Die Rosenkreuzer, ein Blick in dunkele Vergangenheit."
22. Eckartshausen, *Aufschlüsse zur Magie,* 2:388–90.
23. Hermes Trismegistus, *Emerald Tablet* 5.
24. See Burggrav, *Biolychnium, seu Lucerna, cum vita eius, cui accensa est Mystice vivens iugiter;* Helmont, *De magnetica vulnerum curatione,* 25–26. [See also Helmont, *A Ternary of Paradoxes,* 12 § 20. —*Trans.*]

CHAPTER A. ACACIA–ASPHODEL

1. *Translator's note:* Entry augmented. See Exodus 25–27; Mackey, *An Encyclopaedia of Freemasonry and Its Kindred Sciences,* 7–9; Papus, *Traité méthodique de science occulte,* 743–44.
2. *Translator's note:* Entry augmented. Cf. Jollivet-Castelot, *La médecine spagyrique,* 11.
3. Dioscorides, *On Medical Materials* 4.41.
4. Serres, *Le théatre d'agriculture et mésnage des champs,* 561.
5. *Translator's note:* Entry augmented. See Pliny, *Natural History* 25.40.
6. *Translator's note:* Cf. Dioscorides, *On Medical Materials* 1.123.
7. *Translator's note:* Entry augmented. See Dioscorides, *On Medical Materials* 3.91.
8. *Translator's note:* Entry augmented. See Homer, *Odyssey* 24.81; Philostratus, *On Heroes* 53.8–13; 1 Peter 5:4.
9. *Translator's note:* Cf. Theophrastus, *Enquiry into Plants* 9.19.2–4.

10. *Translator's note:* Entry augmented. See Pausanias, *Descriptions of Greece* 1.44.3.
11. *Translator's note:* Entry augmented. See Papus, "Premiers éléments d'homéopathie pratique," 102, 109.
12. *Translator's note:* Entry augmented. See Mavéric, *La médecine hermétique des plantes,* 27, 178 [English edition: *Hermetic Herbalism,* 21, 143].
13. *Translator's note:* Cf. Hartmann, *With the Adepts,* 9.
14. *Translator's note:* Entry augmented. See Theophrastus, *Enquiry into Plants* 3.11.3.
15. *Translator's note:* Entry augmented. See Dioscorides, *On Medical Materials* 2.125; cf. Agrippa, *De occulta philosophia libri tres,* liber I, caput XXXII, xxxvii [English edition: *Three Books of Occult Philosophy,* bk. 1, chap. 32, 107].
16. *Translator's note:* Entry augmented. See Homer, *Odyssey* 11.539 and 24.13–14; cf. Gubernatis, *La mythologie des plantes,* 2:28–29 s.v. *asphodèle.*

CHAPTER B.
BACHELOR'S BUTTON–BUTTERCUP

1. *Translator's note:* Entry augmented. See Cazin, *Traité pratique et raisonné des plantes médicinales indigènes,* 201.
2. Paracelsus, *The Hermetic and Alchemical Writings,* 2:218.
3. *Translator's note:* See Guaïta, *Le serpent de la Genèse,* 1:372; Pseudo-Albertus, *Liber aggregationis seu secretorum,* 4v–5r (Chaldean herb no. 14) = *Les admirables secrets d'Albert le Grand,* 83; cf. Papus, *Traité élémentaire de magie pratique,* 546 s.v. *mélisse.*
4. *Translator's note:* Cf. Pernety, *Dictionnaire mytho-hermétique,* 291 s.vv. *meliphyllum, melissophyllum.*
5. Guaïta, *Le serpent de la Genèse,* 1:337–38.
6. *Translator's note:* Entry augmented. See Bhagavad Gita 15.1; cf. Sédir's comments in chapter 6.
7. *Translator's note:* Entry augmented. See Dioscorides, *On Medical Materials* 3.12.
8. *Translator's note:* Entry augmented. See Hippocrates, *Use of Liquids* 5; Pliny, *Natural History* 29.11.
9. *Translator's note:* Entry augmented. See Jollivet-Castelot, *La médecine spagyrique,* 13.
10. *Translator's note:* Entry augmented. See Papus, "Premiers éléments d'homéopathie pratique," 103.
11. *Translator's note:* Cf. Huysmans, *La cathédrale,* 290 [English edition: *The Cathedral,* 201].
12. *Translator's note:* Cf. Collin de Plancy, *Dictionnaire infernal,* 1:121–22.
13. *Translator's note:* Entry augmented. See Guaïta, *Le serpent de la Genèse,* 1:372;

Pseudo-Albertus, *Liber aggregationis seu secretorum,* 5r (Chaldean herb no. 16) = *Les admirables secrets d'Albert le Grand,* 84–85.

14. *Translator's note:* See, for example, Virgil, *Aeneid* 9.619–20.
15. *Translator's note:* Entry augmented. Cf. the entry ⋆white bryony and the corresponding footnote.
16. Columella, *On Agriculture* 10.250.
17. *Translator's note:* Cf. Huysmans, *La cathédrale,* 290–91 [English edition: *The Cathedral,* 201–2].
18. *Translator's note:* Entry revised and augmented. See Wecker, *De secretis libri XVII,* 667; cf. Pierre Meyssonier's French translation (Wecker, *Les secrets et merveilles de nature,* 982), which reads *semence de pourpié, ou des fueilles du grand Lappa ou Bardane*; and Porta, *Magiae naturalis,* liber II, capitula XXII, 91; cf. François Arnoullet's French translation (Porta, *La magie naturelle,* 209), which reads *semence de pourcelaine, ou des fueilles de glouteron* ("seeds of ⋆purslane or leaves of ⋆greater burdock").
19. *Translator's note:* Entry augmented. See Cazin, *Traité pratique et raisonné des plantes médicinales,* 901–6.

CHAPTER C. CABBAGE–CYPRESS

1. *Translator's note:* Entry augmented. See Cazin, *Traité pratique et raisonné des plantes médicinales,* 302–4.
2. *Translator's note:* Cf. Pernety, *Dictionnaire mytho-hermétique,* 372 s.v. *pentadactylon.*
3. *Translator's note:* Entry augmented. See Guaïta, *Le serpent de la Genèse,* 1:372; Pseudo-Albertus, *Liber aggregationis seu secretorum,* 3r (Chaldean herb no. 6) = *Les admirables secrets d'Albert le Grand,* 78; cf. Papus, *Traité élémentaire de magie pratique,* 546 s.v. *nepte.*
4. *Translator's note:* Entry augmented. See Culpeper, *Culpeper's English Physician and Complete Herbal,* 144–46; Jollivet-Castelot, *La médecine spagyrique,* 271–72.
5. *Translator's note:* Entry augmented. See Ezekiel 31:10.
6. *Translator's note:* Entry augmented. See Mavéric, *La médecine hermétique des plantes,* 53 [English edition: *Hermetic Herbalism,* 45].
7. *Translator's note:* See Guaïta, *Le serpent de la Genèse,* 1:372; Pseudo-Albertus, *Liber aggregationis seu secretorum,* 2v–3r (Chaldean herb no. 4) = *Les admirables secrets d'Albert le Grand,* 77; cf. Papus, *Traité élémentaire de magie pratique,* 539 s.v. *chélidoine.*
8. *Translator's note:* Cf. Maurin, *Formulaire de l'herboristerie,* 150.

9. *Translator's note:* Entry augmented. See Mavéric, *La médecine hermétique des plantes,* 42 [English edition: *Hermetic Herbalism,* 35].
10. *Translator's note:* Cf. Plutarch, *Timoleon* 26.
11. *Translator's note:* Cf. Pernety, *Dictionnaire mytho-hermétique,* 359 s.v. *oscieum.*
12. *Translator's note:* Cf. Pliny, *Natural History* 25.30.
13. *Translator's note:* Entry revised. See Forsyth, *Demonologia,* 71–72; Forsyth, *The New Domestic Medical Manual,* 324; cf. the entry *meadow saffron and the corresponding footnote.
14. *Translator's note:* See Guaïta, *Le serpent de la Genèse,* 1:372; Pseudo-Albertus, *Liber aggregationis seu secretorum,* 4r (Chaldean herb no. 11) = *Les admirables secrets d'Albert le Grand,* 81; cf. Agrippa, *De occulta philosophia libri tres,* liber I, caput XLIX, xxxvii [English edition: *Three Books of Occult Philosophy,* bk. 1, chap. 49, 148].
15. *Translator's note:* Cf. Pliny, *Natural History* 24.38.
16. *Translator's note:* Entry augmented. See Papus, *Traité élémentaire de magie pratique,* 545 s.v. *léthargie.*
17. *Translator's note:* Cf. Pernety, *Dictionnaire mytho-hermétique,* 539 s.v. *zatanea,* 546 s.vv. *zuccaiar ou zuccar.*
18. *Translator's note:* Entry revised and augmented. Cf. the entry *sorb and the corresponding footnote.
19. See Pictorius, *Pantopōlion,* 71; Pictorius, "De speciebvs magiae ceremonialis, quam Goetiam vocant, epitome," in *Sermonum convivalium libri decem deque ebrietate Iusus quidam,* 77; cf. Thiers, *Traité des superstitions,* 1:153.
20. *Translator's note:* Entry augmented. See Candolle, *Essai sur les propriétés médicinales des plantes,* 101.
21. *Translator's note:* Pseudo-Albertus, *Liber aggregationis seu secretorum,* 6r–v (planetary herb no. 5) = *Les admirables secrets d'Albert le Grand,* 87–88; cf. Papus, *Traité élémentaire de magie pratique,* 239, 540 s.v. *dents,* 548 s.v. *plaies,* 550 s.v. *quintefeuille.*
22. *Translator's note:* Cf. Pernety, *Dictionnaire mytho-hermétique,* 377 s.v. *philadelphus.*
23. *Translator's note:* Entry augmented. See Cazin, *Traité pratique et raisonné des plantes médicinales,* 623–24.
24. *Translator's note:* Cf. Pliny, *Natural History* 21.29.
25. Guaïta, *Le serpent de la Genèse,* 346. [Sédir's entry has been augmented with recourse to Guaïta's discussion of the *coca plant. See further appendix 3. —*Trans.*]
26. *Translator's note:* Entry augmented. See Candolle, *Essai sur les propriétés médicinales des plantes,* 101.

27. *Translator's note:* Cf. Pernety, *Dictionnaire mytho-hermétique,* 185 s.vv. *handal & handel.*
28. *Translator's note:* Entry augmented. See Dioscorides, *On Medical Materials* 3.112.
29. *Translator's note:* Cf. Pernety, *Dictionnaire mytho-hermétique,* 396 s.v. *populago.*
30. *Translator's note:* Cf. Hartmann, *With the Adepts,* 9.
31. *Translator's note:* Entry revised. See Papus, *Traité élémentaire de magie pratique,* 549 s.v. *punaises;* cf. the entry *larkspur and the corresponding footnote.
32. *Translator's note:* Entry augmented. See Hildegard of Bingen, *Physica* 1.209; cf. Huysmans, *La cathédrale,* 278 [English edition: *The Cathedral,* 193].
33. *Translator's note:* Cf. Huysmans, *La cathédrale,* 290 [English edition: *The Cathedral,* 201].
34. *Translator's note:* Cf. Pernety, *Dictionnaire mytho-hermétique,* 367 s.vv. *paralysis herba ou paralytica.*
35. *Translator's note:* Entry augmented. See Huysmans, *La cathédrale,* 292 [English edition: *The Cathedral,* 202].
36. *Translator's note:* Entry augmented. See Dierbach, *Flora mythologica,* 200.
37. *Translator's note:* Cf. Pernety, *Dictionnaire mytho-hermétique,* 448 s.vv. *sanguinalis, sanguinaria.*
38. *Translator's note:* Entry augmented. See Papus, *Traité élémentaire de magie pratique,* 542 s.v. *grossesse.*
39. *Translator's note:* Entry augmented. See Rodin, *Les plantes médicinales,* 426–49.
40. *Translator's note:* Entry augmented. See Mavéric, *La médecine hermétique des plantes,* 42 [English edition: *Hermetic Herbalism,* 36]; cf. Pernety, *Dictionnaire mytho-hermétique,* 463 s.vv. *sicyos & sicys.*
41. *Translator's note:* Entry augmented. See Papus, *Traité élémentaire de magie pratique,* 549 s.v. *punaises.*
42. Cf. Piemontese, *The Secrets of Alexis,* 139.
43. *Translator's note:* Cf. Pernety, *Dictionnaire mytho-hermétique,* 477 s.v. *suffo,* 531 s.vv. *umbilicus terræ.*

CHAPTER D. DAFFODIL–DITTANY OF CRETE

1. *Translator's note:* Entry augmented. Cf. Pernety, *Dictionnaire mytho-hermétique,* 233 s.vv. *keiri ou keirim.*
2. *Translator's note:* Entry augmented. See Papus, *Traité élémentaire de magie pratique,* 544 s.v. *insomnie.*

CHAPTER E.
EDELWEISS–EYEBRIGHT

1. *Translator's note:* Cf. Hartmann, *With the Adepts,* 9.
2. Cf. Piemontese, *The Secrets of Alexis,* 139.
3. *Translator's note:* Entry augmented. See Jollivet-Castelot, *La médecine spagyrique,* 17.
4. *Translator's note:* Entry augmented. See Pliny, *Natural History* 21.33.
5. *Translator's note:* Entry augmented. This is Papus's version of a love charm from the *Petit Albert.* See Papus, *Traité élémentaire de magie pratique,* 525; cf. Pseudo-Albertus, *Secrets merveilleux de la magie naturelle et cabalistique du Petit Albert,* 149.
6. *Translator's note:* Entry augmented. See Rodin, *Les plantes médicinales,* 43.

CHAPTER F.
FAVA BEAN–FUMITORY

1. *Translator's note:* Entry augmented. See Mavéric, *La médecine hermétique des plantes,* 42 [English edition: *Hermetic Herbalism,* 35].
2. *Translator's note:* Entry augmented. See Theophrastus, *Enquiry into Plants* 7.14.1; Pliny, *Natural History* 22.30.
3. *Translator's note:* Entry augmented. See Theophrastus, *Enquiry into Plants* 9.22; Pliny, *Natural History* 27.9; and Dioscorides, *On Medical Materials* 4.184.
4. *Translator's note:* Cf. Galen, *On the Powers and Mixtures of Simple Remedies* 9.39.
5. Thiers, *Traité des superstitions,* 1:314.
6. *Translator's note:* Cf. Pernety, *Dictionnaire mytho-hermétique,* 417 s.v. *pteris.*
7. *Translator's note:* Entry augmented. See Dioscorides, *On Medical Materials* 4.94; cf. Cazin, *Traité pratique et raisonné des plantes médicinales,* 975–76.
8. Cf. Tuchmann, "La fascination," 241 s.v. *Acorus calamus.*
9. *Translator's note:* Entry augmented. See Mario, "La flore mystérieuse," *La vie mystérieuse* 38, 220.
10. *Translator's note:* Cf. Hartmann, *With the Adepts,* 71.

CHAPTER G.
GARLIC–GRAPEVINE

1. *Translator's note:* Cf. Athenaeus, *Deipnosophists* 10.422d.
2. *Translator's note:* Cf. Huysmans, *La cathédrale,* 290 [English edition: *The Cathedral,* 201].

3. *Translator's note:* Cf. Hartmann, *With the Adepts,* 9.
4. *Translator's note:* Entry augmented. See Cazin, *Traité pratique et raisonné des plantes médicinales,* 480–82.
5. *Translator's note:* Cf. Pernety, *Dictionnaire mytho-hermétique,* 537 s.vv. *xiphidium, xiphium.*
6. *Translator's note:* Entry augmented. Cf. Dioscorides, *On Medical Materials* 4.20.
7. *Translator's note:* Cf. Pernety, *Dictionnaire mytho-hermétique,* 225 s.v. *ipcacidos;* cf. Duchanteau, *La grand livre de la nature,* 73 s.v. *ipcacidos.*
8. *Translator's note:* Papus, *Traité élémentaire de magie pratique,* 553 s.v. *vigne.*

CHAPTER H. HAZEL–HYACINTH

1. *Translator's note:* Entry augmented. See Guaïta, *Le serpent de la Genèse,* 1:372; Pseudo-Albertus, *Liber aggregationis seu secretorum,* 2r (Chaldean herb no. 1) = *Les admirables secrets d'Albert le Grand,* 74–75; cf. Papus, *Traité élémentaire de magie pratique,* 237, 543 s.v. *héliotrope.*
2. *Translator's note:* Cf. Hartmann, *With the Adepts,* 71.
3. *Translator's note:* Entry augmented. See Pseudo-Albertus, *Liber aggregationis seu secretorum,* 5v (planetary herb no. 1) = *Les admirables secrets d'Albert le Grand,* 85; cf. Papus, *Traité élémentaire de magie pratique,* 234–35.
4. *Translator's note:* Cf. Pernety, *Dictionnaire mytho-hermétique,* 265 s.v. *mandella.*
5. *Translator's note:* Cf. Pernety, *Dictionnaire mytho-hermétique,* 188 s.v. *helebria.*
6. *Translator's note:* Cf. Pliny, *Natural History* 25.86.
7. *Translator's note:* Entry augmented. See Bosc, *Traité théorique et pratique du haschich, des substances psychiques et des plantes magiques,* 75–76.
8. *Translator's note:* Cf. Papus, *Traité élémentaire de magie pratique,* 547 s.v. *oiseaux.*
9. See further Baudelaire, *Les paradis artificiels;* Bosc, *Traité théorique et pratique du haschich,* 17–73; Guaïta, *Le serpent de la Genèse,* 1:360–62; and Pandorac, *Testament d'un haschischéen.* [Sédir also refers to an unspecified work by Matgïoi, the pseudonym of Albert de Pouvourville. Most likely this is a reference to his "La clef orientale des faux paradis," but this work is concerned almost exclusively with opium. See further Sédir's essay on opium in appendix 3. —*Trans.*]
10. *Translator's note:* See Guaïta, *Le serpent de la Genèse,* 1:372; Pseudo-Albertus, *Liber aggregationis seu secretorum,* 3v (Chaldean herb no. 8) and 6v (planetary herb no. 6) = *Les admirables secrets d'Albert le Grand,* 79 and 88; cf. Papus, *Traité élémentaire de magie pratique,* 235–36, 544 s.v. *jusquiame.*
11. *Translator's note:* Thiriat, "Croyances, superstitions, préjugés, usages et coutumes dans le Département des Vosges," *Mélusine* 21, 499.

12. *Translator's note:* Entry augmented. See Gessmann, *Die Pflanze im Zauberglauben,* 29 s.v. *Christdorn.*
13. *Translator's note:* Entry augmented. See Pernety, *Dictionnaire mytho-hermétique,* 268 s.v. *matersylva.*
14. *Translator's note:* Cf. Huysmans, *La cathédrale,* 290 [English edition: *The Cathedral,* 201].
15. *Translator's note:* Entry augmented. See Celsus, *On Medicine* 4.9.
16. *Translator's note:* Entry augmented. See Pseudo-Albertus, *Les admirables secrets d'Albert le Grand,* 167.
17. *Translator's note:* Entry augmented. See Guaïta, *Le serpent de la Genèse,* 1:372; Pseudo-Albertus, *Liber aggregationis seu secretorum,* 3r–v (Chaldean herb no. 7) = *Les admirables secrets d'Albert le Grand,* 78–79; cf. Papus, *Traité élémentaire de magie pratique,* 544 s.v. *langue de chien.*
18. Thiers, *Traité des superstitions,* 1:152.
19. *Translator's note:* Entry augmented. Cf. Hartmann, *With the Adepts,* 71.

CHAPTER I. IRIS–IVY

1. *Translator's note:* Entry augmented. Cazin, *Traité pratique et raisonné des plantes médicinales,* 526–28.
2. Anonymous, *Le bâtiment des receptes, ou les vertus et propriétés des secrets,* 11–12.

CHAPTER J. JIMSONWEED–JUNIPER

1. *Translator's note:* Entry augmented. See Cazin, *Traité pratique et raisonné des plantes médicinales,* 1023–32.
2. *Translator's note:* Cf. Pernety, *Dictionnaire mytho-hermétique,* 105 s.vv. *datel ou tatel.*
3. *Translator's note:* Cf. Pliny, *Natural History* 24.36; Gubernatis, *La mythologie des plantes,* 2:152–55 s.v. *genévrier.*
4. *Translator's note:* Cf. Pernety, *Dictionnaire mytho-hermétique,* 185 s.v. *hara,* 289 s.vv. *mel juniperinum.*

CHAPTER K. KNOTWEED–KUSHA GRASS

1. *Translator's note:* Entry augmented. See Guaïta, *Le serpent de la Genèse,* 1:372; Pseudo-Albertus, *Liber aggregationis seu secretorum,* 5v–6r (planetary herb no. 2)

= *Les admirables secrets d'Albert le Grand,* 78–79; cf. Papus, *Traité élémentaire de magie pratique,* 237, 550 s.v. *renouée.*

2. *Translator's note:* Cf. Pernety, *Dictionnaire mytho-hermétique,* 313 s.v. *molybdæna,* 408 s.v. *proserpinaca,* 458 s.v. *seminalis.*
3. *Translator's note:* Entry augmented. See Monier-Williams, *A Sanskrit-English Dictionary,* 242 s.v. कुश, 403 s.v. दर्भ.
4. *Translator's note:* Entry augmented. See Bhagavad Gita 6.11; cf. Jacolliot, *La genèse de l'humanité,* 343–44.

CHAPTER L. LARKSPUR–LUNGWORT

1. *Translator's note:* Cf. Papus, *Traité élémentaire de magie pratique,* 552 s.v. *taies.*
2. *Translator's note:* Cf. Pernety, *Dictionnaire mytho-hermétique,* 39 s.v. *aquilena.*
3. *Translator's note:* Entry augmented. See, for example, Rozin, "L'apperçu des plantes usuelles," 353.
4. *Translator's note:* Pseudo-Albertus, *Les admirables secrets d'Albert le Grand,* 146.
5. *Translator's note:* Entry augmented. See Cazin, *Traité pratique et raisonné des plantes médicinales,* 558–61.
6. *Translator's note:* Entry augmented. See Maurin, *Formulaire de l'herboristerie,* 150; cf. Cazin, *Traité pratique et raisonné des plantes médicinales,* 831–32.
7. Hildegard of Bingen, *Physica* 1.23.
8. *Translator's note:* Entry augmented. See Guaïta, *Le serpent de la Genèse,* 1:372; Pseudo-Albertus, *Liber aggregationis seu secretorum,* 3v–4r (Chaldean herb no. 9) and 6r (planetary herb no. 3) = *Les admirables secrets d'Albert le Grand,* 80 and 86–87; cf. Papus, *Traité élémentaire de magie pratique,* 239–240, 539 s.v. *chrynostates,* 544 s.v. *lis.*
9. *Translator's note:* Entry augmented. See Blavatsky, *The Secret Doctrine,* 1:379; cf. the passage on the lotus flower extracted from the so-called *Agrouchada-Parikchai* in Jacolliot's *La genèse de l'humanité,* 341–42.

CHAPTER M. MADDER–MYRTLE

1. *Translator's note:* Entry augmented. See Cazin, *Traité pratique et raisonné des plantes médicinales,* 622–23.
2. *Translator's note:* Entry augmented. See Lenglet-Mortier and Vandamme, *Nouvelles et véritables étymologies,* 105–8.
3. *Translator's note:* Cf. Papus, *Traité élémentaire de magie pratique,* 543 s.v. *guimauve,* 546 s.v. *matrice,* 552 s.v. *urines.*

4. *Translator's note:* Here Sédir appears to be referencing Reclus, "Études sur les origines magiques de la médecine"; cf. Pseudo-Albertus, *Secrets merveilleux de la magie naturelle et cabalistique du Petit Albert,* 55–57.
5. *Translator's note:* Entry augmented. See Pernety, *Dictionnaire mytho-hermétique,* 45 s.v. *aroph,* 119 s.v. *dudaim* (cf. Genesis 30:14–16 and Song of Songs 7:13), and 207 s.v. *jabora.*
6. *Translator's note:* Entry augmented. Cf. Hartmann, *With the Adepts,* 9.
7. *Translator's note:* Cf. Dioscorides, *On Medical Materials* 4.189.
8. *Translator's note:* Cf. Colossians 4:14.
9. *Translator's note:* Entry augmented. See Huysmans, *La cathédrale,* 294 [English edition: *The Cathedral,* 204].
10. *Translator's note:* Entry augmented. See Jollivet-Castelot, *La médecine spagyrique,* 11.
11. *Translator's note:* Cf. Pliny, *Natural History* 16.95.
12. *Translator's note:* See Guaïta, *Le serpent de la Genèse,* 1:372; Pseudo-Albertus, *Liber aggregationis seu secretorum,* 4r (Chaldean herb no. 10) = *Les admirables secrets d'Albert le Grand,* 80–81; cf. Papus, *Traité élémentaire de magie pratique,* 542 s.vv. *gui de chêne.*
13. *Translator's note:* Cf. Davidson, *The Mistletoe and Its Philosophy,* 14.
14. *Translator's note:* Cf. Pernety, *Dictionnaire mytho-hermétique,* 103 s.v. *dabat,* 188 s.vv. *hele ou helle.*
15. *Translator's note:* Cf. Pernety, *Dictionnaire mytho-hermétique,* 462 s.v. *serpigo.*
16. Tuchmann, "La fascination," 243–44 s.v. *armoise.*
17. *Translator's note:* Entry augmented. See Papus, *Traité élémentaire de magie pratique,* 538 s.v. *armoise.*
18. *Translator's note:* Cf. Pseudo-Apuleius, *Herbarium* 12.1; cf. Tuchmann, "La fascination," 243.
19. *Translator's note:* Cf. Huysmans, *La cathédrale,* 292 [English edition: *The Cathedral,* 202].
20. *Translator's note:* Cf. Pernety, *Dictionnaire mytho-hermétique,* 390 s.v. *ploma.*
21. *Translator's note:* Cf. Matthew 13:31–32; Mark 4:30–32; Luke 13:18–19.
22. *Translator's note:* Entry augmented. See Berkeley, *Siris,* 24.
23. *Translator's note:* Cf. Papus, *Traité élémentaire de magie pratique,* 546 s.v. *migraines,* 553 s.v. *vomissements.*

CHAPTER N. NEEDLELEAF–NUTMEG

1. Cf. Dioscorides, *On Medical Materials* 3.94.
2. *Translator's note:* See Pseudo-Albertus, *Secrets merveilleux de la magie naturelle et*

cabalistique du Petit Albert, 141–42; cf. Papus, *Traité élémentaire de magie pratique,* 427.

3. *Translator's note:* See Guaïta, *Le serpent de la Genèse,* 1:372; Pseudo-Albertus, *Liber aggregationis seu secretorum,* 2r–v (Chaldean herb no. 2) = *Les admirables secrets d'Albert le Grand,* 75–76; cf. Pseudo-Albertus, *Secrets merveilleux de la magie naturelle et cabalistique du Petit Albert,* 10; Papus, *Traité élémentaire de magie pratique,* 547 s.v. *ortie.*

CHAPTER O. OAK–OREGANO

1. *Translator's note:* Cf. Papus, *Traité élémentaire de magie pratique,* 541 s.v. *epilepsie.*
2. *Translator's note:* Cf. Papus, *Traité élémentaire de magie pratique,* 531.
3. *Translator's note:* Entry augmented. See Lenain, *La science cabalistique,* 137; Jollivet-Castelot, *La médecine spagyrique,* 8.
4. *Translator's note:* Papus, *Traité élémentaire de magie pratique,* 546 s.v. *métrorrhagie.*
5. *Translator's note:* Entry augmented. Cf. Pseudo-Apuleius, *Herbarium* 15.3.
6. *Translator's note:* Entry augmented. See Theophrastus, *Enquiry into Plants* 9.18.9; Pliny, *Natural History* 26.63; cf. Dioscorides, *On Medical Materials* 3.128. The passage from Theophrastus is censored from Arthur Hort's edition and English translation but may be found in Friedrich Wimmer's edition of the Greek text.
7. Kircher, *Ars magna lucis et umbrae in X. libros digesta,* 44.
8. *Translator's note:* Entry augmented. See Gessmann, *Die Pflanze im Zauberglauben,* 98 s.v. *Wohlgemuth.*

CHAPTER P. PALM–PURSLANE

1. *Translator's note:* Cf. 1 Kings 6:29.
2. *Translator's note:* See Pernety, *Dictionnaire mytho-hermétique,* 490 s.v. *thamar;* cf. Exodus 15:27; Leviticus 23:40; Numbers 33:9; Deuteronomy 34:3; Judges 1:16 and 3:13; 2 Chronicles 28:15; Nehemiah 8:15; Psalm 92:12; Song of Songs 7:7–8; Joel 1:12.
3. *Translator's note:* Papus, *Traité élémentaire de magie pratique,* 553 s.v. *vers.*
4. *Translator's note:* Entry augmented. See Plutarch, *On Isis and Osiris* 68.
5. *Translator's note:* Cf. Huysmans, *La cathédrale,* 290 [English edition: *The Cathedral,* 201].
6. *Translator's note:* Entry augmented. See *Homeric Hymn to Demeter* 210; cf. the entry ⋆elecampane and the corresponding footnote.

7. *Translator's note:* Cf. Papus, *Traité élémentaire de magie pratique,* 264.
8. *Translator's note:* Entry augmented. See Rodin, *Les plantes médicinales,* 158–59.
9. *Translator's note:* Entry augmented. See Cazin, *Traité pratique et raisonné des plantes médicinales,* 819–20.
10. *Translator's note:* See Guaïta, *Le serpent de la Genèse,* 1:372; Pseudo-Albertus, *Liber aggregationis seu secretorum,* 2v–3r (Chaldean herb no. 5) = *Les admirables secrets d'Albert le Grand,* 77–78; cf. Papus, *Traité élémentaire de magie pratique,* 548 s.v. *pervenche.*
11. *Translator's note:* Entry augmented. See Papus, *Traité élémentaire de magie pratique,* 549 s.v. *peste.*
12. *Translator's note:* Entry augmented. Cf. Bjerregaard, *Lectures on Mysticism and Nature Worship,* 96.
13. *Translator's note:* Cf. Pernety, *Dictionnaire mytho-hermétique,* 375 s.v. *peucé,* 388 s.v. *pitys.*
14. *Translator's note:* Entry augmented. See Dioscorides, *On Medical Materials* 1.79; cf. Cazin, *Traité pratique et raisonné des plantes médicinales,* 861–62.
15. *Translator's note:* Cf. Pseudo-Albertus, *Liber aggregationis seu secretorum,* 6r (planetary herb no. 4) = *Les admirables secrets d'Albert le Grand,* 77–78.
16. *Translator's note:* Cf. Papus, *Traité élémentaire de magie pratique,* 549 s.v. *plantain.*
17. *Translator's note:* Entry augmented. Cf. Pernety, *Dictionnaire mytho-hermétique,* 394 s.v. *polyneuron.*
18. *Translator's note:* Entry augmented. See Jollivet-Castelot, *La médecine spagyrique,* 7.
19. *Translator's note:* Entry augmented. See Homer, *Iliad* 13.389 and 14.482.
20. *Translator's note:* Entry augmented. See Porta, *Magiae naturalis,* liber II, caput XXVI, 101; cf. Culpeper, *Culpeper's English Physician and Complete Herbal,* 321.
21. *Translator's note:* Entry augmented. See Pliny, *Natural History* 20.78; Cazin, *Traité pratique et raisonné des plantes médicinales,* 803.
22. Cf. Piemontese, *The Secrets of Alexis,* 139. [See further appendix 3. —*Trans.*]
23. *Translator's note:* Entry augmented. See Cazin, *Traité pratique et raisonné des plantes médicinales,* 1074–75.
24. Wecker, *De secretis libri XVII,* 667; cf. Porta, *Magiae naturalis,* 91. [Cf. the entry *greater burdock and the corresponding footnote. —*Trans.*]
25. *Translator's note:* Cf. Papus, *Traité élémentaire de magie pratique,* 544 s.v. *jusquiame,* 549 s.v. *pourpier,* 550 s.v. *pourpier,* 553 s.v. *visions.*

CHAPTER Q. QUINCE

1. *Translator's note:* Entry augmented. See Cazin, *Traité pratique et raisonné des plantes médicinales,* 332–33.

CHAPTER R. RAISIN TREE–RUE

1. *Translator's note:* Entry augmented. See Culpeper, *Culpeper's Complete Herbal,* 289 s.vv. rampion (hairy sheep's).
2. *Translator's note:* Entry augmented. See Gubernatis, *La mythologie des plantes,* 2:307–9 s.v. *raiponce.*
3. *Translator's note:* Cf. Pernety, *Dictionnaire mytho-hermétique,* 233 s.v. *kanech.*
4. *Translator's note:* Cf. Pernety, *Dictionnaire mytho-hermétique,* 425 s.v. *ramed,* 426 s.v. *raved,* 540 s.v. *zipar.*
5. *Translator's note:* Cf. Pernety, *Dictionnaire mytho-hermétique,* 316 s.vv. *mucarum & mucharum.*
6. *Translator's note:* See Guaïta, *Le serpent de la Genèse,* 1:372; cf. Apuleius, *Metamorphoses* 11.
7. *Translator's note:* Pseudo-Albertus, *Liber aggregationis seu secretorum,* 5r (Chaldean herb no. 15) = *Les admirables secrets d'Albert le Grand,* 83–84; cf. Papus, *Traité élémentaire de magie pratique,* 550 s.v. *rose.*
8. *Translator's note:* Cf. Hartmann, *With the Adepts,* 71.
9. *Translator's note:* Entry augmented. See Pliny, *Natural History* 25.6.
10. *Translator's note:* Cf. Pernety, *Dictionnaire mytho-hermétique,* 36 s.v. *anthos,* 248 s.v. *libanotis.*
11. *Translator's note:* Entry augmented. See Duchesne, *Répertoire des plantes utiles et des plantes vénéneuses du globe,* 119; cf. Dupuis, "Les andromèdes," 299.
12. Thiers, *Traité des superstitions,* 1:185.
13. Pliny, *Natural History* 20.51.
14. *Translator's note:* Cf. Homer, *Odyssey* 10.390–94.
15. *Translator's note:* Cf. Pernety, *Dictionnaire mytho-hermétique,* 185 s.v. *harmel.*

CHAPTER S. SAFFRON–STRAWBERRY

1. *Translator's note:* Entry augmented. See Dioscorides, *On Medical Materials* 1.26.
2. *Translator's note:* Cf. Cazin, *Traité pratique et raisonné des plantes médicinales,* 334–35.
3. *Translator's note:* Cf. Forsyth, *Demonologia,* 72; Forsyth, *The New Domestic Medical Manual,* 324.
4. *Translator's note:* Cf. Lenglet-Mortier and Vandamme, *Nouvelles et véritables étymologies,* 2–3, 126.
5. *Translator's note:* See Guaïta, *Le serpent de la Genèse,* 1:372; Pseudo-Albertus,

Liber aggregationis seu secretorum, 4r–v (Chaldean herb no. 12) = *Les admirables secrets d'Albert le Grand,* 81–82.

6. Pseudo-Llull, *De secretis natvrae, sev de quinta essentia liber vnus, in tres distinctiones diuisus, omnibus iam partibus absolutus,* 127–29.
7. *Translator's note:* Cf. Pernety, *Dictionnaire mytho-hermétique,* 396 s.v. *porrosa.*
8. *Translator's note:* Cf. Hartmann, *With the Adepts,* 9.
9. Cf. Piemontese, *The Secrets of Alexis,* 139.
10. *Translator's note:* Entry augmented. See Jollivet-Castelot, *La médecine spagyrique,* 14.
11. *Translator's note:* Entry augmented. See Maurin, *Formulaire de l'herboristerie,* 340; cf. Cazin, *Traité pratique et raisonné des plantes médicinales,* 966–67.
12. *Translator's note:* Entry augmented. See Gessmann, *Die Pflanze im Zauberglauben,* 90–91 s.v. *Teufelsabbiss;* cf. Rodin, *Les plantes médicinales,* 42–43.
13. *Translator's note:* Entry augmented. See Candolle, *Essai sur les propriétés médicinales des plantes,* 132–33; cf. Maurin, *Formulaire de l'herboristerie,* 128–29.
14. *Translator's note:* Entry augmented. See Cazin, *Traité pratique et raisonné des plantes médicinales,* 121–22.
15. *Translator's note:* Entry augmented. See Guaïta, *Le serpent de la Genèse,* 1:372; Pseudo-Albertus, *Liber aggregationis seu secretorum,* 2v (Chaldean herb no. 3) = *Les admirables secrets d'Albert le Grand,* 76–77; cf. Papus, *Traité élémentaire de magie pratique,* 552 s.vv. *vierge de pasteur.*
16. *Translator's note:* Cf. Huysmans, *La cathédrale,* 290 [English edition: *The Cathedral,* 201].
17. *Translator's note:* Entry augmented. See Gessmann, *Die Pflanze im Zauberglauben,* 96 s.v. *Weisswurz.*
18. *Translator's note:* Entry augmented. Cf. Pernety, *Dictionnaire mytho-hermétique,* 454 s.v. *secacul.*
19. *Translator's note:* Entry augmented. See Cazin, *Traité pratique et raisonné des plantes médicinales,* 1014–16.
20. *Translator's note:* Entry augmented. See Gessmann, *Die Pflanze im Zauberglauben,* 31 s.v. *Eberesche.*
21. *Translator's note:* Entry augmented. See Dioscorides, *On Medical Materials* 2.114.
22. *Translator's note:* Entry augmented. See Papus, *Traité élémentaire de magie pratique,* 549 s.v. *peste.*
23. *Translator's note:* Cf. Pernety, *Dictionnaire mytho-hermétique,* 362 s.v. *oxus.*
24. *Translator's note:* Entry augmented. See Linnaeus, *Species plantarum,* 1:9–14.
25. *Translator's note:* Entry augmented. See Dioscorides, *On Medical Materials* 4.164; cf. Cazin, *Traité pratique et raisonné des plantes médicinales,* 429–30.
26. *Translator's note:* Entry augmented. See Pseudo-Apuleius, *Herbarium* 109.

27. *Translator's note:* Entry augmented. See Rodin, *Les plantes médicinales,* 292–94; cf. Cazin, *Traité pratique et raisonné des plantes médicinales,* 424–28.
28. *Translator's note:* Cf. Papus, *Traité élémentaire de magie pratique,* 542 s.v. *fraisier.*

CHAPTER T. TAMARIND–TURNIP

1. *Translator's note:* See further Ferrand, *Les musulmans à Madagascar et aux îles Comores,* 1:73–101.
2. *Translator's note:* Entry augmented. See Cazin, *Traité pratique et raisonné des plantes médicinales,* 1062–63.
3. *Translator's note:* Entry augmented. See Papus, *Traité élémentaire de magie pratique,* 550 s.v. *règles.*
4. *Translator's note:* Cf. Papus, *Traité élémentaire de magie pratique,* 540 s.v. *chardons.*
5. *Translator's note:* Cf. Pernety, *Dictionnaire mytho-hermétique,* 231 s.v. *ixia.*
6. *Translator's note:* Entry augmented. See Tuchmann, "La fascination," 282 s.v. *thym.*
7. *Translator's note:* Cf. Pseudo-Apuleius, *Herbarium* 130.
8. *Translator's note:* Entry augmented. See Pliny, *Natural History* 20.90.
9. *Translator's note:* Entry augmented. See Papus, *Traité élémentaire de magie pratique,* 550 s.v. *boutons.*
10. *Translator's note:* Entry augmented. See Rodin, *Les plantes médicinales,* 186; cf. Cazin, *Traité pratique et raisonné des plantes médicinales,* 1071–73.
11. *Translator's note:* Entry augmented. See Cazin, *Traité pratique et raisonné des plantes médicinales,* 677.

CHAPTER U. USNEA

1. *Translator's note:* See Porta, *Natural Magick,* 228–29; cf. Pseudo-Paracelsus, *Of the Supreme Mysteries of Nature,* 118.

CHAPTER V. VALERIAN–VIOLET

1. *Translator's note:* See, for example, Rochas, *Les frontières de la science,* 102–3; cf. Bourru and Burot, *La suggestion mentale et l'action à distance,* 61–62, 106–7, 283–84.
2. *Translator's note:* Cf. Huysmans, *La cathédrale,* 292 [English edition: *The Cathedral,* 202].

3. *Translator's note:* Entry augmented. See Pseudo-Albertus, *Les admirables secrets d'Albert le Grand,* 82; cf. Papus, *Traité élémentaire de magie pratique,* 221, 552 s.v. *verveine.*
4. *Translator's note:* Entry augmented. See Guaïta, *Le serpent de la Genèse,* 1:372; Pseudo-Albertus, *Liber aggregationis seu secretorum,* 4v (Chaldean herb no. 13) and 6v–7r (planetary herb no. 7) = Pseudo-Albertus, *Les admirables secrets d'Albert le Grand,* 82–83, 89–90; cf. Papus, *Traité élémentaire de magie pratique,* 238, 539 s.v. *convives,* 546 s.v. *maladie,* 550 s.v. *rage,* 552–53 s.v. *verveine.*
5. *Translator's note:* Entry augmented. See Pernety, *Dictionnaire mytho-hermétique,* 374 s.v. *peristeron,* 513 s.vv. *veine de Vénus*; cf. Ruland, *Lexicon alchemiae,* 478 s.vv. *vena Veneris.*
6. *Translator's note:* Entry augmented. See Théis, *Glossaire de botanique,* 481 s.v. *vicia;* cf. Lenglet-Mortier and Vandamme, *Nouvelles et véritables étymologies,* 179–80.
7. *Translator's note:* Entry augmented. See Cazin, *Traité pratique et raisonné des plantes médicinales,* 1071–73; cf. Rodin, *Les plantes médicinales,* 263–64.
8. *Translator's note:* Pernety, *Dictionnaire mytho-hermétique,* 282 s.vv. *matrionalis flos.*

CHAPTER W. WALLFLOWER–WORMWOOD

1. *Translator's note:* Entry augmented. See Théis, *Glossaire de botanique,* 107; cf. Pernety, *Dictionnaire mytho-hermétique,* 233 s.vv. *keiri ou keirim.*
2. *Translator's note:* Cf. Pernety, *Dictionnaire mytho-hermétique,* 248 s.vv. *ligni Heraclei.*
3. *Translator's note:* Entry augmented. See Cazin, *Traité pratique et raisonné des plantes médicinales,* 354–56.
4. *Translator's note:* Cf. Pernety, *Dictionnaire mytho-hermétique,* 452 s.v. *saure.*
5. *Translator's note:* Entry augmented. See Cazin, *Traité pratique et raisonné des plantes médicinales,* 959–65.
6. *Translator's note:* Entry augmented. See *Homeric Hymn to Hermes* 407–14.
7. *Translator's note:* Cf. Ovid, *Metamorphoses* 6.29; Pliny, *Natural History* 27.4.
8. *Translator's note:* Entry augmented. See Hartmann, *With the Adepts,* 9.

CHAPTER Y. YARROW

1. *Translator's note:* Entry augmented. See Cazin, *Traité pratique et raisonné des plantes médicinales,* 637–43.

2. *Translator's note:* Cf. Pernety, *Dictionnaire mytho-hermétique,* 306–7 s.vv. *militaris ou stratiotes.*

APPENDIX 1.
OCCULT MEDICINE

1. *Translator's note:* Appendix 1 is a translation of Sédir, "La médecine occulte."
2. *Translator's note:* Cf. Sédir's discussion of the transplantation of diseases in chapter 5.
3. See Fabre d'Olivet, *Caïn, mystère dramatique en trois actes de Lord Byron;* cf. Fabre d'Olivet, *Lettre de Fabre d'Olivet à Lord Byron exposant les motifs et le but de son ouvrage sur Caïn.*
4. *Translator's note:* See further Sédir's monograph bearing the same title, *La médecine occulte.*

APPENDIX 2.
PARACELSIAN PHYSIOLOGY

1. *Translator's note:* Appendix 2 is a translation of Sédir, "La physiologie de Paracelse."
2. See Lévi, *La clef des grands mystères suivant Hénoch, Abraham, Hermès Trismégiste, et Salomon,* 112–14, 126–28, 135, 139–40, *passim.*
3. *Translator's note:* Cf. Sédir's discussion of Louis Claude de Saint-Martin's theory of binary signatures in chapter 3.
4. These are the main lines of Paracelsus's teaching on the organic functioning of the human machine. We advise readers who would like to pursue this matter further and completely reassemble the spirit of his system to reduce the data contained in these pages into tables of correspondences. For those who cannot read Paracelsus in Latin or German, we recommend the following French works: Dariot, *La grand chirurgie de Philippe Aoreole Theophraste Paracelse, grand médecin et philosophe entre les Alemans;* Colonna, *Abrégé de la doctrine de Paracelse et de ses Archidoxes;* Lévi, *La clef des grands mystères,* 378–88; Franck, "Paracelse et l'alchimie au XVIe siècle"; and especially Durey, *Étude sur l'œuvre de Paracelse.*

APPENDIX 3.
ON OPIUM USE

1. *Translator's note:* Appendix 3 is a translation of Sédir, "Sur l'emploi de l'opium."
2. *Translator's note:* Matgioï, the pseudonym of Albert de Pouvourville (1861–1939), otherwise known as Tau Simon in the bishopric of the Église gnostique de France,

was one of the most prolific authors on the subject of opium use. See especially Matgioï, *L'opium;* Matgioï, *L'opium et l'alcool en Indochine;* and Matgioï, *Physique et psychique de l'opium.* Numa Pandorac, the pseudonym of Jules Giraud, published numerous articles on the subject of hashish. See further five articles by Pandorac published in *La Voie*: "Influence du haschisch sur la santé," "Influence du haschisch sur le corp et sur l'esprit," "Influence du haschisch sur nos facultés," "Manuel du haschischéen," and "Prédictions d'un haschischéen"; and one in *L'Initiation:* "Testament d'un haschischéen." All of which culminated in the monograph *Testament d'un haschischéen.* Gabriel de Lautrec (1867–1938), who is best known today as a translator of Mark Twain, published original poems and prose allegedly composed under the influence of opium and hashish. See, for example, Lautrec, *Le serpent de mer.*

3. Maeterlinck, *Le trésor des humbles* [English edition: *The Treasure of the Humble*].

Bibliography

ANCIENT SOURCES

Apuleius, *Metamorphoses*

Metamorphoses. 2 vols. Edited and translated by J. Arthur Hanson. Cambridge, Mass.: Harvard University Press, 1989.

Athenaeus, *Deipnosophists*

The Deipnosophists. 7 vols. Translated by Charles Burton Gulick. Cambridge, Mass.: Harvard University Press, 1927–1941.

Cassianus Bassus, *Agricultural Pursuits*

Geoponica sive Cassiani Bassi scholastici De re rustica eclogae. Edited by Henricus Beckh. Leipzig: Teubner, 1895.

Celsus, *On Medicine*

On Medicine. 3 vols. Translated by W. G. Spencer. Cambridge, Mass.: Harvard University Press, 1935–1938.

Clement of Alexandria, *Miscellanies*

Clemens Alexandrinus: II. Stromata, Buch I–VI. Edited by Otto Stählin. Leipzig, Germany: Hinrichs, 1906.

Columella, *On Agriculture*

On Agriculture. 3 vols. Translated by Harrison Boyd Ash, E. S. Forster, and Edward H. Heffner. Cambridge, Mass.: Harvard University Press, 1941–1955.

Demotic Magical Papyri (PDM)

The Greek Magical Papyri in Translation: Including the Demotic Spells. Edited by Hans Dieter Betz. 2nd ed. Chicago: University of Chicago Press, 1992.

Diogenes Laertius, *Lives of Eminent Philosophers*

Lives of Eminent Philosophers. 2 vols. Translated by R. D. Hicks. Cambridge, Mass.: Harvard University Press, 1925.

Dioscorides, *On Medical Materials*

Pedanii Dioscuridis Anazarbei De materia medica libri quinque. 3 vols. Edited by Max Wellmann. Berlin: Weidmann, 1906–1914.

Eusebius, *Preparation for the Gospel*

Die Praeparatio Evangelica. 2 vols. Edited by Karl Mras. Berlin: Akademie-Verlag, 1954–1956.

Galen, *On Antidotes*

Claudii Galeni opera omnia. Vol. 14.1. Edited by Karl Gottlob Kühn. Leipzig, Germany: Cnoblochius, 1827.

Galen, *On the Composition of Local Remedies*

Claudii Galeni opera omnia. Vol. 13.1. Edited by Karl Gottlob Kühn. Leipzig, Germany: Cnoblochius, 1826.

Galen, *On the Powers and Mixtures of Simple Remedies*

Claudii Galeni opera omnia. Vols. 11–12. Edited by Karl Gottlob Kühn. Leipzig, Germany: Cnoblochius, 1826.

Galen, *On Temperaments*

Claudii Galeni opera omnia. Vol. 1.2. Edited by Karl Gottlob Kühn. Leipzig, Germany: Cnoblochius, 1821.

Greek Magical Papyri (PGM)

The Greek Magical Papyri in Translation: Including the Demotic Spells. Edited by Hans Dieter Betz. 2nd ed. Chicago: University of Chicago Press, 1992.

Hermes Trismegistus, *Emerald Tablet*

Tabula smaragdina: Ein Beitrag zur Geschichte der hermetischen Literatur. Edited by Julius Ruska. Heidelberg, Germany: C. Winter's Universitätsbuchhandlung, 1926.

Hermes Trismegistus, *On Plants of the Seven Planets*

Catalogus codicum astrologorum graecorum: VII. Codices Germanicos. Edited by Franz Boll. Brussels: Henricus Lamertin, 1908.

Hesiod, *Theogony*

Theogony; Works and Days; Testimonia. Edited and translated by Glenn W. Most. Cambridge, Mass.: Harvard University Press, 2018.

Hippocrates, *Nature of Women and Barrenness*

Generation; Nature of the Child; Diseases 4; Nature of Women and Barrenness. Hippocrates vol. 10. Edited and translated by Paul Potter. Cambridge, Mass.: Harvard University Press, 2012.

Hippocrates, *Use of Liquids*

Places in Man; Glands; Fleshes; Prorrhetic 1–2; Physician; Use of Liquids; Ulcers; Haemorrhoids and Fistulas. Hippocrates vol. 8. Edited and translated by Paul Potter. Cambridge, Mass.: Harvard University Press, 1995.

Homer, *Iliad*

The Iliad. 2 vols. Translated by A. T. Murray. Cambridge, Mass.: Harvard University Press, 1944.

Homer, *Odyssey*

The Odyssey. 2 vols. Translated by A. T. Murray. Cambridge, Mass.: Harvard University Press, 1938.

Homeric Hymns

Homeric Hymns, Homeric Apocrypha, Lives of Homer. Edited and translated by Martin L. West. Cambridge, Mass.: Harvard University Press, 2003.

Ovid, *Metamorphoses*

Metamorphoses. 2 vols. Translated by Frank Justus Miller. Cambridge, Mass.: Harvard University Press, 1916.

Pausanias, *Descriptions of Greece*

Description of Greece. 4 vols. Translated by W. H. S. Jones and H. A. Ormerod. Cambridge, Mass.: Harvard University Press, 1918–1935.

Philostratus, *On Heroes*

On Heroes. Translated by Jennifer K. Berenson Maclean and Ellen Bradshaw Aitken. Atlanta, Ga.: Society of Biblical Literature, 2001.

Plato, *Charmides*

Charmides; Alcibiades 1–2; Hipparchus; The Lovers; Theages; Minos; Epinomis. Plato vol. 12. Translated by W. R. M. Lamb. Cambridge, Mass.: Harvard University Press, 1964.

Pliny, *Natural History*

Natural History. 10 vols. Translated by Harris Rackham, W. H. S. Jones, and D. E. Eichholz. Cambridge, Mass.: Harvard University Press, 1938–1962.

Plutarch, *Causes of Natural Phenomena*

Plutarch's Moralia. Vol. 11. Translated by Lionel Pearson, F. H. Sandbach, and Phillip H. De Lacy. Cambridge, Mass.: Harvard University Press, 1965.

Plutarch, *On Isis and Osiris*

Plutarch's Moralia. Vol. 5. Translated by Frank Cole Babbitt. Cambridge, Mass.: Harvard University Press, 1936.

Plutarch, *Timoleon*

Plutarch's Lives. Vol. 6. Translated by Bernadotte Perrin. Cambridge, Mass.: Harvard University Press, 1965.

Porphyry, *Life of Pythagoras*

Porphyrii philosophi Platonici opuscula tria. Edited by August Nauck. Leipzig, Germany: Teubner, 1860.

Proclus, *Commentaries on Plato's Timaeus*

Procli Diadochi In Platonis Timaeum commentaria. 8 vols. Edited by Hermann Diels. Leipzig, Germany: Teubner, 1903.

Pseudo-Apuleius, *Herbarium*

Antonii Musae De herba vettonica liber, Pseudo-Apulei Herbarius, Anonymi De taxone liber, Sexti Placiti Liber medicinae ex animalibus. Edited by Ernst Howald and Henry E. Sigerist. Leipzig, Germany: Teubner, 1927.

Pseudo-Aristotle, *On Plants*

Aristotle, Minor Works: On Colours; On Things Heard; Physiognomics; On Plants; On Marvelous Things Heard; Mechanical Problems; On Indivisible Lines; The Situations and Names of Winds; On Melissus; Xenophanes; Gorgias. Aristotle vol. 14. Translated by W. S. Hett. Cambridge, Mass.: Harvard University Press, 1955.

Pseudo-Clement, *Clementine Recognitions*

Die Pseudoklementinen: II. Rekognitionen in Rufins Übersetzung. Edited by Bernhard Rehm and Georg Strecker. Berlin: Akademie Verlag, 1965.

Testament of Solomon

The Testament of Solomon: Edited from Manuscripts at Mount Athos, Bologna, Holkham Hall, Jerusalem, London, Milan, Paris and Vienna. Edited by C. C. McCown. Leipzig, Germany: Hinrichs, 1922.

Theophrastus, *Enquiry into Plants*

Enquiry into Plants and Minor Works on Odours and Weather Signs. 2 vols. Translated by Arthur F. Hort. Cambridge, Mass.: Harvard University Press, 1916–1926.

Theophrasti Eresii opera quae supersunt omnia. 3 vols. Edited by Friedrich Wimmer. Leipzig, Germany: Teubner, 1854–1862.

Virgil, *Aeneid*

Aeneid, Books 7–12; Appendix Vergiliana. Virgil vol. 2. Translated by H. Rushton Fairclough. Cambridge, Mass.: Harvard University Press, 1918.

EARLY MODERN AND MODERN WORKS

Agricola, Georg Andreas. *The Experimental Husbandman and Gardener.* Translated by Richard Bradley. 2nd ed. London: Mears and Clay, 1726.

———. *Versuch der universal Vermehrung aller Bäume, Stauden und Blumen-Gewächse.* 2 vols. Regensburg, Germany: Petz, 1716–1717.

Agrippa, Heinrich Cornelius. *De occulta philosophia libri tres.* Cologne, Germany: Soter, 1533.

———. *Henrici Cornelii Agrippae ab Nettesheym . . . opera.* 2 vols. Lyon, France: Beringos fratres, 1550.

———. *La philosophie occulte de Henr. Corn. Agrippa.* 2 vols. The Hague, Netherlands: Alberts, 1727.

———. *Three Books of Occult Philosophy or Magic.* Edited by Willis F. Whitehead. Chicago: Hahn and Whitehead, 1898.

Albertus Magnus. *De animalibus libri XXVI, nach der Cölner Urschrift.* 2 vols. Edited by Herman Stadler. Münster, Germany: Aschendorff, 1920.

Andreä, Johann Valentin. *Chymische Hochzeit Christiani Rosencreutz anno 1459.* Strasbourg, France: Zetzner, 1616.

Anonymous. *Künstliche Auferweckung der Pflanzen, Menschen, Thiere aus ihrer Asche: Nebst einem kurtzen Unterricht allerhand Farben auf Glas zu brennen.* Frankfurt, Germany, 1785.

Anonymous. *Le bâtiment des receptes, ou les vertus et propriétés des secrets.* Lyon, France: Lions, 1693.

Ardoynis, Santes de. *Opus de venenis.* Edited by Ferdinando Ponzetti. Basel, Switzerland: Petri et Perna, 1562.

Astruc, Jean. *Traité des maladies des femmes.* 6 vols. Paris: Cavelier, 1763–1770.

Azaïs, Gabriel. *Catalogue botanique: Synonymie languedocienne, provençale, gasconne, quercinoise.* Béziers, France: Malinas, 1871.

B., E. R. "Haunted Trees and Stones." *The Theosophist* 16, no. 4 (November 1894): 98–102.

Babbit, Edwin Dwight. *The Principles of Light and Color: Including among Other Things the Harmonic Laws of the Universe, the Etherio-Atomic Philosophy of Force, Chromo Chemistry, Chromo Therapeutics, and the General Philosophy of the Fine Forces.* 2nd ed. London: Paternoster House, 1896.

Bailey, Liberty Hyde. *Cyclopedia of American Horticulture.* 4 vols. London: Macmillan, 1900.

———. *Standard Cyclopedia of Horticulture.* 2nd ed. 6 vols. London: Macmillan, 1917.

Baillon, Henri. *Cours élémentaire de botanique.* Paris: Hachette, 1882.

Barlet, F.-C. "Université libre des hautes études." *L'Initiation* 13, no. 1 (October 1891): 3–11.

Barrère, Albert. *Argot and Slang: A New French and English Dictionary of the Cant Words.* Rev. ed. London: Whittaker, 1889.

Barthés, Melchior. *Glossaire botanique languedocien, français, latin de l'arrondissement de Saint-Pons (Hérault): Précédé d'un étude du dialecte languedocien.* Montpellier, France: Imprimerie centrale du midi, 1873.

Baudelaire, Charles. *Les paradis artificiels: Opium et haschich.* Paris: Poulet-Malassis et de Broise, 1860.

Bauderon, Brice. *La pharmacopée de Bauderon, à laquelle outre les corrections et augmentations de toutes les precedentes editions sont adjoustée de nouveau.* Edited by François Verny. Lyon, France: Rivière, 1663.

Becher, Johann Joachim. *Chymischer Glücks-Hafen, oder grosse chymische Concordantz und Collection von funffzehen hundert chymischen Processen.* Frankfurt, Germany: Schiele, 1682.

Becke, David von der. *Experimenta et meditationes circa naturalium rerum principia.* Hamburg, Germany: Schultze, 1674.

Béguin, Jean. *Tyrocinium chymicum.* Venice: Baleonius, 1643.

———. *Tyrocinium chymicum, or Chymical Essays: Acquired from the Fountain of Nature and Manual Experience.* London: Passenger, 1669.

Bélus, Jean [Jean Mavéric]. *Traité des recherches pour la découverte des personnes disparues, des enfants, animaux et objets perdus ou volés: Moyens certains pour connaître le lieu où ils se trouvent, ainsi que le signalement des voleurs et l'endroit où ils se cachent, chapitre spécial pour découvrir la provenance des lettres anonymes, étude sur la recherche des trésors cachés, méthode magique rattionnelle.* Paris: Ficker, 1911.

Berkeley, George. *Siris: A Chain of Philosophical Reflexions and Inquiries Concerning the Virtues of Tar Water.* Rev. ed. London: Innys and Hitch, 1747.

Besant, Annie. *Vegetarianism in the Light of Theosophy.* London: Theosophical Publishing Society, 1894.

Bjerregaard, Carl H. A. "The Elementals, the Elementary Spirits, and the Relationship between Them and Human Beings." *The Path* 1, no. 10 (January 1887): 289–300; no. 11 (February 1887): 321–31.

———. *Lectures on Mysticism and Nature Worship.* Chicago: Kent, 1897.

Blagrave, Joseph. *Blagrave's Astrological Practice of Physick Discovering the True Way to Cure All Kinds of Diseases.* London: S.G. and B.G., 1671.

Blankaart, Steven. *Lexicon medicum renovatum: In quo totius artis medicae termini, in anatome, chirurgia, pharmacia, chymia, re botanica etc. usitati, dilucide & breviter exponuntur.* Leiden, Netherlands: Luchtmans, 1756.

———. *The Physical Dictionary: Wherein the Terms of Anatomy, the Names and Causes of Diseases, Chirurgical Instruments, and Their Use, Are Accurately Described.* 7th ed. London: Sprint, 1726.

Blavatsky, Helena Petrovna. *Isis Unveiled: A Master-Key to the Mysteries of Ancient and Modern Science and Theology.* 2nd ed. 2 vols. New York: Bouton, 1887.

———. *The Secret Doctrine: The Synthesis of Science, Religion, and Philosophy.* 2 vols. London: Theosophical Publishing, 1888.

Boehme, Jacob. *Concerning the Election of Grace, or of God's Will Towards Man, Commonly Called Predestination.* Translated by John Sparrow. London: Streater, 1655. Translation of *De electione gratiae, oder, Von der Gnaden-Wahl* (1623).

———. *De la signature des choses ou l'engendrement et de la définition de tous les êtres.* Translated by Paul Sédir. Paris: Charcornac, 1908. Translation of *De signatura rerum, oder, Von der Geburt und Bezeichnung aller Wesen* (1622).

———. *The High and Deep Searching Out of the Three-fold Life of Man through (or according to) the Three Principles.* Translated by John Sparrow. London: Lloyd,

1656. Translation of *De triplicivita hominis, oder Vom dreifachen Leben des Menschen* (1620).

Boerhaave, Herman. *Elementa chemiae.* 2 vols. Leiden, Netherlands: Severinus, 1732.

Bonnejoy, Ernest. *Principes d'alimentation rationnelle hygiénique et économique avec des recettes de cuisine végétarienne.* Paris: Berthier, 1884.

———. *Le végétarisme et le régime végétarien rationnel: Dogmatisme, histoire, pratique.* Paris: Baillière, 1891.

Bonnet, Charles. *Contemplation de la Nature.* 2 vols. Amsterdam: Rey, 1764.

Borel, Pierre. *Historiarum, et observationum medico-physicarum, centuriae IV.* Paris: Billaine, 1656.

Bosc, Ernest. "Égrégore." *La Voile d'Isis,* troisième série, 19 (July 1911): 151–53.

———. *Traité de la longévité ou l'art de devenir centenaire.* Paris: Daragon, 1908.

———. *Traité théorique et pratique du haschich et autre substances psychiques.* 3rd ed. Paris: Édition des curiosités, 1907.

Boscowitz, Arnold. *L'âme de la plante.* Paris: Ducrocq, 1867.

Bourru, Henri, and Ferdinand Burot. *La suggestion mentale et l'action à distance des substances toxiques et médicamenteuses.* Paris: Baillière, 1887.

Bradley, Richard. *A Philosophical Account of the Works of Nature.* London: Mears, 1721.

Britton, Nathaniel Lord. *An Illustrated Flora of the Northern United States, Canada, and the British Possessions.* 3 vols. New York: Scribner's, 1896–1898.

Brosse, Guy de la. *De la nature, vertu, et utilité des plantes.* Paris: Baragnes, 1678.

Brummitt, R. K., and C. E. Powell. *Authors of Plant Names.* 2nd ed. Surrey, UK: Royal Botanic Gardens, Kew, 2004.

Burgoyne, Thomas H. *The Light of Egypt, or the Science of the Soul and the Stars.* 5th ed. 2 vols. Denver, Colo.: Astro-Philosophical, 1903.

Burggrav, Johann Ernst. *Biolychnium, seu Lucerna, cum vita eius, cui accensa est Mystice vivens iugiter, cum morte eiusdem expirans omnesque affectus graviores prodens.* Franeker, Netherlands: Balck, 1611.

Candolle, Augustin Pyramus de. *Essai sur les propriétés médicinales des plantes.* Paris: Crochard, 1816.

Campanella, Tommaso. *De sensu rerum et magia.* Paris: Béchet, 1638.

Cazin, François-Joseph. *Traité pratique et raisonné des plantes médicinales indigènes.* 3rd ed. Paris: Asselin, 1868.

Chambers, Robert, and William Chambers. *Chambers's Encyclopædia: A Dictionary of Universal Knowledge for the People.* Rev. ed. 8 vols. London: Chambers, 1878.

Charas, Moyse. *Pharmacopée royale galénique et chymique.* 2 vols. Paris: Charas, 1676.

———. *The Royal Pharmacopœa, Galenical and Chymical.* London: Starkey, 1678.

Claves, Gaston LeDoux de. *Philosophia chymica tribus tractatibus comprehensa.* Geneva: Vignon, 1612.

Collin de Plancy, Jacques. *Dictionnaire infernal, ou bibliothèque universelle.* 2nd ed. 4 vols. Paris: La librairie universelle, 1825–1826.

Colonna, Francesco Maria Pompeo [Crosset de la Haumerie]. *Abrégé de la doctrine de Paracelse et de ses Archidoxes: Avec une explication de la nature des principes de chymie.* Paris: D'Houry fils, 1724.

Culpeper, Nicholas. *Culpeper's Complete Herbal: Consisting of a Comprehensive Description of Nearly All Herbs with Their Medicinal Properties and Directions for Compounding the Medicines Extracted from Them.* London: Foulsham, 1880.

———. *Culpeper's English Physician and Complete Herbal: To Which Are Now First Added, Upwards of One Hundred Additional Herbs, with a Display of Their Medicinal and Occult Properties, Physically Applied to Cure All Disorders Incident to Mankind.* Edited by Ebenezer Sibly. London: Green, 1789.

Dariot, Claude. *La grand chirurgie de Philippe Aoreole Theophraste Paracelse, grand médecin et philosophe entre les Alemans.* Lyon, France: Harsy, 1589.

Darwin, Erasmus. *The Botanic Garden: A Poem in Two Parts.* New York: Swords, 1798.

———. *Zoonomia, or the Laws of Organic Life.* 2 vols. London: Johnson, 1794.

Davidson, Peter. *La gui et sa philosophie.* Translated by Paul Sédir. Paris: Chamuel, 1896.

———. *The Mistletoe and Its Philosophy.* 2nd ed. Glasgow: Goodwin, 1898.

Delvau, Alfred. *Dictionnaire de la langue verte: Argots parisiens comparés.* 2nd ed. Paris: Dentu, 1867.

D'Espérance, Madame [Elizabeth Hope]. *Shadow Land or Light from the Other Side.* London: Redway, 1897.

De Quincey, Thomas. *Confessions of an English Opium-Eater.* 2nd ed. London: Taylor and Hessey, 1823.

Dierbach, Johann Heinrich. *Flora mythologica, oder Pflanzenkunde in Bezug auf Mythologie und Symbolik der Griechen und Römer.* Frankfurt, Germany: Sauerländer, 1833.

Digby, Kelnem. *Dissertatio de plantarum vegetatione.* Amsterdam: Pluymert, 1678.

Dodoens, Rembert. *A New Herball, or Historie of Plants.* Translated by Henry Lyte. London: Newton, 1586.

Doinel, Jules. "Rituel du Consolamentum." *L'Initiation* 22, no. 6 (March 1894): 209–12.

Duchanteau, Touzay. *La grand livre de la nature, ou l'apocalypse philosophique et hermétique.* Paris: L'imprimerie de la vérité, 1790.

———. *La grand livre de la nature, ou l'apocalypse philosophique et hermétique.* 2nd ed. Introduction by Oswald Wirth. Paris: Librairie du merveilleux, 1910.

Duchesne, E. A. *Répertoire des plantes utiles et des plantes vénéneuses du globe.* Paris: Renouard, 1836.

Duchesne, Joseph. *Ad veritatem hermeticae medicinae.* Frankfurt, Germany: Richter, 1605.

———. *The Practise of Chymicall, and Hermeticall Physicke, for the Preservation of Health.* Translated by Thomas Timme. London: Creede, 1605.

Dupuis, Aristide. "Les andromèdes." *Revue horticole* 32 (1860): 298–301.

Durville, Hector. *Les actions psychiques à distance: Quatre observations personelles.* Paris: Perthuis, 1915.

Durey, Louis. *Étude sur l'œuvre de Paracelse, médecin hermétiste, astrologue, alchimiste, et sur quelques autres médecins hermétistes, Arnauld de Villeneuve, J. Cardan, Cornélius Agrippa.* Paris: Vigot frères, 1900.

Eckartshausen, Karl von. *Aufschlüsse zur Magie aus geprüften Erfahrungen über verborgene philosophische Wissenschaften und verdeckte Geheimnisse der Natur.* 4 vols. Munich, Germany: Lentner, 1788–1792.

Eléazar, "Faculté des sciences hermétiques." *L'Initiation* 38, no. 4 (January 1898): 97–101.

Estienne, Robert. *Dictionarium seu latinae linguae thesaurus, cum gallica fere interpretatione.* Paris: Estienne, 1531.

Fabre d'Olivet, Antoine. *Caïn, mystère dramatique en trois actes de lord Byron, traduit en vers français et réfuté dans une suite de remarques philosophiques et critiques.* Paris: Servier, 1823.

———. *Lettre de Fabre d'Olivet à Lord Byron exposant les motifs et le but de son ouvrage sur Caïn.* Paris: Éditions de l'Hyperchimie, 1901.

Fechner, Gustav Theodor. *Nanna oder über das Seelenleben der Pflanzen.* Leipzig, Germany: Voß, 1848.

Ferrand, Gabriel. *Les musulmans à Madagascar et aux îles Comores.* 3 vols. Paris: Leroux, 1891–1902.

Flammarion, Camille. "Les radiations solaires et les couleurs." *Bulletin de la Société astronomique de France* 8 (August 1897): 305–17.

Forsyth, J. S. *Demonologia, or Natural Knowledge Revealed.* London: Newman, 1831.

———. *The New Domestic Medical Manual: Being a Practical and Familiar Guide to the Treatment of Diseases Generally.* London: Sherwood, Jones, 1824.

Franck, Adolphe. "Paracelse et l'alchimie au XVIe siècle." In *Collection d'ouvrages relatifs aux sciences hermétiques,* ed. Jules Lermina, 1–32. Paris: Chacornac, 1889.

Franck von Franckenau, Georg. *De palingenesia sive resuscitatione artificali plantarum, hominum et animalium.* Halle (Saale), Germany: Serre, 1717.

———. *Flora Francica aucta, oder, Vollständiges Kräuter-Lexikon.* Edited by Johann Gottfried Thilo. Leipzig, Germany: Waysenhaus und Frommann, 1766.

Fuchs, Leonhart. *De historia stirpium commentarii insignes.* Basel, Switzerland, 1542.

———. *Plantarum effigies.* Lyons, France: Arnoullet, 1551.

Fulcanelli. *Le mystère des cathédrales: Esoteric Interpretation of the Hermetic Symbols of the Great Work.* Translated by Mary Sworder. Las Vegas, Nev.: Brotherhood of Life, 1986.

Gerard, John. *The Herball or Generall Historie of Plantes.* London: Norton, 1597.

Gessmann, Gustav Wilhelm. *Die Pflanze im Zauberglauben: Ein Katechismus der Zauberbotanik.* Vienna: Hartleben, 1899.

Görres, Johann Joseph von. *Die christliche Mystik.* 4 vols. Regensburg, Germany: Manz, 1836–1842.

Guaïta, Stanislas de. *Le serpent de la Genèse.* 2 vols. Paris: Chamuel, 1891–1897.

Gubernatis, Angelo de. *La mythologie des plantes: Les légends du règne végétal.* 2 vols. Paris: Reinwald, 1878–1882.

Haatan, Abel. *Traité d'astrologie judiciaire: Influences planétaires, signes du zodiaque, mystères de la naissance, détermination de l'horoscope, domification du ciel, interprétation du thème généthliaque, clef générale des prophéties, astrologiques.* Paris: Chamuel, 1895.

Haeckel, Ernst. *Anthropogenie oder Entwickelungsgeschichte des Menschen: Gemeinverständliche wissenshaftliche Vorträge über die Grundzüge der menschlichen Keimes- und Stammes-Geschichte.* Leipzig, Germany: Engelmann, 1874.

Haller, Abrecht von. *Onomatologia medica completa oder Medicinisches Lexicon das alle Benennungen und Kunstwörter welche der Arzeywissenschaft und Apoteckerkunst eigen sind deutlich und vollständig erkäret zu allegemeinem Gebrauch.* Frankfurt, Germany: Gaum, 1755.

Hartmann, Eduard von. *Der Spiritismus.* Leipzig, Germany: Friedrich, 1885.

———. *Spiritism.* Translated by Charles C. Massey. London: Psychological Press, 1885.

Hartmann, Franz. *The Life and Doctrines of Jacob Boehme: The God-Taught Philosopher.* Boston: Occult Publishing Company, 1891.

———. *The Life and the Doctrines of Philippus Theophrastus, Bombast of Hohenheim, Known by the Name of Paracelsus.* London: Paul, Trench, Trübner, 1896.

———. *With the Adepts: An Adventure among the Rosicrucians.* 2nd. ed. New York: Theosophical Publishing, 1910.

Hellwig, Otto von. *Curiosa physica, oder Lehre von unterschiedlichen Natur-Geheimnissen.* Sondershausen, Germany: Käyser, 1700.

Helmont, Jan Baptist van. *De magnetica vulnerum curatione.* Paris: Le Roy, 1621.

———. *A Ternary of Paradoxes: The Magnetick Cure of Wounds, the Nativity of Tartar in Wine, the Image of God in Man.* Translated by Walter Charleton. London: Flesher, 1650.

Hildegard of Bingen. *Physica: Liber subtilitatum diversarum naturarum creaturarum.* Berlin: De Gruyter, 2010.

Hingston, James. *The Australian Abroad: Branches from the Main Routes round the World.* 2 vols. London: Low, 1879–1880.

Hollandus, Johann Isaac. *Opera vegetabilia: In welchen gelehret wird, wie aus dem Weine und anderen Vegetabilien der Lapis Vegetabilis praepariret, oder alle Dinge in ihre höchste Exaltation gebracht werden müssen.* Vienna: Krauss, 1746.

Hooker, William Jackson, "Echinocactus Williamsii." *Curtis's Botanical Magazine* 3 (1847): Tab. 4296.

Hooper, Robert. *Lexicon medicum or Medical Dictionary: Containing an Explanation of the Terms in Anatomy, Human and Comparative, Botany, Chemistry, Forensic Medicine, Materia Medica, Obstetrics, Pharmacy, Physiology, Practice of Physic, Surgery, Toxicology, and the Different Branches of Natural Sciences Connected with Medicine.* 8th rev. ed. Edited by Klein Grant. London: Longman, Brown, Green, and Longmans, 1848.

Horst, Georg Conrad. *Zauber-Bibliothek oder von Zauberei, Theurgie und Mantik, Zauberern, Heren, und Herenprocessen, Dämonen, Gespenstern, und Geistererscheinungen.* 6 vols. Mainz, Germany: Kupferberg, 1821–1826.

Huysmans, Joris-Karl. *La cathédrale.* 6th ed. Paris: Stock, 1898.

———. *The Cathedral.* Translated by Clara Bell. London: Paul, Trench, Trubner, 1922.

Jacolliot, Louis. *La genèse de l'humanité.* Paris: Lacroix, 1879.

———. *Le spiritisme dans le monde: L'initiation et les sciences occultes dans l'Inde et chez tous les peuples de l'antiquité.* Paris: Lacroix, 1879.

———. *Occult Science in India and among the Ancients: Mystic Initiations, and the History of Spiritism.* Translated by Willard L. Felt. New York: Lovell, 1884.

Jollivet-Castelot, François. *La médecine spagyrique: Oswald Crollius, Joseph du Chesne, Jean d'Aubry, avec la réédition intégrale du traité des signatures et correspondances de Crollius.* Paris: Durville, 1912.

Kiesewetter, Karl. "Die Palingenesie in ihrer Geschichte und Praxis geschildert." *Sphinx* 7, no. 46 (May 1889): 207–16.

———. "Die Rosenkreuzer, ein Blick in dunkele Vergangenheit." *Sphinx* 1, no. 1 (January 1886): 45–54.

———. "Magishe Räucherungen." *Sphinx* 1, no. 3 (March 1886): 220–21.

———. "La palingénésie historique et pratique." Translated by L. Desvignes. *L'Initiation* 31, no. 7 (April 1896): 41–64.

Kingsford, Anna. *The Perfect Way in Diet: A Treatise Advocating a Return to the Natural and Ancient Food of Our Race.* London: Paul, Trench, 1881.

Kircher, Athanasius. *Ars magna lucis et umbrae in X. libros digesta.* Amsterdam: Janssonius, 1671.

———. *Mundus subterraneus in XII libros digestus.* 2 vols. Amsterdam: Janssonius et Weyerstraet, 1665.

Kostka, Jean [Jules Doinel]. *Lucifer démasqué.* Paris: Delhomme et Briguet, 1895.

Lamarck, Jean-Baptiste. *Encyclopédie méthodique: Botanique.* 8 vols. Paris: Panckoucke, 1783–1808.

Langhans, Christoph. *Neue Ost-Indische Reise.* Leipzig: Rohrlachs, 1705.

L'Aulnaye, François-Henri-Stanislas de. *Histoire générale et particulière des religions et du culte de tous les peuples du monde, tant anciens que modernes.* 2 vols. Paris: Fournier, 1791–1792.

Lautrec, Gabriel de. *Le serpent de mer.* Paris: Éditions du siècle, 1925.

Leadbeater, C. W. *Vegetarianism and Occultism.* Adyar Pamphlets 33. Chicago: Theosophical Publishing House, 1913.

Lémery, Nicolas. *A Course of Chymistry: Containing an Easie Method of Preparing Those Chymical Medicines Which Are Used in Physick.* 4th ed. London: Bell, 1720.

———. *Cours de chymie: Contenant la manière de faire les opérations qui sont en usage dans la médecine, par une méthode facile.* 7th ed. Paris: Michallet, 1690.

———. *Pharmacopée universelle: Contenant toutes les compositions de pharmacie qui sont en usage dans la médecine, tant en France que par toute l'Europe.* Paris: D'Houry, 1698.

———. *Pharmacopoeia Lemeriana contracta: Lemery's Universal Pharmacopoeia.* London: Kettilby, 1700.

———. *Traité universel des drogues simples: Mises en ordre alphabétique.* Paris: D'Houry, 1699.

Lenain, Lazare. *La science cabalistique, ou l'art de connaître les bons génies qui influent sur la destinée des hommes.* Amiens, France, 1823.

Lenglet-Mortier, and Diogène Vandamme. *Nouvelles et véritables étymologies medicales tirées du gaulois.* Paris: Le Quesnoy, 1857.

Lévi, Éliphas. *La clef des grands mystères suivant Hénoch, Abraham, Hermès Trismégiste, et Salomon.* Paris: Baillière, 1861.

———. *Dogme et rituel de la haute magie.* 2 vols. Paris: Baillière, 1861.

———. *Le grand arcane ou l'occultisme dévoilé.* Paris: Chamuel, 1898.

Linnaeus, Carl. *Des Ritters Carl von Linné vollständiges Pflanzensystem.* 14 vols. Edited by Gottlieb Friedrich Christmann and Georg Wolfgang Franz Panzer. Nuremburg, Germany: Raspe, 1777–1778.

———. *Species plantarum.* 2 vols. Stockholm: Salvius, 1753.

Linocier, Geoffroy. *L'histoire des plantes.* Paris: Macé, 1620.

Lloyd, John Thomas. "The Mombreros (Coca Users) of Columbia." *Drug Treatise* 27 (1913): 5–14.

Luys, Jules Bernard. "Sur l'action des médicaments à distance chez les sujects hypnotisés." *Comptes rendus hebdomadaires des séances et mémoires de la Société de biologie* 30, 8th series (August 1886): 426–28.

Mackey, Albert. *An Encyclopaedia of Freemasonry and Its Kindred Sciences.* Philadelphia: Moss, 1874.

Maeterlinck, Maurice. *Le trésor des humbles.* Paris: Société du Mercure de France, 1896.

———. *The Treasure of the Humble.* Translated by Alfred Sutro. New York: Dodd, Mead, 1912.

Mattioli, Pietro Andrea. *Discorsi di M. Pietro Andrea Mattioli sanese, medico cesareo, ne' sei libri di Pedacio Dioscoride anazarbeo Della materia medicinale.* Venice: Pezzana, 1774.

Maldigny Clever de. "Confession spiritualiste: Quatrième et dernier article." *Journal du magnétisme* 16 (1857): 393–418.

Mario, Marc (Maurice Jogand). "La flore mystérieuse: Les fleurs porte-bonheur, les plantes maléfiques." *La vie mystérieuse* 29 (1910): 69; 30 (1910): 91–92; 31 (1910): 105–6; 32 (1910): 125; 34 (1910): 150; 35 (1910): 171–72; 36 (1910): 182–83; 38 (1910): 220–21; 39 (1910): 228–29; 41 (1910): 266–67; 45 (1910): 324–25; 47 (1910): 363–64; 50 (1911): 24–25.

Martin-Lauzer, A. "De la grande consoude." *Revue de thérapeutique médico-chirurgicale* 1, no. 13 (1853): 353–55.

Martius, Karl Friedrich Philipp von. *Die Pflanzen und Thiere des tropischen America, ein Naturgemälde.* Leipzig, Germany: Fleischer, 1831.

———. "Die Unsterblichkeit der Pflanze: Ein Typus." In *Reden und Vorträge über Gegenstände aus dem Gebiete der Naturforschung.* Stuttgart, Germany: Cotta'scher Verlag, 1838.

———. *Reise in Brasilien.* 3 vols. Munich, Germany: Lindauer, 1823–1831.

Matgioï [Albert de Pouvourville]. "La clef orientale des faux paradis." *L'Initiation* 55, no. 7 (April 1902): 6–41; 55 no. 8 (May 1902): 149–62.

———. *L'opium et l'alcool en Indochine.* Brussels: Établissements généraux d'imprimerie, 1909.

———. *L'opium: Sa pratique.* Paris: Société d'étitions littéraires et artistiques, 1902.

———. *Physique et psychique de l'opium.* Paris: Figuière, 1914.

Mattei, Cesare. *Electro-Homœopathy: The Principles of a New Science Discovered.* 2nd ed. Bologna, Italy: Mareggiani, 1880.

———. *Elettromiopatia: Scienza nuova che cura il sangue e sana l'organismo.* Casale Monferrato, Italy: Bertero, 1878.

Mattioli, Pietro Andrea. *Les commentaires de M. P. André Matthiole, medecin sienois, sur les six livres de la matiere madecinale de Pedacius Dioscoride, anazarbéen.* Lyon, France: De Ville, 1680.

Maurer, Felix. *Amphitheatrum magiae universae theoreticae et practicae, oder gründlicher, ausführlicher Bericht und Unterricht von denen grössesten, geheimesten*

Wunder-Machten Gottes, der Natur, der Engel, der Teufel, der Menschen. Nuremburg, Germany: Rüdiger, 1714.

Mavéric, Jean. *Hermetic Herbalism: The Art of Extracting Spagyric Essences.* Edited and translated by R. Bailey. Rochester, Vt.: Inner Traditions, 2020.

———. *La médecine hermétique des plantes ou l'extraction des quintessences par art spagyrique.* Paris: Dorbon-aîné, 1911.

Mavéric, Jean, and Rip Monfloride. *La magie rurale: Révélations de la magie campagnarde, villageoise, champêtre, sylvestre, fluviale et cynégétique.* Paris: Durville, 1913.

Maxwell, William. *De medicina magnetica libri III in quibus tam theoria quam praxis continetur.* Frankfurt, Germany: Zubrodt, 1679.

Mercuriale, Girolamo. *De morbis muliebribus praelectiones.* Venice: Giunti, 1601.

Michelet, Victor-Émile. *Les compagnons de la hiérophanie: Souvenirs du mouvement hermétiste à la fin du XIXe siècle.* Paris: Dorbon-aîné, 1937.

Monginot, François. *Traité de la conservation et prolognation de la santé.* Paris: Bourdin et Perier, 1635.

Monier-Williams, Sir Monier. *A Sanskrit-English Dictionary: Etymologically and Philologically Arranged with Special Reference to Greek, Latin, Gothic, German, Anglo-Saxon, and Other Cognate Indo-European Languages.* Oxford, UK: Clarendon, 1872.

Morley, Henry. *The Life of Henry Cornelius Agrippa von Nettesheim.* 2 vols. London: Chapman and Hall, 1856.

Mousseaux, Roger Gougenot des. *Les hauts phénomènes de la magie, précédés du spiritisme antique.* Paris: Plon, 1864.

Nynauld, Jean de. *De la lycanthropie, transformation, et extase des sorciers.* Paris: Millot, 1615.

Oetinger, Friedrich Christoph. *Die Philosophie der Alten widerkommend in der güldenen Zeit.* 2 vols. Frankfurt, Germany, 1762.

———. *Gedanken über die Zeugung und Geburten der Dinge: Aus Gelegenheit der Bonnetischen Palingenesie.* Frankfurt, Germany, 1774.

Oken, Lorenz. *Esquisse du système d'anatomie et d'histoire naturelle.* Paris: Béchet jeune, 1821.

Oliver, Daniel. *First Book of Indian Botany.* London: Macmillan, 1869.

Oxley, William. "Fac-simile of Plant with Flower (*Ixora crocata*) Produced by a Materialised Spirit Form (Yolande), at Newcastle-on-Tyne, August 4th, 1880." *Herald of Progress* 1, no. 8 (September 1880): 105–7.

Pandorac, Numa [Jules Giraud]. "Influence du haschisch sur la santé." *La Voie* 26 (June 1906): 75–82.

———. "Influence du haschisch sur le corp et sur l'esprit." *La Voie* 31 (December 1906): 71–84.

———. "Influence du haschisch sur nos facultés." *La Voie* 28 (September 1906): 63–74.

———. "Manuel du haschischéen." *La Voie* 19 (November 1905): 62–75.

———. "Prédictions d'un haschischéen: Sur le haschisch." *La Voie* 15 (June 1905): 66–76; 16 (July–August 1905): 69–82; 17 (September 1905): 72–89; 18 (October 1905): 71–79.

———. "Testament d'un haschischéen." *L'Initiation* 2 no. 4 (January 1889): 59–70.

———. *Testament d'un haschischéen.* Paris: Durville, 1913.

Papus [Gérard Encausse]. *Anatomie philosophique et ses divisions: Précéde d'un essai de classification méthodique des sciences anatomiques.* Paris: Chamuel, 1894.

———. "Échos." *L'Initiation* 31, no. 9 (June 1896): 291–93.

———. "L'École hermétique." *L'Initiation* 53, no. 1 (October 1901): 39–58.

———. "Les facultés occultes de l'homme." *Rosa alchemica* 8, no. 3 (March 1903): 121–29.

———. "Premiers éléments d'homéopathie pratique: Cours professé à l'École supérieure libre de sciences médicales appliquées, 1912–1913." *Mysteria* 1, no. 2 (February 1913): 98–122.

———. *Traité élémentaire de magie pratique: Adaptation, réalisation, théorie de la magie avec appendice sur l'histoire et la biographie de l'evocation magique.* 2nd ed. Paris: Charcornac, 1906.

———. *Traité élémentaire de science occulte.* 9th ed. Paris: Société d'éditions littéraires et artistiques, 1903.

———. *Traité méthodique de science occulte.* Paris: Dorbon-aîné, 1891.

Paracelsus [Theophrast von Hohenheim]. *De natura rerum: IX Bücher.* Strasbourg, France: Jobin, 1584.

———. *Paracelsus, His Archidoxis, Comprised in Ten Books, Disclosing the Genuine Way of Making Quintessences, Arcanums, Magisteries, Elixirs, etc., Together with His Books, Of Renovation and Restauration, Of the Tincture of the Philosophers, Of the Manual of the Philosophical Medicinal Stone, Of the Virtues of the Members, Of the Three Principles, and Finally His Seven Books, Of the Degrees and Compositions of Receipts and Natural Things.* Translated by John Harding. London: Brewster, 1660.

———. *The Hermetic and Alchemical Writings of Paracelsus the Great.* 2 vols. Edited and translated by A. E. Waite. London: Elliott, 1894.

Parkinson, John. *Theatrum botanicum.* London: Cotes, 1640.

Pernety, Antoine-Joseph. *Dictionnaire mytho-hermétique.* Paris: Bauche, 1758.

Pezoldt, Adam Friedrich. "De palingenesia seu artificiali rerum ex regno vegetabili, animali et minerali, igne combustarum, e suis cineribus resuscitatione et repraesentatione." *Academiae Caesaro-Leopoldinae naturae curiosorum ephemerides sive observationum medico physicarum centuria* 7 (1719): 31–37.

Philipon, René. *Stanislas de Guaita et sa bibliothèque occulte.* Paris: Librairie Dorbon, 1899.

Pictorius, Georg. *Pantopōlion: Continens omnivm fermè quadrupedum, auium, piscium, sepentium, radicum, herbarum, seminum, fructuum, aromatum, metallorum & gemmarum naturas.* Basel, Switzerland: Petri, 1563.

———. *Sermonum convivalium libri decem deque ebrietate lusus quidam.* Basel, Switzerland: [Petri], 1571.

Piemontese, Alessio [Girolamo Ruscelli]. *The Secrets of Alexis: Containing Many Excellent Remedies against Divers Diseases, Wounds, and Other Accidents.* London: Stansby, 1615.

Piobb, Pierre. *Formulaire de haute magie: Recettes et formules pour fabriquer soi-même les philtres d'amour, talismans, etc., clefs absolues des sciences occultes, esprits, invocations, évocations, tables tournantes, etc.* Paris: Daragon, 1907.

Pivion, Edmond. *Étude sur le régime de Pythagore: Le végétarisme et ses avantages.* Paris: Berthier, 1885.

Pomet, Pierre. *Histoire générale des drogues: Traitant des plantes, des animaux, & des mineraux.* Paris: Loyson, 1694.

Porta, Giambattista della. *La magie naturelle: Qui est, Les secrets & miracles de nature.* Translated by François Arnoullet. Lyon, France, 1615.

———. *Magiae naturalis, sive de miraculis rerum naturalium libri IIII.* Naples, Italy: Cancer, 1558.

———. *Natural Magick by John Baptista Porta.* London: Young and Speed, 1658.

Prel, Carl du. "Der Pflanzenphönix." *Sphinx* 7, no. 40 (April 1889): 193–202.

———. *Die Philosophie der Mystik.* Leipzig, Germany: Günther, 1885.

———. "Forciertes Pflanzenwachstum." *Sphinx* 7, no. 39 (March 1889): 145–58.

———. "Pflanzenmystik." *Sphinx* 7, no. 37 (January 1889): 17–24.

Pseudo-Agrippa, *Fourth Book of Occult Philosophy.* Translated by Robert Turner. London: Rooks, 1665.

Pseudo-Albertus. *Les admirables secrets d'Albert le Grand, contenant pluseiurs traittez sur la conception des femmes, & les vertus des herbes, des pierres precieuses, & des animaux.* Cologne, Germany: Dispensateur des secrets, 1703.

———. *Les solide tresor des merveilleux secrets de la magie naturelle et cabalistique du Petit Albert.* Geneva: Aux dépens de la Compagnie, 1704.

———. *Liber aggregationis seu secretorum Alberti Magni de virtutibus herbarum, lapidum et animalium quorundam.* [Venice?], ca. 1493.

———. *Secrets merveilleux de la magie naturelle et cabalistique du Petit Albert.* Lyon, France: Les héritiers de Béringos fratres, 1722.

Pseudo-Aquinas. *Secreta alchimiae magnalia.* 3rd ed. Leiden, Netherlands: Basson, 1602.

Pseudo-Llull. *De secretis natvrae, sev de quinta essentia liber vnus, in tres distinctiones diuisus, omnibus iam partibus absolutus.* Cologne, Germany: Birckmann, 1567.

Pseudo-Paracelsus. *Of the Supreme Mysteries of Nature.* Translated by Robert Turner. London: Brook and Harison, 1656.

Rabelais, *Tiers liure des faictz et dictz heroïques du noble Pantagruel.* Paris: Chrestien Wechel, 1546.

Ragon, Jean-Marie. *Maçonnerie occulte suivie de l'initiation hermétique.* Paris: Dentu, 1853.

———. *Orthodoxie maçonnique, suivie de la maçonnerie occulte et de l'initiation hermétique.* Paris: Dentu, 1853.

Reclus, Élie. "Études sur les origines magiques de la médecine: La mandragore." *L'Humanité nouvelle* 5 (1899): 560–71.

Reichenbach, Karl. *Die Pflanzenwelt in ihren Beziehungen zur Sensitivitat und zum Ode: Eine physiologische Skizze.* Vienna: Braumüller, 1858.

———. *Physico-Physiological Researches on the Dynamics of Magnetism, Electricity, Heat, Light, Crystallization, and Chemism in Their Relations to Vital Force.* Translated by John Ashburner. London: Baillière, 1851.

———. *Reichenbach's Letters on Od and Magnetism.* Translated by F. D. O'Byrne. London: Hutchinson, 1926.

Rochas, Albert de. *Les frontières de la science.* Paris: Libraire des sciences psychologiques, 1904.

Rodin, Hippolyte. *Les plantes médicinales et usuelles des champs—jardins—forêts.* 3rd ed. Paris: Rothschild, 1876.

Rozin, A. "L'apperçu des plantes usuelles." *L'esprit des journaux, francais et etrangers* 2 (1794): 343–64.

Ruland, Martin. *Lexicon alchemiae, siue, dictionarium alchemisticum: Cum obscuriorum verborum, et rerum hermeticarum, tum Theophrast-Paracelsicarum phrasium, planam explicationem continens.* Frankfurt, Germany: Palthenius, 1612.

Şăineanu, Lăzar. *L'histoire naturelle et les branches connexes dans l'œuvre de Rabelais.* Paris: Champion, 1921.

Saint-Denys, Marquis d'Hervey de. *Les rêves et les moyens de les diriger: Observations pratiques.* Paris: Amyot, 1867.

Saint-Martin, Louis Claude de. *De l'esprit des choses, ou coup d'oeil philosophique sur la nature des natures et sur l'objet de leur existence.* 2 vols. Paris: Laran, 1797.

Schröder, Johann. *La pharmacopée raisonnée de Schroder: Commentée par Michel Ettmuller.* 2 vols. Lyon, France: Amaulry, 1698.

Schuré, Edouard. *The Great Initiates: Sketch of the Secret History of Religions.* Translated by Fred Rothwell. 2 vols. London: Rider and Son, 1922.

Sédir, Paul [Yvon Le Loup]. *Histoire et doctrines des Rose-Croix.* Bihorel, France: Bibliothèques des Amitiés spirituelles, 1932.

———. “La médecine occulte.” In *Les sciences maudites.* Edited by François Jollivet-Castelot, Paul Ferniot, and Paul Redonnel, 135–40. Paris: La maison d’art, 1900.

———. *La médecine occulte.* Paris: Bibliothèque universelle Beaudelot, 1910.

———. “La physiologie de Paracelse.” *L’Initiation* 52 no. 12 (September 1901): 215–22.

———. *Le bienheureux Jacob Boehme: Le cordonnier-philosophe.* 2nd ed. Paris: Éditions de l’Initiation, 1901.

———. *L’enfance du Christ (Conférences sur l’Évangile).* 2nd ed. Paris: Bibliothèque universelle Beaudelot, 1914.

———. “L’esotérisme indou: Les avatars.” *La Rose + Croix: Revue synthétique des sciences d’Hermès* 9 (September 1902): 282–97.

———. *Les plantes magiques: Botanique occulte, constitution sècrete des végétaux vertus des simples, médecine hermétique, philtres, onguents, breuvages magiques, teintures, arcanes, élixirs spagyriques.* Paris: Bibliothèque Chacornac, 1902.

———. “Sur l’emploi de l’opium.” *L’Initiation* 62, no. 4 (January 1904): 66–76.

Sennert, Daniel. *De chymicorum cum Aristotelicis et Galenicis consensu ac dissensu.* Wittenberg, Germany: Schürer, 1619.

Serres, Oliver de. *Le théatre d’agriculture et mésnage des champs.* 2nd ed. Paris: Saugrain, 1617.

Störck, Anton von. *Observations nouvelles sur l’usage de la cigüe.* Paris: Didot, 1762.

Sylburg, Friedrich, ed. *Etymologicon magnum seu Magnum grammaticae penu.* Heidelberg, Germany: Commelin, 1594.

Tachenius, Otto. *Hippocrates chymicus.* Paris: D’Houry, 1674.

———. *Otto Tachenius, His Hippocrates chymicus, Which Discovers the Ancient Foundations of the Late Viperine Salt.* London: James, 1677.

Tau Théophane [Léon Champrenaud], and Tau Simon [Albert de Pouvourville]. *Les enseignements secrets de la Gnose.* Paris: Société d’éditions contemporaines, 1907.

Tavernier, Jean-Baptiste. *Les six voyages en Turquie, en Perse, et aux Indes.* 2 vols. Paris, 1679.

Théis, Alexandre de. *Glossaire de botanique, ou dictionnaire étymologique de tous les noms et termes relatifs à cette science.* Paris: Dufour, 1810.

Thiers, Jean-Baptiste. *Traité des superstitions selon l’écriture sainte, les décrets des conciles et les sentimens des saints pères et des théologiens.* 2 vols. Paris: Dezallier, 1679.

Thiriat, Xavier. “Croyances, superstitions, préjugés, usages et coutumes dans le Département des Vosges.” *Mélusine* 19 (October 1877): 451–58; 20 (October 1877): 477–79; 21 (November 1877): 498–502.

Trevisan, Bernard. *Le texte d’alchymie, et le songe-verd.* Paris: D’Houry, 1695.

Tuchmann, Jules. “La fascination: 4. Les fascinés.” *Mélusine* 7 (1894–1895): 15–21, 41–45, 64–69, 114–18, 137–41, 159–65, 170–82, 205–14, 231–38, 241–55, 273–85.

Turland, Nicholas J. *The Code Decoded: A User's Guide to the International Code of Nomenclature for Algae, Fungi, and Plants.* 2nd ed. Sofia, Bulgaria: Pensoft, 2019.

Turland, Nicholas J., John H. Wiersema, Fred R. Barrie, Werner Greuter, David L. Hawksworth, Patrick S. Herendeen, Sandra Knapp et al., eds. *International Code of Nomenclature for Algae, Fungi, and Plants (Shenzhen Code) Adopted by the Nineteenth International Botanical Congress, Shenzhen, China, July 2017.* Glashütten, Germany: Koeltz Botanical Books, 2018.

Vergnes, J.-P. "De la transplantation des maladies." *Le Voile d'Isis* 26, no. 13 (January 1921): 33–40; no. 14 (February 1921): 118–21; no. 15 (March 1921): 192–96; no. 16 (April 1921): 263–71.

Wecker, Johann Jacob. *De secretis libri XVII.* Basel, Switzerland: Ludovicus Rex, 1629.

———. *Les secrets et merveilles de nature.* Revised, corrected, and edited by Pierre Meyssonier. Rouen, France: Reinsart, 1608.

Wilson, Andrew. *Chapters on Evolution.* Rev. ed. New York: G. P. Putnam's Sons, 1883.

Wolfram, E. *The Occult Causes of Disease: Being a Compendium of the Teachings Laid Down in His "Volumen Paramirum," by Bombastus von Hohenheim, Better Known as Paracelsus.* Translated by Agnes Blake. London: Rider, 1912.

Index of Ancient Sources

Index of Plant Names

Index of Authors and Subjects